Commission of the European Communities

The Climate of Europe: Past, Present and Future

Natural and Man-Induced Climatic Changes: A European Perspective

edited by

HERMANN FLOHN
Meteorological Institute, University of Bonn, F.R.G.

and

ROBERTO FANTECHI
Commission of the European Communities, Brussels, Belgium

D. Reidel Publishing Company

A MEMBER OF THE KLUWER ACADEMIC PUBLISHERS GROUP 

Dordrecht / Boston / Lancaster

Library of Congress Cataloging in Publication Data
Main entry under title:

The Climate of Europe: past, present and future.

At head of title: Commission of the European Communities.
Includes bibliographical references.
1. Europe–Climate. 2. Paleoclimatology. 3. Climatic changes.
4. Man–Influence on nature. I. Flohn, Hermann. II. Fantechi, Roberto.
III. Commission of the European Communities.
QC989.A1C58 1984 551.694 83-24776
ISBN 90-277-1745-1

Publication arrangements by
Commission of the European Communities
Directorate-General Information Market and Innovation, Luxembourg

EUR 8038

Published by D. Reidel Publishing Company
P.O. Box 17, 3300 AA Dordrecht, Holland

Sold and distributed in the U.S.A. and Canada
by Kluwer Academic Publishers,
190 Old Derby Street, Hingham, MA 02043, U.S.A.

In all other countries, sold and distributed
by Kluwer Academic Publishers Group,
P.O. Box 322, 3300 AH Dordrecht, Holland

Printed in The Netherlands

TABLE OF CONTENTS

EDITORS' PREFACE

The present book originated from an idea of one of us (R.F.). As the responsible person for the Climatology Research Programme of the Commission of the European Communities, he was of the opinion that such a programme should be heralded or accompanied by a book looking at climate from the standpoint of Europe and at Europe in a climatic perspective. For this reason he found encouragement and approval from Dr. Ph. Bourdeau, Director, Environmental and Raw Material Research Programmes, and from Prof. P. Fasella, Director General for Science, Research and Development. Over all, the Commission of the European Communities has to be acknowledged for its financial and logistic support, which has made both the preparation and the publication of the volume possible.

The contributors to the book have to be acknowledged first of all for their acceptance of a task which was by no means an easy one. They were not simply requested to write chapters just to be put together; they had to write a book representing at the same time both different disciplines and a synthesis of conclusions; both varied opinions and a consensus on conclusions: both Europe as a well defined unit and the global climate as an indivisible process. The Authors' difficulties were increased by the fact that they were working in separate places away from each other, so that part of the book has actually been written by correspondence and through the telephone.

This partially explains the delay alluded to below and at the same time shows the patience and dedication of all the people involved in the preparation of the book.

Unfortunately the finalisation and printing of the report was delayed, for various reasons, until late Summer 1982. At that time, most authors reviewed their individual chapters again. It was found that the substance of the book including the General Summary could be maintained, without any major change of the conclusions, only with some amendments derived from recent references. Indeed, the settlement of some controversies an increasing consensus among scientists seems to evolve. This is also evident in a quite recent comprehensive report "Carbon Dioxide Review: 1982" (edited by W. C. Clark) which was published during the last revision, and which could not be taken into account.

In contrast to this, the book deals primarily with climate, focussing mainly, but not exclusively, on its European aspects. Furthermore it discusses in much greater detail our rapidly increasing evidence on past climatic evolution on two different time-scales: that of written history (10^3 years) and that of geology (10^4 - 10^6 years).

The individual chapters have been written by the authors as given below. In several extended discussions held at Brussels, the authors had taken the greatest of care to reach consensus about their views and to produce a concerted manuscript for publication. The General Summary was one of the final results of these discussions.

A. BERGER	Chapter 4.2
A. BOURKE	Chapter 6 (except 6.4)
W. DANSGAARD	Chapter 5.3 - 5.4
J.C. DUPLESSY	Chapter 4.1 - 5.3
H. FLOHN	Chapter 1, 3.2, 5.1-2, 5.5
H.H. LAMB	Chapter 2, 3.14
E. ROSINI	Chapter 6.4
C.J.E. SCHUURMANS	Chapter 3.1 (except 3.14)

The burdensome task of improving the style of the whole volume has been accomplished by A. Bourke and H. H. Lamb.

The result of the whole effort is a truly European achievement which will hopefully represent a lasting witness of what such a cooperation can bring about.

GENERAL SUMMARY

The problem

The evidence of the last ice age is sufficient of itself to show that the world's climate today is very different from what it was at times in the past. It is equally clear that nowadays seasonal conditions can fluctuate widely from one year to another: thus seasonal mean temperatures in Europe vary over a range of 8°-13°C in winter and 5°-7°C in summer, while the rainfall in particular seasons can deviate by 80% or more from long-term averages.

Less obvious is the fact that even if one eliminates the year-to-year fluctuations by taking the means of temperature, rainfall and other climatic elements over periods of 30 years or more, the long-term averages will be found to vary from one period to another. In Europe, the 30-year mean temperature can vary by 1°-2°C, and the rainfall by 40% or more in the Mediterranean region, and by some 15-30% elsewhere.

The picture is further complicated by the fact that the degree of variability changes from time to time. During the late 17th and early 19th centuries, climatic variability was much greater than in the remarkable decades between 1920 and 1960, which combined the warmest mean temperatures in the last five to seven centuries with a low level of variation from year to year. Again there is often a tendency for the short-term fluctuations to cluster, as occurred in much of Europe with the three severe winters of 1939-42 or the seven mild winters of 1971-78. Such deviations in one region are usually accompanied by fluctuations in the opposite direction in neighbouring areas; thus, for example the remarkable summer drought of 1976 in Western Europe was accompanied by a cool and very wet season in the European U.S.S.R.

The implications for human activities of climatic changes and fluctuations need no stressing; their economic impact is strongly felt in such fields as water supplies, agriculture, fisheries, transport, energy use, etc. Planning for the economic future of a region such as Western Europe would benefit appreciably from even a broad indication of the direction in which the climate is likely to move in the next 50 or 100 years. For this reason the Commission of the European Community assigned to a group of scientists the task of producing the present report on the relevant aspects of past climates, on natural and man-made influences on the atmosphere, and on possible paths that the climate might reasonably be expected to follow in the coming years.

Structure of the report

The main body of the report consists of six chapters prepared by individual specialists. Each chapter (after the first) and each important section ends with a summary and/or set of conclusions, as indicated in the table of contents; these provide a rapid means of scanning the report in rather greater detail than is provided here.

A broad view of the complex background to the problem of climatic variation and the different factors involved is given in Chapter 1. The changes which, almost exclusively under the impact of natural causes, have taken place in the world's climate over the past thousand years or more are reviewed in Chapter 2. The following chapter analyses in greater detail the recent centuries for which instrumental weather data are available, the number and reliability of which rose considerably with the founding of State Meteorological Services during the nineteenth century.

Chapter 4 deals with a subject which has caused growing concern in recent years: the extent to which man himself may affect the climate by his activities, and particularly the risk that a continuing rise of carbon dioxide in the atmosphere will cause a significant rise of world temperature and other consequent changes.

Chapter 5 looks into the past to find analogues to what appear to be the more likely lines of climatic development in the future, and uses these as indicators of the consequences of a probable temperature rise by carbon dioxide and other trace gases, and also of a possible interlude which could be triggered, for example, by a spell of enhanced volcanic outbreaks.

How European agriculture has reacted to climate change and fluctuations over the past two hundred years or so is described in Chapter 6, which also considers the extent to which food production is likely to be affected in the future.

An outline of the main findings and recommendations

To assist the busy and hurried reader the more important conclusions of the report are summarized under separate headings in the following paragraphs, and a list is appended of some of the main areas of uncertainty which call urgently for further research.

A. *The climatic system and its internal interactions*

Climate and weather are associated in the atmosphere with a complex interacting hierarchy of processes from small fine-weather clouds to fully developed tropical hurricanes and mid-latitude cyclones. But this fast-living system is only part of a larger entity: the climatic system. It embraces atmosphere, ocean, floating and continental ice, soil and vegetation: all these subsystems vary on quite different time-scales and interact in an even more complex way - partly damping, partly amplifying the motions and variations of the other subsystems. Recent experience has shown that the slower subsystems (the upper mixed layer of the ocean, the floating sea-ice) are driven by the combined effects of atmospheric motions. In turn they react, on longer time-scales, upon the behaviour of the atmosphere.

The climatic system as a whole is capable of undergoing fluctuations on may time-scales. One can outline a variance spectrum ranging over all possible periods of variations, many of which have in fact been observed, from hundreds of million years to an hour or less in the case of weather changes. Estimates of the amplitude of the fluctuations of the globally averaged temperature in the past give values of the order of 0.1°C per decade; in very rare cases this value may be significantly higher, even under purely natural conditions. Looking ahead we should concentrate our main interest on the likely evolution within the next 100 years, without neglecting longer time-scales with stronger variations.

B. *Some vulnerable aspects of the climatic systems*

One aspect of the climatic system which must be stressed is its sensitivity to some internal interactions. Two key areas-each of a size of 10-12 million km^2-are well-known:

a) The varying extent of the shallow sea-ice on the Arctic and Subantarctic Oceans: about 70 percent (in the Arctic) or 15 percent (Subantarctic) are multi-year ice-floes with a thickness up to 6 m, while the remainder consists of seasonal ice with a thickness of only 50-150 cm.

b) The regions of coastal and equatorial upwelling of cool nutrient-rich water from deeper layers of the ocean. In the equatorial Pacific and Atlantic in particular upwelling and downwelling change simultaneously over very large regions, with a dramatic effect on water temperature, evaporation, rainfall, CO_2-exchange and hence on fisheries.

C. *Is climatic variation predictable?*

The art of predicting climatic variation is still in its infancy. However, with increasing understanding of the internal interactions and of the sensitive areas it may be hoped that, by combining theoretical and empirical methods, fairly reliable prediction techniques, on a geophysical basis, for up to six months or perhaps one year can be developed.

Since there appears to be little hope of predicting the natural climatic changes over the next century or two, we may attempt a sort of conditional forecast, limited to these man-made effects. Such a forecast may be based on the assumption that natural effects remain within the same limits as during recent centuries. In particular, we assume:

a) no clustering of major volcanic eruptions,
b) no major change of solar radiation,
c) no major change of cloudiness.

Assumption (c) is an allusion to our incomplete knowledge of the relationship between radiation fluxes and clouds: this relationship need not, however, be overemphasized, as has been shown by recent investigations.

D. *External influences of a natural kind*

Obviously a distinction must be made between non-systematic variations within the climatic system ("internal") and more systematic variations forced from outside ("external" impacts). The latter include changes in the composition of the atmosphere, in the geometry of the Earth's orbit, possible changes in the Sun's radiation (the great motor of the climatic system) or great volcanic eruptions. Unfortunately the climatic system is so complicated that it is not yet possible to give a satisfactory explanation of the climatic changes observed during the last centuries, except to say that their causes are overwhelmingly natural. Factors such as volcanic activity and possible changes in solar radiation are at present unpredictable: this limits, in principle, any attempts to forecast climatic changes over the next 100 years or so.

Because of the scale of atmospheric motions, any significant climatic change in our region is likely to be accompanied by similar (or opposite) changes in other regions of the globe. Thus, the European climate cannot be seen in isolation; it must be considered within its large-scale natural framework.

E. *External influences due to man*

The great challenge with which we are confronted today is the growing impact of man on the climatic system. This interference started at least 8000 years ago, during the neolithic revolution to pastoralism and agriculture, when the natural vegetation cover was increasingly converted into arable land and pasture.
Now, with a rapidly increasing population already near 4500 millions, the rate of this conversion has led to serious consequences: deforestation, desertification, overgrazing, bushfires and soil degradation, all of which change the climatic conditions near the surface; so also on a smaller scale do irrigation and reservoirs. The quantity of dust particles produced by many of these operations exceed that produced by industrial combus-

tion and urbanization. Contrary to common belief, the climatic effect of these man-made particles is to cause a slight warming of the lower atmosphere.

The particles remain in the air for a short time, from a few days to about one month; the atmospheric residence time of some man-made trace gases is much longer. Within the atmospheric processes responsible for climate, the balance between the solar radiation (with highest intensity in the visible part of the spectrum) and terrestrial radiation (with highest intensity in the infrared) plays a dominant role. Like glass, some gases are transparent with respect to solar radiation, but absorb terrestrial radiation and thereby heat the air: this is the so-called "greenhouse effect".

F. *The role of carbon dioxide and other man-made gases*

The most important of these gases is carbon dioxide (CO_2), which has increased since the end of the 19th century by about 15 percent and is increasing still by 0.3-0.5 percent per year. But some other man-made gases add to the effect, notably N_2O as the final decay product of nitrogen fertilizers, and the chlorofluoromethanes (CFM's) used as propellants in spray cans.

One may arrive at a convenient first approximation to the total contribution of these complex effects and interactions by introducing the concept of a "combined greenhouse effect" which takes account of all the gases involved in terms of the "equivalent CO_2 content" of the atmosphere.

The budget of carbon in the climate system is by no means fully understood, since the fate of CO_2 in the oceans and in the vegetation and the soil has not yet been quantitatively determined with sufficiently accuracy. In addition to the input of CO_2 from fossil fuel (coal, oil, natural gas), the destruction of vegetation and soil probably releases further CO_2. However, this latter contribution and the transport processes in the ocean are far from being certain quantitatively; thus it is virtually impossible to predict the future behaviour of atmospheric CO_2 beyond, say, 30 years even if we knew precisely the growth rate of fossil fuel consumption during the next 100 years. But all model calculations suggest that, with unlimited use of coal and oil shales, the CO_2 content of the atmosphere may rise by a factor of 4-8 during the next few centuries.

G. *Climatic models*

A climatic model is a mathematical description of the climate based on physical principles. The present position in model development can be summarized as follows: Modelling of the interactions between atmosphere, ocean and drifting sea-ice has just started, and the dynamic processes in the ocean are much less well understood than those in the atmosphere. No model is as yet available which adequately simulates the seasonal and diurnal variations of weather and climate with sufficient accuracy and horizontal resolution. But this is only a first test; a more stringent test would be the successful simulation of a true climatic change, e.g. the transition from the present climate into a full glaciation. Any responsible forecasting of future climatic evolution should be based, in the future, on a family of comprehensive, successfully tested models.

It should be realized, however, that our present knowledge of the climatogenic processes is still fragmentary. For example it is still possible that a minor (in fact indetectable) increase of the global cloud cover could counterbalance most of the greenhouse effect produced by increased CO_2. If so, the climate might continue to be dominated in the future by natural impacts. However several models now available predict instead a decrease of cloudiness linked with a growing instability of the troposphere.

H. *Present climatic trends*

Using available climatic models, sensitivity studies show that an increasing CO_2 content of the atmosphere should lead to warming. On the basis of the observed increase of the CO_2 content over the last two decades the estimated warming amounts only to a few tenths of a degree Celsius. Contrary to this, the observed changes of hemispheric average temperature have shown a small decrease over this timespan. Even at higher northern latitudes where the anthropogenic warming should be largest, observations show no clear sign as yet of the predicted effect.

However, this lack of clear evidence of the anthropogenic warming effect cannot be used to discount the projected future warming. According to present knowledge the predicted CO_2 -induced warming is up to now much too small to be detected with certainty; it remains within our experience of the natural range of climatic variability. This probably will continue to be the case until about the end of this century.

I. *The use of past climates as clues to the future*

In the meantime we can investigate past climates which may serve as examples, as models of future climatic evolution. Thus we may select warm phases in the past as parts of a scenario of what might happen with an uncontrolled increase of CO_2. Furthermore we may look into past cool phases and into transitions toward earlier glacials as a scenario of what could happen, if natural cooling effects (possibly much stronger than during the last 1000 or 10000 years) were to override the man-made warming effect.

Under the assumptions mentioned in paragraph (C) we are able to use one of the existing CO_2 -temperature models to estimate the global temperature change equivalent to a given CO_2-level in the atmosphere, as in Table 1. The percentage error indicates the range of values derived from two equally realistic versions of this model; e.g. the equivalent CO_2- level for an expected warming of 4°C is estimated to lie between 630 and 880 ppm.

Table 1: Past climates, change of global temperature ΔT and equivalent CO_2-level (with probable error in percent).

Time	Climatic Phase	ΔT(C)	Equivalent CO_2	Content
1000 AD	Early Medieval Optimum	+1.0	408 ppm	± 6%
6000 BP	Holocene Warm Phase	+1.5	455 ppm	± 8%
120000 BP	Last Interglacial (Eem)	{+2.0	508 ppm	± 9%
		+2.5	555 ppm	±10%
14-3.5 million years BP	Late Tertiary	+4.0	755 ppm	±16%

AD = anno domini (after Christ); BP = years before the present; ppm = one millionth part per volume.

Most recent discussions indicate that the role of other "greenhouse gases" (as mentioned in paragraph F) will increase with time, mostly because of their greater residence time in the atmosphere. In this case the equivalent CO_2 contents of Table 1, last three lines, should be reduced by 10-15%; the value equivalent to a 4°C warming would be near 620 ppm.

J. *The earlier stages of a warm scenario*

In the first phase of a warm scenario, the sea ice around Greenland would retreat towards its northern coast, accompanied by more northerly cyclone tracks than now and frequent dry anticyclonic flow patterns over Europe. During the second phase, illustrated by the Holocene warm period 6000 years ago, most places were somewhat more humid than today. In the arid belt between West Africa and Northwest India/Pakistan the desert was replaced by grassland populated by cattle-raising nomads. However, studies of the role of natural and man-triggered changes of the surface boundary conditions lead to a similar conclusion, that since the final retreat of the North American ice-sheet came after that phase, as also did the beginning of man-made desertification, any return to such a beneficial climate is very unlikely.

Increased warming, equivalent to a CO_2-rise to above 450-500 ppm and to a climate such as occurred during the peak of the last interglacial, should lead to marked shifts of the climate and vegetation belts, e.g. to a retreat of the permafrost in Siberia and Canada and of the Arctic sea ice to its core in the Central Arctic between Canada and Siberia. A significant consequence for Europe would be a northward extension (or displacement) of the Mediterranean climatic belt with dry summers and rainy winters. Apart from such shifts the climate overall would probably also be slightly more humid.

K. *The later stages of a warm scenario*

If the CO_2-level were allowed to rise above a level near 750 ppm (or even 620 ppm, see under (I)), the risk of quite important climatic changes would increase sharply. According to model studies and to paleoclimatic experience, this value is close to the critical threshold for disappearance of the drifting Arctic sea ice. This event would be initiated first by a seasonal melting between July and October, while the newly formed winter ice should be much thinner; it seems likely that this stage would be only a brief interlude (of a few decades of even less) leading to a prolonged melting and ultimately to complete disappearance of the floating ice except during winter along narrow coastal belts. The large volumes of the enormous ice-sheets in Greenland and Antarctica hinder rapid melting; they would remain nearly unaffected for several centuries at least, even in such a warm scenario.

Such a situation - one pole ice-free, the other still highly glaciated - is difficult to imagine. However, according to conclusive paleoclimatic investigations, this asymmetry existed once in the past from about 14 to 3 million years ago, and led to a northward displacement of all climatic zones by 400-800 km in the Northern Hemisphere and probably somewhat less in the Southern Hemisphere. Since the climatic role of an open Arctic Ocean during winter is much greater than during summer, one should expect a marked reduction of the subtropical winter rain belt, which would be replaced by more or less arid conditions. During parts of this time the climatic conditions near Vienna resembled those now existing in the Chotts of Tunisia.

L. *The shift of climatic zones and regional changes in a warm scenario*

This warm scenario indicates a risk of increasing climatic changes. To a first approximation, they may be expressed as meridional shifts of the climatic zones, but they would certainly be accompanied by other changes of a regional nature. The equivalent CO_2 values given in Table 1 assume that the greenhouse effect of CO_2 is increased by 50 percent because of other man-made trace gases. If this is correct, this risk of major climatic changes rises increasingly above a CO_2 level near 450 ppm. All these climatic consequences of the last-mentioned stage become still more serious in view of the probability of a doubling (or even tripling) of the world's population by the time of this event. A shift of the great climatic belts of the northern hemisphere by 400-800 km would in itself affect mankind as a whole: beneficially in some areas, probably including northern Europe and Greenland, but destructively in other areas, drastically changing water supplies and agricultural productivity.

Since no comprehensive climate model of this pattern has been designed so far, no detailed prediction of the regional climates under these boundary conditions is possible. It is also not possible to simply transfer the local climatic evidence from the Late Tertiary to the present or the predicted situation, because the distribution of land and sea was slightly different from today, and the present mountains, even the Alps, existed only in a quite rudimentary form. One of the more serious threats might be desiccation of the Mediterranean countries, especially south of about Latitude 42°N; but the frequency of summertime droughts should increase even well to the north of this latitude. However, a drying up of the Mediterranean Sea, as during the climax of the asymmetric global environment in the Late Tertiary about 6 million years ago can safely be excluded because of the present depth of the Strait of Gibraltar.

M. *Possible rise of sea level in a warm scenario*

The question of a possible rise of the sea level by 5-7 m, as during the last interglacial, which might be caused, for example, by a collapse of the West Antarctic Ice Sheet, is extremely important, but it cannot be answered as yet. The present suppositions are largely speculative. No conclusive investigations of the mechanism, place and time-scale of such a scenario exist. At present there are no signs of any abrupt change in the ice shelves around Antarctica, either now or developing in the immediate future. Any such development would be detectable with Satellite observations.

N. *A cool scenario as might arise from increased volcanic activity*

A scenario for cooling, due to unpredictable natural effects such as a clustering of heavy volcanic eruptions, similar to those of the period 1810-1840, could lead rather rapidly to an expansion of the Arctic sea ice and to increased frequency of meridional circulation types with all kinds of extreme weather. The example of the Little Ice Age, a complex period of frequent cold years and other climatic extremes, mainly between 1550 and 1850, demonstrates the high variability, with bad hervests in central and western Europe and effective cooling in northern Europe and in the mountain regions. Under such conditions, increasing rainfall in the Mediterranean region could be expected, but varying both from year to year and from area to area. While such an evolution might be locally beneficial, the consequent deterioration of agricultural productivity in northern Europe and in upland areas would be serious. The probability of such an evolution seems, on the whole, to be smaller than the probability of the first stages of our warm scenario.

A particularly severe cooling of this kind, if it lasted some decades, could perhaps initiate the nucleus of a new glaciation, probably in some areas of Arctic Canada. Any suggestion of transition toward a full glaciation during the next 100 years seems quite unlikely. However, as indicated by the astronomical theory, the probability of such a transition should increase after a few thousand years. It is indeed probable that a cold climate will peak around 5000 years from now, but research into the details of such an evolution is still needed.

O. *Implications of climatic change for European agriculture*

Clearly it is in the northern half of Europe, where the thermal growing season is often critically short, that agriculture would be most radically affected by changes in the temperature regime. In particular, any sustained trend towards late springs and cold autumns would cause major problems. Even further south, in the maritime areas of western Europe where the annual range of temperature is less, quite small temperature changes in the critical seasons would significantly affect the length of the growing seasons. A cooling trend here could have grave consequences, particularly for upland farming. Higher temperatures would, on balance, be favourable outside the already warm south.

Historically, wet seasons have represented the biggest problem for farmers outside the Mediterranean region. Shortage of water however may be the recurrent headache of the future; it seems likely that European agriculture in general will become progressively more vulnerable to dry seasons, such as that of 1976.

Increased variability of climate between one growing season and the next, and the consequent fluctuations in yield, would create special problems in an economy seeking to rationalise agricultural production and marketing.

Modern agriculture is far from free of the risk of sharp

sudden loss due to plant disease or pest. Climatic change may critically alter the impact of particular pathogens.

P. *An approach to prognosis*

In the foregoing analysis of climatic change, and in approaching a prognosis of the climatic tendency to be expected in the future, one must distinguish between effects operating in very different time-scales:

(1) Short term climatic change attributable to natural causes.
This is apparently compounded of a number of processes operating on different time-scales. Between about 1950 and 1970 the net effect on northern latitudes has been in the direction of cooling, albeit with fluctuations from year to year; in the Tropics and the Southern Hemisphere a weak warming trend prevails. Not all the natural processes are thought to operate in the direction of cooling at the present time, but other things being equal, the net effect may continue to be in that direction for some decades to come.

(2) Long-term climatic change due to natural causes.
Our diagnostic of the longer-term tendency of the climate is more positive. On a time-scale of millennia instead of centuries a decline, fluctuating because of the shorter-term variations to a glacial or near-glacial climatic regime peaking some 5000 to 10000 years from now can be expected.

(3) Changes due to side-effects of Man's activity.
The main element is thought to be the effects of an increase of carbon dioxide in the atmosphere, added to by contributions from other man-made pollutants. These are expected to produce a net warming of the global climate, perhaps by as much as 2 to 4°C, giving the rise to a regime warmer than any in the last two or three million years, following the cool interlude (1) and preceding the frigid regime (2). Accompanied by large-scale shifts of climatic and hydrological regimes, this effect could dominate the situation over the next few centuries.

Consequently, there is no contradiction in saying that the greatest immediate climatic menace to the balance of human society is a considerable warming in the foreseeable future and that looking much further ahead, we are approaching a new glaciation.

The need for further research

A lot of scientific effort has been devoted in recent years to better understanding of the climatic system and to unravelling the past history of the world's climate, to seeking clues as to its future development and the consequences for mankind of any likely change. Although much progress has been made, there still remains a number of grey areas of uncertainty and controversy, on which it is desirable that future research should be concentrated. While this group is not entitled to formulate specific recommendations, some "grey areas" in climatic research may be outlined. Among the topics to which priority should be given are the following:

(a) As background to the study of the impact of man on climate, it is essential to understand the natural climatic variations. In addition to simple (energy-balance) and sophisticated (general ciculation, interactive atmosphere-ocean-ice) models, models with a modest degree of sophistication should be developed, such as generalized advective-radiative-convective schemes.

(b) There is a need of careful and detailed reconstructions in space and time of past climates and time-series analysis, based on meteorological observations and proxy data obtained from deep-sea, ice and lake-bed cores, carefully checked historical manuscripts and tree-ring data.

(c) Case studies of extreme climatic events and anomalies (e.g. 1815-17, 1916-18, 1920-22) are also helpful, especially when accompanied by an assessment of their agricultural and economic impact.

(d) Teleconnections between climatic anomalies over the European area and other regions, especially areas with oceanic upwelling, including possible time-lags call for further investigations, which should be accompanied by parallel studies of the agricultural and economic impacts of such anomalies and their interdependence.

(e) A detailed reconstruction of the land-sea, ice and vegetation distribution during the last (Eemian) interglacial with its sea-level 5-7 m above the present would be of great value.

(f) There is a similar need for a detailed study of the termination of the Eemian and the transition to the short, intense glacial period which followed. Here the relations between variations of sea-level, global ice volume and regional surface temperatures and their time-scale, as well as vegetation changes on land are of particular interest in view of the possibility of Antarctic ice-surges.

(g) Quantitative assessment of the climate of the early Pliocene 3.5 million years ago, especially in the Europe-Africa region, should be linked with the development of vegetation maps for this period. Atlantic deep-sea cores should form the basis of a reconstruction of the surface temperature distribution.

(h) The role of oceanic upwelling for evaporation and CO_2 exchange, including isotope studies (hydrogen, carbon), needs further quantitative study.

(i) As regards the fundamental question of how future climatic change is likely to affect agriculture, there is need for improved models of the influence of seasonal weather anomalies on the growth and yield of important food crops. Such models would help to identify the specific kind of climatic variations, in terms of temperature and precipitation over particular periods of the year to which a crop is most sensitive and use the inter-seasonal variation in a particular growing season to predict regional yields.

(j) A more immediate agroclimatic problem which is ripe for further research is how the economics of hill farming in the maritime countries of western Europe are already affected and may be further influenced in the future, by changes in the length of the growing season brought about by quite small temperature variations in spring and autumn, and by consequent changes in the accumulated warm-season which is vital for growth.

(k) Climatic variation could intensify the dangerous vulnerability of today's food crops, particularly under monoculture, to attack by disease and pests. A detailed study of the last historic outbreak of potato blight in Europe in the 1840s and its relation to weather conditions would provide valuable insights into this danger.

Another field for intensified research is the impact of climatic changes on marine fauna, particularly in polar waters, for example around Greenland, where the fishing industry was badly affected by the general cooling in the 1960.

Chapter 1

INTRODUCTION

Climate is the time-integral of weather over a period of years. The usual length of the "normal" or reference period, as standardised by the World Meteorological Organization, is 30 years. Climate comprehends not only monthly, seasonal and annual averages of temperature, precipitation, wind and other elements, but also variability and extremes of these parameters. While weather and climate occur in the atmosphere, their seasonal and interannual variations are largely controlled by other systems, e.g. oceans, soils and vegetation, snow and ice cover. Together, these subsystems form the climatic system (Bolin 1975) (Fig. 1.1), with all its spatial and time-scales, with a multitude of interactions including positive and negative feedbacks, i.e. intensifying and damping effects. These subsystems have characteristic time-scales varying between a few days in the weather-producing troposphere (below about 10 km) and more than 1000 years in the abyssal depth of the ocean or even some tens or hundreds of thousands of years in continental ice-sheets (e.g. Antarctica).

Climatic variability embraces shorter time-scales: from one month to 10 years. No generally accepted definition of climate change -as distinguishable from the shorter climatic fluctuations consisting of the anomalies of a given month, season or year- exists, and it is hardly useful to discuss here semantic differences at length. Longer time-scales of climatic change, as verified between 10000 and 100000 years (see Section 5.4.2) may seem to be of no more than academic interest when considering the climate of the next few centuries, but increasing evidence seems to show that the transition between quite different climatic "modes" on this time scale may be quite short, of the order of centuries or even less.

The physical background of climatic fluctuations and change consists, in the first place, of internal interactions between the above-mentioned sub-systems of the climatic system. The most effective processes are changes of the extension (and thickness) of drifting sea-ice in both polar oceans, and also changes of the water temperature of the oceanic upper mixing layer, with a characteristic depth of the order of 100 m, which may

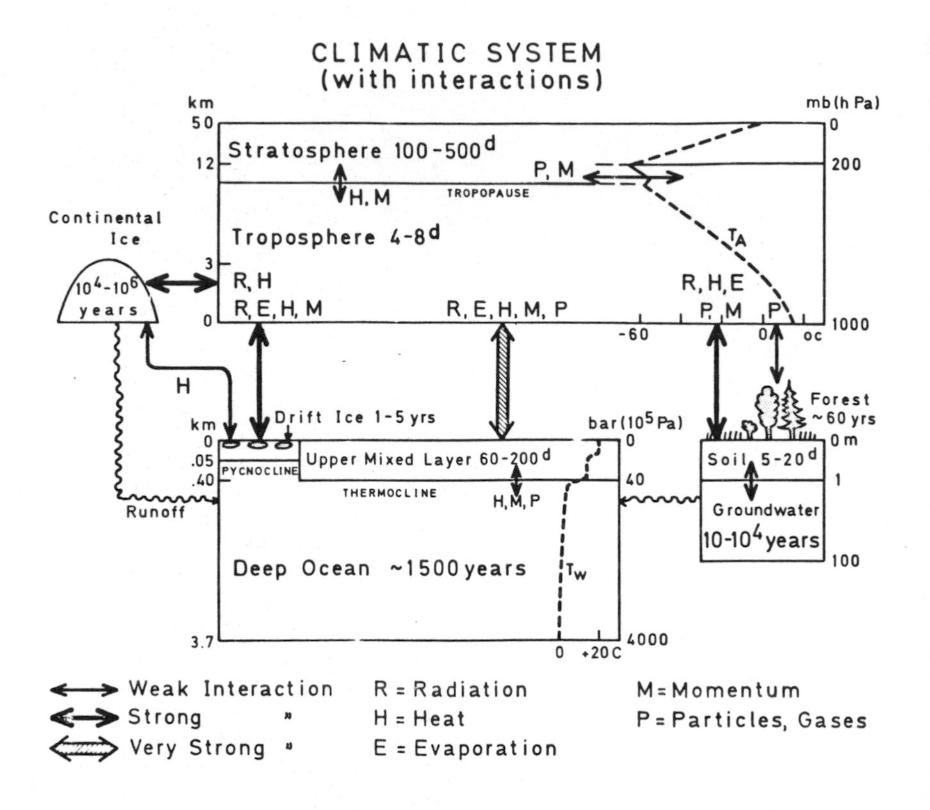

Fig. 1.1. Climatic system with its subsystems and interactions between them (double arrows). The characteristic timescale is defined either by decay of autocorrelation coefficients below the significance level (troposphere, upper mixed ocean layer) or the average residence time of characteristic substances, e.g. water (deep ocean, ice, soil and groundwater) or particles (stratosphere).

persist, with some horizontal displacement, for 6-12 months or even longer. The average life-time of an Arctic multi-year ice floe varies between 5 and 10 years, while 30 percent of the area of Arctic drift-ice (85 percent of the sub-Antarctic) are purely seasonal, with a life of no more than 8-10 months. Since snow (on the continents) and ice tend to cool the atmosphere further, a positive feedback effect enlarges global or hemispheric temperature changes in polar regions - there such changes are greater by a factor of about 3 than in lower latitudes. Drastic changes of water temperatures occur, at some coasts and along the Pacific and Atlantic equator, through vertical motion of the water caused by the stress of surface winds. In equatorial regions, this vertical component -nearly simultaneously over large distances- changes its sign irregularly on a time-scale of a few years; upwelling cool, nutrient-rich deep water is then replaced by warm water or vice-versa, with a large-scale feedback on the atmospheric circulation, evaporation and precipitation.

In addition to these internal effects within the climatic system, external changes contribute to the variability. Most significant in this context are heavy volcanic eruptions (Lamb 1970): their climatic consequences are described in Section 5.4.2.1 . The volcanogenic particles tend to concentrate in polar regions, absorb and scatter the incoming radiation of the sun and produce, in the global average, a general slight cooling of the atmosphere near the surface, but a warming in the stratosphere. Their average residence time in the stratosphere is of the order of 1-2 years. Frequency and intensity of volcanic eruptions vary greatly with time; while some lead to outbreaks of fluid lava with only minor amounts of gases, the climatically most effective types are the "explosive" eruptions which blow large masses of hot gases and debris to high altitudes (to 30 km and more). After pronounced maxima (Fig. 1.2.) around 1780, 1810-40 and 1880 a marked minimum of large eruptions happened between 1912 and 1948; some experts consider that this minimum was responsible for the period of global warming centred during this time.

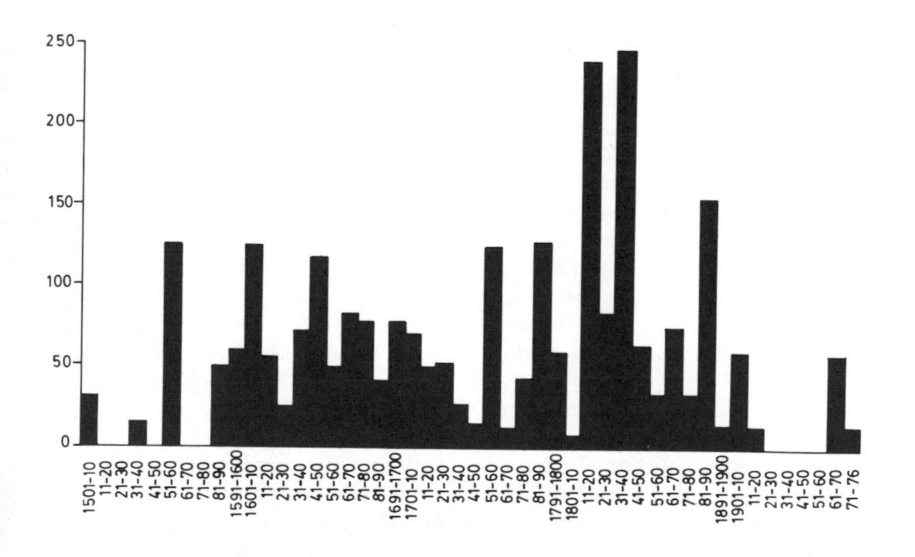

Fig. 1.2. Decadal averages of volcanic activity (dust veil index) in the northern hemisphere only (Lamb 1970, amended 1977).

Although some authors believe that variations in solar activity contributed to the cooling of the 'Little Ice Age' and to the 20th century warm period, the possible role of fluctuations of solar radiation in producing climatic changes is still controversial. The true variations of the "solar constant" due to sunspots, faculae and other features, in the visible and near-infrared part of the spectrum (0.3-3 μm) are quite small, in the order of 0.1 percent. Strong variations exist in the particle (proton) emission of the sun, as well as in the far ultraviolet or Roentgen rays; however, their energy contribution is very small (of the order of 10^{-5} of the total solar radiation), their possible role for weather and climate in the troposphere remains doubtful and shall not be discussed here.

Other external effects are produced by man's increasing activity; they can easily be perceived on a local scale, i.e. in cities and industrial centers, which are, by day and night, during summer and winter, warmer than the surrounding areas. This is partly due to direct heating ("waste heat") and to atmospheric pollutants, such as dust particles. In addition to combustion processes, agricultural operations trigger large amounts of mineral particles from the soil over very large areas. The climatic effect of aerosol particles shall not be overrated, because of their short residence time in the atmosphere (few days or some weeks); for details cf. section 4.2.

More important are the man-produced trace gases in the atmosphere, such as CO_2 and others (see section 4.2.), in so far as they absorb the infrared terrestrial rays; most of these gases remain for many years in the atmosphere. On an even longer time-scale are effects of land use since the begin of the neolithic era 6-8000 years ago, especially deforestation and overgrazing leading to an increasing degradation of tropical and subtropical soils.

Estimates from U.N. agencies give the annual loss of (mainly tropical) forests by deforestation at 110,000 km^2, the loss due to desertification as 60,000 km^2 per year: these figures are truly alarming, as one of the most serious consequences of the increase of world's population. Their climatic effect appears in increasing albedo (reflectivity of the Earth's surface), depletion of soil moisture etc.; it cannot be expressed in simple terms. On the other hand, at least 400,000 km^2 are now occupied by reservoirs, and more than two million km^2 are under irrigation; both features affect strongly the continental water budget (see section 4.2.).

Up to now, natural causes appear in general to be more important in their climatic effects than the man-made processes, most of which in their majority (except probably for the slow cooling produced by the increase of the surface albedo due to destruction of vegetation) lead to a slight warming of the lower atmosphere near the surface (section 4.2.). Therefore, the recent cooling of the northern hemisphere (Chapter 2) -which may be partly due to the recent increase of moderate volcanic activity -was a strong argument against any current predominance of man-made effects. Since about 1972, this cooling trend had ceased, but it is not possible to take the most recent warming as a signal for reversal. However if present rate of increase of the atmospheric CO_2 -content and other trace gases is maintained or further increased, then man-made effects will reach, and later exceed, the natural fluctuations of average global temperature, which are of the order of $\pm$ 0.6°C over the last 100 years; this may happen around the turn of the century, i.e. after about 20 years. After that date the consequences of a man-made global warming may step well outside the climatic experience obtained during the little more than 300 years of instrumental observations, and even beyond the climatic changes of the last 10000 years.

The present upsurge of interest in the climate question has been aroused by some dramatic events of the recent past: the Sahel drought between 1968 and 1973 recurring during 1976/78, the 1976 drought in Western Europe which occurred simultaneously with a moist-cool summer in the European part of the U.S.S.R., the series of seven unusually mild winters 1971/2-1977/8 immediately followed by the severe winter of 1978/9 in Western Europe in contrast to the sequence of three very severe winters 1976/7-1978/9 in the eastern and central U.S.A. are only a few examples. Indeed: the very weak, apparently negligible changes of 30-year climatic averages -of the order of 1°C or 10 percent of rainfall (see sections 3.1. and 3.2.)- are in fact mainly produced by the varying frequency and intensity of extreme anomalies which tend to cluster in adjacent seasons and years. It is their economic and social impact which is most deeply felt; if these anomalies follow each other at a short time-scale, their effect may amplify itself.

Furthermore: such extreme anomalies never occur in isolation, confined to one area only: due to the scale of atmospheric motions and of oceanic response they are accompanied, simultaneously or with short delay, by similar or opposite extremes in distant areas. Such teleconnections are quite frequent, but still not very well understood; they contribute substantially to the economic impact on a global scale. A remarkable example was the year 1972: droughts in the Sahel, in India and N.E. Brazil, unusual cold in Arctic Canada, drought in the grain-procucing areas of the Soviet-Union, but abundant rains (and floods) in Australia, cessation of equatorial upwelling ("El Niño") in the Pacific and at the coast of Ecuador-Peru, reducing fish catch there to about 10 percent, dramatic increase in the prices of cereal, soy-beans and other crops. All this happened simultaneously, not at random, but more or less regularly interconnected, as many other examples in the past have demonstrated. The examples given in Chapter 6, although primarily concerned with Europe, have often been accompanied by extreme climatic anomalies at other areas of the globe.

Experience shows that not infrequently social unrest and revolutions are triggered by such clusters of climatic anomalies: mass emigrations to the U.S.A. especially from western Germany after 1816/7, from Ireland after 1845, the sequence of European revolutions of 1848, even the French revolution of 1789 have all been preceded by such extreme climatic anomalies with marked economic consequences (See Chapter 6). A recent example was the deposition of the Emperor of Ethiopia after the Sahel drought. Certainly, the occurrence of climatic anomalies is rarely the only or even the predominant cause: the climatic role has normally been to trigger historical events which had ripened in the aftermath of a multitude of interacting economic and social effects. These examples may indicate that the innocent-looking apparently negligible numerical changes of climatic averages hide a much more important fact: the varying frequency and intensity of climatic anomalies, which are the more effective when they repeat in clusters.

In this way, the variability of climate is an economic factor of high importance, to which unfortunately due attention has not always been given in earlier investigations into climatic change and, more important, into economic variability. The variability of long records of precipitation or temperature can easily be checked as a function of time: for examples, see sections 3.1. and 3.2. such studies had revealed -as a preliminary result- that the variability of climatic elements was substantially higher during parts of the 19th century than during the first half of the 20th century. The impression of a recent increase of variability is not everywhere justified. A worldwide study of statistical parameters of climatic variability is urgently needed, together with the teleconnections between far distant climatological anomalies (e.g. Hastenrath and Heller 1977, 1978).

Climatologists who have studied the daily vagaries of weather as shown on hemispheric weather maps for several decades have pointed out, as a general impression, that in Europe great variability, i.e. high frequency and intensity of extreme weather, is usually correlated with a well-known circulation type characterized by quasi-stationary anticyclones in latitudes 50-65°, centered somewhere near the British Isles or Scandinavia, defined as "blocking anticyclones". On their eastern sides, cold air is continuously transported across Europe into subtropical latitudes, at higher levels penetrating even into the tropics, accompanied by heavy showers and great frequency of rainfall or snow. On their western sides, warm air flows into subpolar and even polar latitudes. In such cases, representative upper-air temperatures at 5 or 8 km may be higher in Greenland than in North Africa. Because of the slow motion of these blocking anticyclones and their tendency of recurrence, intense rainfall anomalies can last much longer than in the usual case of fast-moving disturbances in a westerly current. In their central area they cause long droughts (as in 1976). But extremely heavy or long-lasting precipitation occurs preferentially at both flanks of the anticyclones. The term "meridional circulation types" describes such patterns with strong meanders of the high-tropospheric flow, in contrast to "zonal circulation types" characterized by a dominant westerly flow accompanied by weak waves only (Fig. 1.3.).

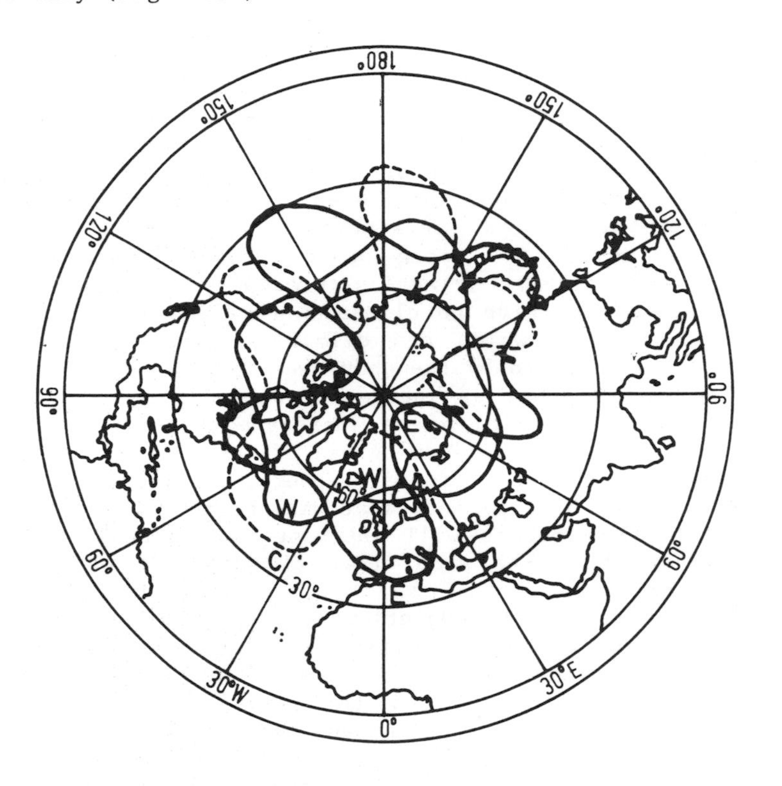

Fig. 1.3. Typical flow patterns at 500 mb level (height ca. 5.5 km) in the northern hemisphere (Wangenheim-Girs types; Lamb 1972, Fig. 7.6.) W (heavy line)= westerly type with more or less zonal flow, E (light line) and C (broken line) alternative meandering patterns with marked meridional flow. Note a blocking anticyclonic ridge above North Atlantic (C) and Eastern Europe (E), a deep trough over the western Mediterranean (E). At the 200 mb level (ca. 12 km) these troughs extend, during the cool seasons, far into the tropics.

In a rather broad sense we may distinguish between two main modes of the atmospheric circulation which characterize two types of climate with different variability:

(A) "Zonal" circulation mode: limited variability, rapid succession of travelling cyclones and anticyclones, of rainy and fine weather;

(B) "Meridional" circulation mode: high variability, episodes (lasting between one week and several months) of very cool fo very warm weather, accomponied by permanently wet or dry sequences, increasing tendency towards all sorts of weather extremes.

The varying predominance of these modes during the last 1000 years will be described in Chapter 2. The latter mode is responsible for many cases of strong impacts of climatic anomalies on agricultural production - even famine (Chapter 6) during the period of wide-spread subsistence agriculture in Europe. Now the same impact is (among other effects) represented in the varying price levels of agricultural products. These levels are still quite sensitive to climatic anomalies in the world's great bread-baskets, which are climatically not independent, but loosely interrelated by the above-mentioned teleconnections.

Taking into account an increasing demand for energy -especially on a global scale, where increasing population and industrialization leads to a growing use of fossil fuel- a further increase of CO_2 content of the atmosphere, of other trace gases and pollutants can be expected. What will be the climatic consequences? Man-made impacts will interact with natural effects, both being of about the same order of magnitude by the turn of the century.

How far are climatic evolutions of the future predictable? The present state of art of climate forecasting does not look promising, to say the least, and only recently systematic investigations have been started to improve our understanding of the basic internal interactions and external impacts. Particularly sobering is our inability to predict, with any degree of skill, such important natural effects as volcanic eruptions. This inability -which cannot be expected to be cured soon- prevents, in the foreseeable future, any rational climatic forecast on a scale of years or decades ahead. Attempts to search for reliable climatic cycles with a period between 1 and 100 years have been frequently made, but with rather limited success: most of these cycles are weak compared with the total variance, tend to disappear suddenly (sometime immediately after their "discovery") or to change their phase. A notable exception is perhaps a recently investigated 20-22 year cycle -perhaps identical with the "Hale cycle" of solar activity- of the occurrence of droughts in the U.S.A. west of the Mississippi (Mitchell et al. 1979) maintained since about 1700. In this case, however, a prediction of the specific area involved is not yet possible, which reduces its usefulness. A weak cycle of 2-2.5 years has also been found and is described in Section 3.1.

Unfortunately, our knowledge of the internal interactions within the climatic system is also insufficient. Recent evidence has demonstrated that in the ocean, cyclonic (cold) and anticyclonic (warm) eddies exist frequently, at a horizontal scale much smaller (100-200 km) but on a time-scale much longer (several months) than in the atmosphere. Due to the lack of an adequate observing network, their role in transporting heat, their interaction with the atmosphere and with the deeper oceanic layers is little known. In addition to the fast development of different types of atmospheric models, several interactive air-sea models -especially those including the polar ice- are now in progress. Looking at the structure and results obtained with models (Smagorinsky 1974; Shutts and Green 1979), one feels that there is no easy way towards models suitable for a conditional climatic prediction, assuming selected levels of man-made impacts. The interaction between solar and terrestrial radiation and clouds is not yet completely understood, and an adequate simulation of the interaction between the upper mixed layer of the ocean, of drift-ice and the atmosphere remains a difficult, but challenging task. Progress is certainly going on (see e.g. Manabe et al,. 1979, Washington et al. 1980), and most specialists agree that increased efforts will be finally successful. However, it would be unrealistic to expect reliable and rigidly verified results, at a regional and national scale, very soon.

In the meanwhile, substitute approaches have to be investigated. One way is to look into the historical evolution of our climate and to select some analogues not only for the expected man-made global warming, but also for possible cooling due to natural events. These analogues can be used as "scenarios", if one takes into account that the physical boundary conditions- e.g. vegetation and soil status, variations of sea-level and continental ice and, on a much longer time-scale, variations of the land-sea distribution and of the altitude of mountains- have changed and may change further. Such limitations can be partly overcome with the use of adequate models; even the best models are also subject to limitations due to many necessary simplifications in their fundamental design. Investigations of past climates as lessons for the future (see Chapters 2, 5 and 6) serve also another purpose: they are needed to provide the background against which predictive climate models must be tested before their results can be trusted.

Such a combination of advanced interactive models of the climatic system, responding to prescribed changes in the atmospheric constituents and of surface boundary conditions, and of investigations of selected stages of the climatic past as analogues, may be considered as a realistic intermediate step towards a conditional prediction of future climatic evolution.

Looking into the past evolution of our climate reveals

that many equilibrium states are possible, much more different from the present situation than might be expected. Historical, climatological and paleo-climatological data now confront us with quite surprising facts, opening a much deeper insight into the geophysical behaviour of the climatic system; Chapter 5 will deal, on a purely scientific basis, with some of the most challenging prospects of future evolution.

In addition to the references quoted in this chapter, a list is appended of selected monographs and published conference reports. Conference documents are normally more up-to-date, but tend to contain much discussion on controversial issues which have not yet been resolved. The rapid progress in many areas makes it difficult to compose a comprehensive monograph without the risk of its being partly out of date even before publication.

Chapter 2

CLIMATE IN THE LAST THOUSAND YEARS: NATURAL CLIMATIC FLUCTUATIONS AND CHANGE

2.1 Introduction

Climate -even under its natural development alone- varies continually. Each year, each decade, each century, each millennium, since long before any question of impact of human activity, has produced a somewhat different record. It is important to gauge the magnitudes and time-scales of these variations, since planning should not be based on expectations of return to some non-existent norm. And the magnitude and extent of any changes attributable to Man's activities -or even whether any such effects are occurring on more than a local scale- cannot be determined without knowing the range, and the likely timing, of changes due to natural causes. Data on past climate are also needed to test and calibrate mathematical models of climate, and are increasingly used as scenarios to describe the probable main features of any future climatic regimes that might be brought into being by a similar change of the overall mean temperature level to certain levels that occurred in the past (see Section 6.1.).

Extension of the record to earlier times by systematic use of the numerous historical documentary reports of weather available in Europe as well as various forms of fossil, or "proxy", data, indicates that the last thousand years saw a particularly great swing of the prevailing temperature level of more or less global extent. And this was accompanied by other changes, e.g. of the wind and rainfall regimes, that are largely determined by the global temperature level and its distribution. There was a long period of more or less sustained higher temperatures in this part of the world between A.D. 900 and 1300 or later, followed by a period of regression and glacier growth until between about 1550 and 1850, when there was probably the greatest extent of mountain glaciers and sea ice in both hemispheres since the last major ice age. This cold period in recent centuries has therefore come to be known as the Little Ice Age.

It was followed by a marked recovery of warmth and recession of the ice, particularly in the present century, between about 1920 and 1960.

There are some points of interest about the geographical extent of the warm and cold climate regimes which respectively affected Europe in the early and middle to late centuries of the last millennium. The medieval warm epoch seems to have extended over nearly all of North America, the North Atlantic and Europe in the twelfth century. At that time information from the highest northern latitudes seems to indicate that the Arctic generally was still enjoying a relatively mild regime, which had reached its peak (or peaks) some centuries earlier, and which did not continue into the thirteenth century as the warmth did in much of Europe.

Evidence from the southern hemisphere at that time is sparser, but there are indications of warmth in New Zealand roughly contemporaneaus with the warmth in Europe; and it has been supposed that the wave of Polynesian (and Maori) sea-migrations between about A.D. 600 and 1300 coincided with a time of warmth and slightly higher sea level (attributable to some melting of glaciers and ice-sheets) giving more water-clearance over the coral bars at the entrances to their island lagoons. There are indications of various sorts that around A.D. 600 a milder regime extended as far south as the coast of Antarctica (Ross Sea), but this coincides rather with a cold phase in Europe.

It is clear from this survey that there is something less than complete phase agreement between the warmest times in different parts of the world (see also Section 5.2).

There is more evidence from the last millennium. From both oxygen isotope work and forest studies, it seems clear that New Zealand underwent a sequence of temperature changes broadly paralleling that in Europe, though with some differences of the order of half a century in timing. From the variations of the glaciers in Chile and elsewhere, including those near the equator in Africa and New Guinea, it seems clear that the colder regime of the Little Ice Age centuries was virtually world-wide in extent. Nevertheless, the sub-Antarctic ocean zone was probably on the whole less cold than today: at least in the late eighteenth century it is known from Cook's voyages that open water extended somewhat farther south than in the present century. This may be attributed to the known fact that the storm belt was displaced somewhat farther south, and this probably broke up more of the Antarctic sea ice. Also, in the 1820s the Weddell Sea was clear of ice, a condition not known to have recurred until the 1970s. From studies (supported by radiocarbon dates) of penguin rookeries on the coast of Antarctica and of the evidence of ice changes in the Ross Sea, there is evidence of an antiphase relationschip between the variations from century to century in the Far South compared with most of the rest of the world. The coldest phases there seem to have been around

A.D. 1450-1650 and 1834-1900 or after, with milder conditions between 1250 and 1450 and between 1650 and 1834 coinciding more or less with the coldest conditions in Europe and the sub-Arctic and the greatest southward extensions of the northern polar sea-ice.

From a variety of evidence, it seems that the colder climate, particularly in the seventeenth and eighteenth centuries, was subject to a wide range of year-to-year variations (and variations between one group of a few years and the next) than the present century, although a tendency to renewed increase of such variability has been noticed in the last two or three decades (see details in Chapter 3). There is a clear need for research aimed at recognizing early symptoms of any longerlasting climatic change such as a return to conditions more like those of recent centuries (including periods of greater variability), or alternatively a shift towards a warmer regime.

The time-scales of the natural changes and fluctuations of climate range over a very wide spectrum, from periods of 10^8 years and more, associated with the life-history of the Earth and the Sun, down to the gusts of wind from second to second. The short-term fluctuations may be treated statistically as random occurrences, but no approaches to forecasting are likely to be satisfactory which do not proceed from recognition of the physical processes involved in the variations considered. A long period of research, such as the 20-year programme proposed in the U.S.A. and as newly urged by the World Meteorological Organization, is likely to be needed before we reach this position. The variations with which we are most concerned for planning purposes must be those on time-scales from a single season up to 100 to 200 years, but it is worth noting in connexion with the century-to-century prospect that a recent survey (Wigley et al. 1979) indicates that each of the last three millennia has produced a period of colder climate and advancing glaciers in its middle centuries in this part of the world (though the hemispheric patterns in each case differed, at least in the Arctic and in the tropics) after a warmer period around the beginning of the millennium. The longer-term variations also are therefore of more than academic interest. For further references to the results presented in this chapter see Lamb (1977), especially Chapter 17.

2.2 Data

The evidence from which the past record of climate can be reconstructed is of the following types:

(i) Regular meteorological observations, including instrument measurements of temperature, rainfall, etc.

(ii) Regular weather observations without instruments (but often including wind direction and a description of its strength).

(iii) Historical documentary reports of the character of individual seasons and some specific events, such as a great flood or a great frost (often with dates).

(iv) Fossil data of various forms, often now called "proxy" climate data, which registers the effects of weather and climate.

The longest series of meteorological instrument records which have been carefully scrutinized, and can be treated as homogeneous within reasonable limits of error, are for places in Europe. The series of temperature measurements which start earliest are the monthly means for central England beginning in 1659, Berlin (research not yet completed) from 1697, and Zwanenburg/De Bilt in the Netherlands from 1735 with some overlapping earlier Netherlands records starting with Hoofddorp from 1706. Observation series in northern Europe begin with Uppsala in Sweden from 1739, St. Petersburg (now Leningrad) from 1726. There is a record, using primitive instruments and units, from Florence (Firenze) from 1654 to 1670, which it may now for the first time be possible to standardize against later records: otherwise the earliest Italian record is the rainfall at Padova from 1725. The longest rainfall series are the monthly values, a composite series representing London (Kew) going back to 1697, area averages for the East Midlands of England from 1726 and for "England and Wales" from 1727, and a Netherlands series starting with observations at Delft in 1715. Histories of the early observations, and the problems of interpreting old units and instrument exposures and rendering the series homogenous, have been given by Manley (1953), Lamb and Johnson (1966), von Rudloff (1967) and Lamb (1977).

Mountain station records begin with a fragment from the St. Gotthard Pass from 1781 and a continuous series from the St. Bernard Pass starting in 1817. The longest record from a high mountain top is that from Sonnblick (3106 m) in Austria from 1886. There was an observatory on Ben Nevis (1343 m) in Scotland which provides a 20-year record from 1883 to 1904.

Some early sea surface temperature observations are given by Lamb (1977), the earliest analysed data being maps of the averages for January and July for 1780 to 1820 in the Atlantic Ocean and the results of an expedition voyage between Scotland, Iceland and Norway in 1789.

Series of regular weather observations without instruments include numerous ships logs in the archives of various nations. Some of the ships -those which are easiest to use for climatological studies- maintaining positions either in a port or patrolling a short stretch of coast, the earliest of which dates between about 1630 and 1680. Other series are found in diaries, some of them being specifically weather diaries or registers. Among the most notable are Merle's diary in eastern England 1337 to 1344, Haller's observations in Zürich from 1546 to 1576,

Tycho Brahe's observations on an island in The Sound (Øresund) between Denmark and Sweden 1582 to 1597, and those of the Landgraf Hermann IV of Hessen from 1621 to 1650; also daily records near Zürich, studied by Pfister (1978), from 1683 onwards. A recent study (Douglas et al. 1978, 1979) has produced synoptic

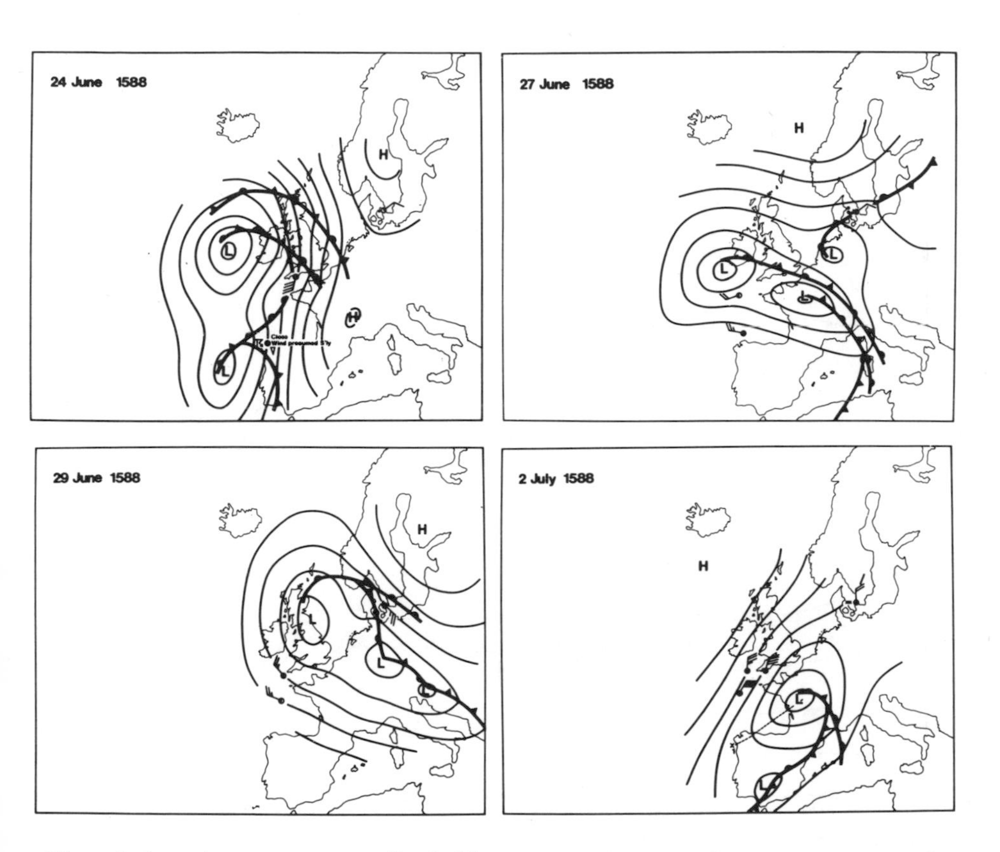

Fig. 2.1. A sequence of daily synoptic weather maps around midsummer 1588, based on reports from the ships of the Spanish Armada and a few other weather observations.

This is part of the earliest series of weather maps for individual days so far available. It is interesting as a sample of a Little Ice Age summer when the glaciers in Europe and the Arctic sea ice were advancing substantially.

The meteorological analysis of part of the series was tested by adding the observations of Tycho Brahe in Denmark afterwards. Great care has been taken with the logical continuity of the analysis and its ability to explain the weather observed also on the intervening days (not mapped).

weather maps for over sixty individual days in the summer of 1588 from the weather reported by the ships of the Spanish Armada. Some examples illustrating a sequence in late June - early July 1588 are seen in Fig. 2.1. The 1588 summer so analysed has considerable interest as a sample of a Little Ice Age summer, during a phase when the European glaciers and the Arctic sea ice were advancing seriously.

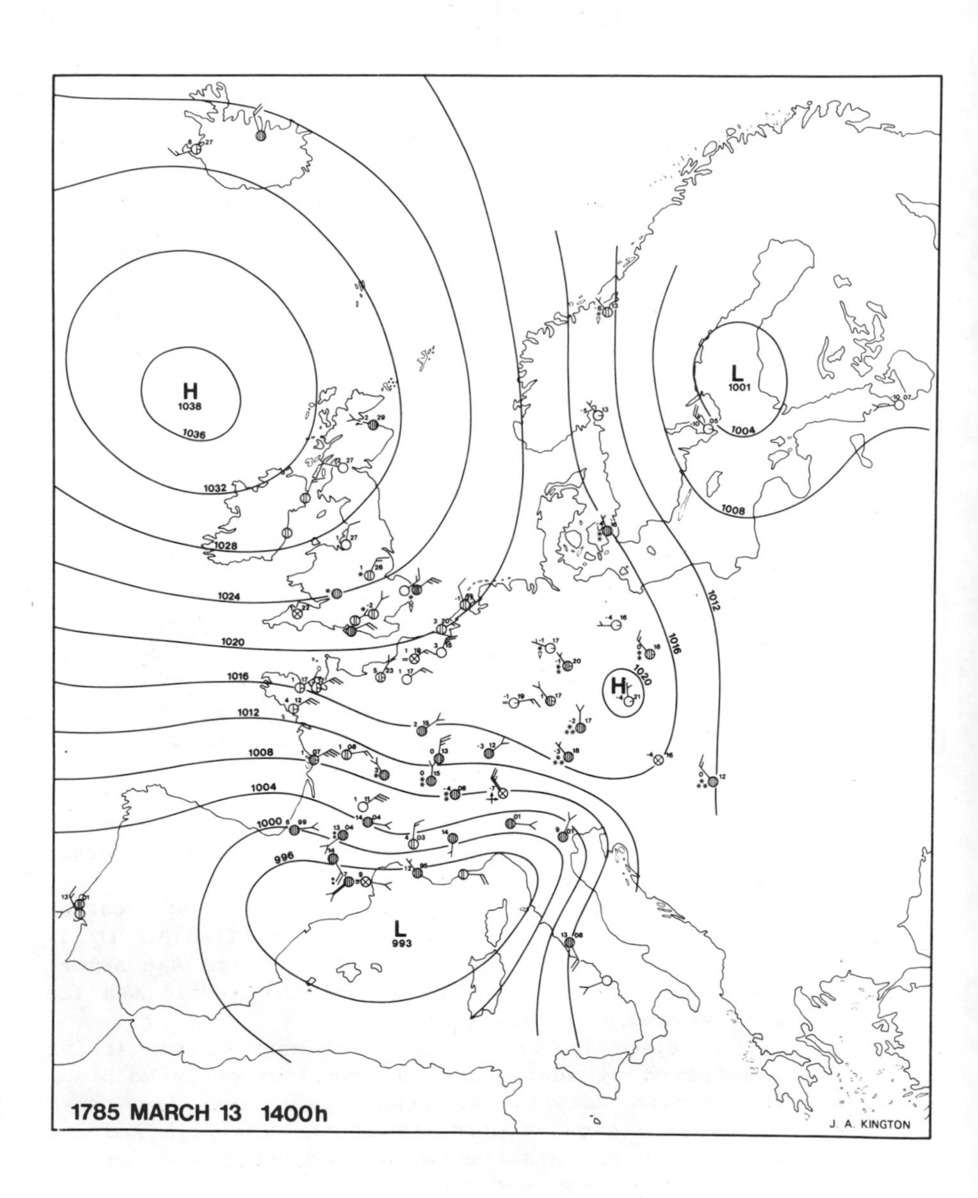

Fig. 2.2 Daily synoptic weather map compiled for March 13, 1785, the coldest March in the thermometer records in many parts of Europe.

Data are entered on the map at each observation point, using the modern meteorological symbols and plotting conventions. Wind directions are shown by the arrows, and the reported wind strength is (roughly) indicated by the number of flecks on the tail of the arrow.

Temperatures in °C are indicated by the figures to the (upper) left-hand side of the circles indicating the observing point's position on the map. Sky-cover is indicated by the amount of shading inside the circle.

The main weather symbols present are: Rain ●
Snow *, Hail shower $\overset{\Delta}{\nabla}$, shower ∇, Thunderstorm ⩘
Fog ≡, Heavy thunderstorm ⩘, Drizzle ❟

The hour of observation was in the afternoon, about 14h. The lines are isobars indicating the barometric pressure, reduced to sea level.

The analysis of the maps is due to J.A. Kington of the Climatic Research Unit, University of East Anglia, Norwich.

The earliest continuous series of daily weather maps so far in existence is that for Europe and the northeast Atlantic as far as Iceland over the years 1781 to 1786, prepared by J.A. Kington in the Climatic Research Unit at the University of East Anglia, Norwich. A sample map from March 1785, the coldest March in the record, is illustrated in Fig. 2.2, showing the network of observation points available.

The available historical documentary reports in Europe are of many types. Those indicating the prevailing character of a single season are most useful. They occur in a miscellany of state, local, monastic and manorial papers, estate and farm management account books, and official chronicles, in the archives of great trading companies and corporations, and incidental to the reporting of battles, famines and other great events. There are also in some places inscriptions in stone recording the height of a great flood, etc. Together, they allow the weather of individual summers, winters, etc. to be mapped and series of index values expressing the relative frequency (over periods of a decade or more) of wet and dry summer months and of mild and severe winter months to be derived (Lamb 1965, 1977; Ingram et al. 1978; Alexandre 1978). A sample map illustrating the winter of 1431-32 is seen in Fig. 2.3.

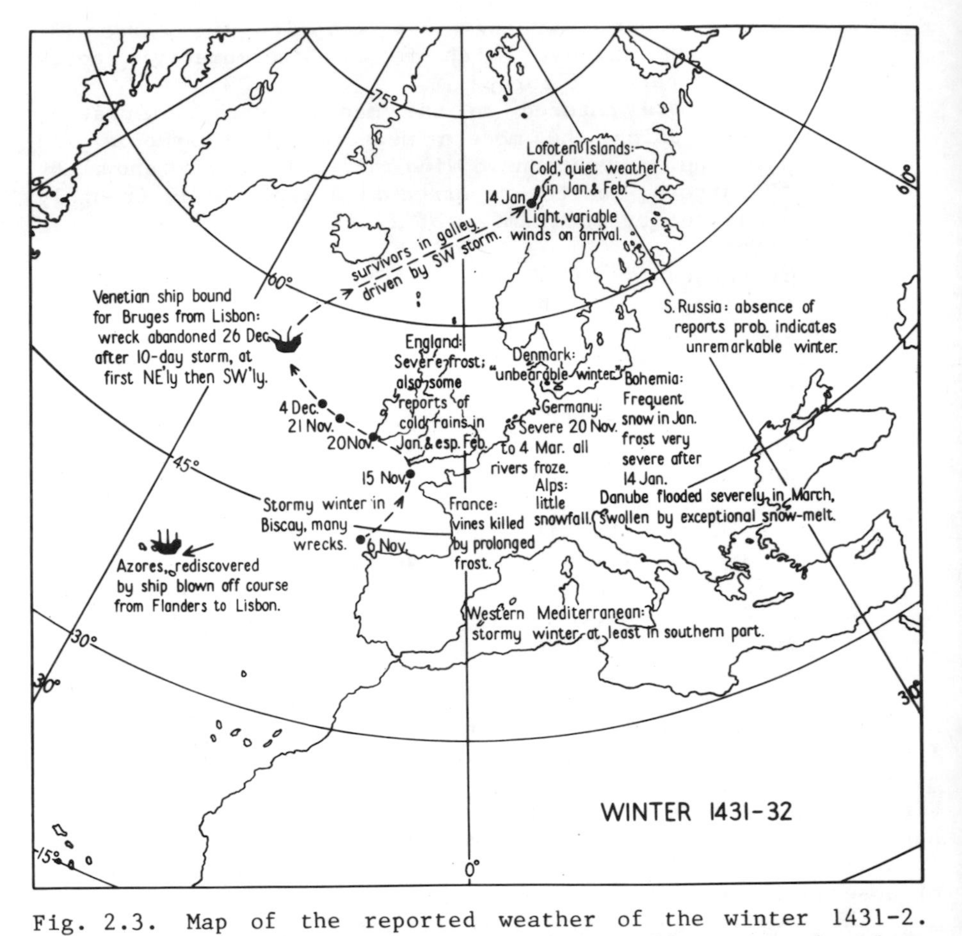

Fig. 2.3. Map of the reported weather of the winter 1431-2. The experiences of ships blown off course make clear the persistence of (mostly, or frequently, strong) easterly winds over Europe and the eastern Atlantic between latitudes 40 and 55°N. This is a classic case of a winter dominated by anticyclones centred between northern Scotland and central Sweden.

Among the various forms of proxy climate data some would include the long records of vintage dates and quality, in some cases going back to the thirteenth or fourteenth century (Lahr 1950; Müller 1953; Ladurie 1967, 1971), and the series of prices of wheat in England since 1316 by Beveridge (1921,1922) and of wheat and rye in various European countries from dates ranging from 1200 to 1500 onwards (Libby 1974; Flohn 1978)(see Fig. 6.7). Under this heading, much use may also be made of strictly fossil data: the results of pollen analysis at various depths in peat bogs and lake sediments, the widths of the yearly growth-rings in trees and the variations of wood-density during the season and from year to year, variations of thickness, physical structure and the proportion of the heavy isotope of oxygen (O^{18}) in the dated ice-layers in the Greenland ice-sheet and, similarly, in the proportions of the stable isotopes of carbon and nitrogen in the wood of tree-rings, thicknesses of the year-layers or "varves" in some lake sediments, variations of the Arctic sea ice at the Iceland coast, dates of lake ice and ice on the Baltic, ice on the Dutch canals, glacier advances and retreats,and so on. Bibliographies for all these items and illustrations of many of them are given in Lamb (1977). The series so far available which seem likely to be most important for reconstructing the climate of Europe since the Middle Ages are the vintage records, the various glacier and ice records referred to, and studies of the longest obtainable series of tree-rings (see, for instance, Dansgaard et al. 1975; Huber and Giertz-Siebenlist 1969; Koch 1945; Schweingruber 1975, 1976: van den Dool et. al. 1978; cf. also Section 5.421 and Figs. 5.10-5.11). However, the relation between tree-rings and climate has not as yet been established under European climatic conditions. Of particular value are density measurements of early and late wood in annual rings, as made by Schweingruber et al. (1979). A comprehensive review of the situation of dendroclimatology with a quite promising outlook has recently been published (Hughes et al. 1982).

The possibilities of using many of these types of record have barely begun to be exploited. In most cases there is need for further research to improve understanding of the climatic events and their effects (the "message" registered in the data which is not infrequently ambiguous). But the manifold types of data which exist offer opportunities of corroboration of the climatic interpretation and ultimately a record that is independent of any oversights and distorted emphasis, prejudices, superstitions and subjective impressions, which may mar some of the descriptive reports of the weather by the people who experienced it. Thus, we are approaching an objective record of the climate over the period here surveyed (and one which cannot be alleged to have been invented to explain the events of human history).

2.3 The climatic record so far as known

To achieve a chronological presentation of the record of Europe's climate over approximately the past 1000 years, as far as at present known, we are obliged to start with the times for which the descriptive accounts which exist are sparsest and in some cases open to most doubts. Figure 2.4 shows the course of indices of the prevailing character of the summers and winters in the British Isles and central Europe, so far as reports exist, by half centuries from the beginning of our era to A.D. 1300 and the reported numbers of wet years in Italy.

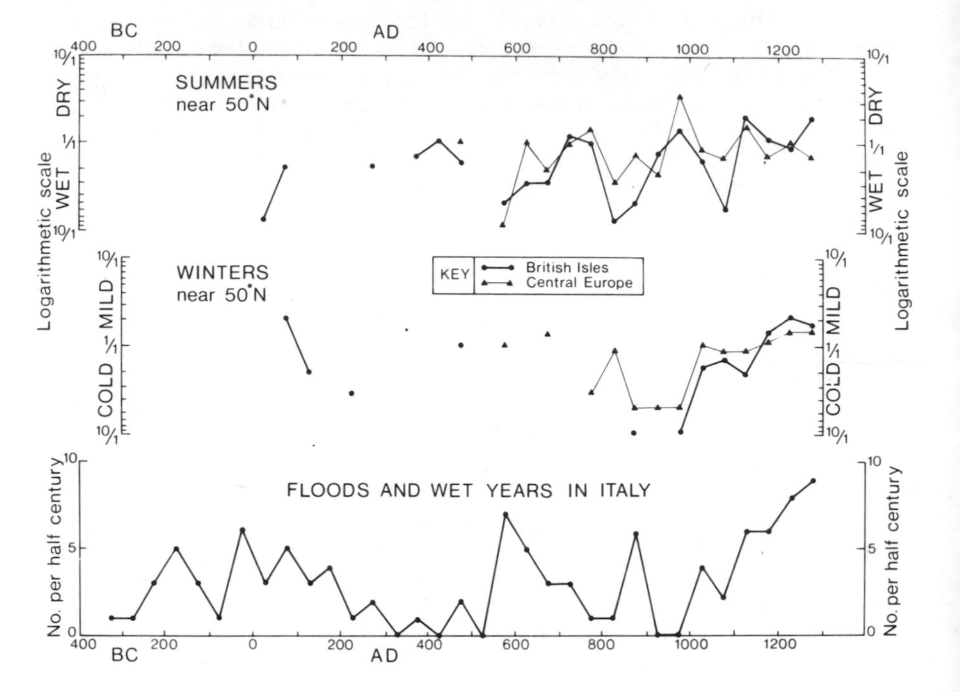

Fig. 2.4. Records of the summers and winters in Europe near latitude 50°N from A.D. 1-1300 and of wet years with flooding in Italy from 350 B.C.

The character of the summers and winters is indicated by ratios of the numbers of years with reports indicating dry/wet summers, mild/cold winters, by half-centuries.

The ratios are plotted on a logarithmic scale. Because of the correlation between summer dryness and warmth in Europe both indices can be read as indicating the prevailing warmth of the seasons.

Although the numbers of reports are too few to give satisfactory statistical reliability to the ratios of dry to wet summers and mild to cold winters, there is probably some ground for confidence in the extent to which the British Isles and central European indices follow parallel courses. There is no reasonable doubt about the trend towards greater warmth in the high Middle Ages after 900 to 1000 A.D. and the increasing wetness in the central Mediterranean about that time. There is good evidence, including substantial glacier advances, also for some wetter and colder times from around A.D. 400 or 415 to the 600s and again around 800 in northwest and central Europe (certainly for some notable cold winters in the 700s and from the later 800s onwards). In central Asia, there is evidence of great drought around A.D. 300 and 800.

2.3.1. *The medieval warm epoch*

The oxygen isotope record from the ice-sheet in the far north-west of Greenland (77°N 56°W), derived by Dansgaard and his co-workers (Fig. 6.42), indicates a somewhat different timing of events there. There seems to have been an early peak of warmth in that area already by A.D. 600-650, after colder conditions in the previous century, and fairly well sustained warmth after A.D. 800 building up to a climax around A.D. 1100. (It should be borne in mind that anticyclones covering northern Greenland, and giving high temperatures there in summer, are at times accompanied by northerly winds and cold weather over Europe. There was a sharp decline in northern Greenland by A.D. 1200, and it is known that already in the 1190s the sea ice was increasing in the East Greenland Current as far south as Iceland. Reports of the early settlers in Iceland regarding the vegetation and the range of their settlements into the interior of the island, the crops they grew, and so on, as well as the lack of sea ice, are taken to indicate a somewhat warmer climate there already from the late 800s, and this seems to have persisted in Iceland until perhaps as late as 1200 to 1250, although there are hints that the storminess of the seas may have been increasing in the thirteenth century, making sea communications more difficult. By about 1340, the sea route from Iceland to the Viking colony in southwest Greenland had been changed to a more southerly track than before because of the ice which made it no longer possible to go so near to the south-east coast of Greenland. And, after 1410, communication with the Greenland colony was lost altogether.

In western, central and northern Europe the climax of the medieval warm epoch seems to have been between about A.D. 1150 and 1300. In eastern Europe, as in north Greenland, the climax seems to have come earlier -in this case about a century earlier; in the 1200s many of the summers there, and apparently in an area extending west to the Swiss Alps, were already wet and the winters were tending to be colder, although neither season seems to have produced as bad a record as in the period between 1550 and 1800. Some at least of the Alpine glaciers advanced during the thirteenth century, and were apparently quiescent or retreating somewhat for a time after that until they produced general advances in the late 1500s and around 1600.

From 1300 onwards, in the course of the changes which ushered in the Little Ice Age, the climate became notably erratic and underwent a number of sharp variations from decade to decade as well as from year to year. The variability and the occurrence of extreme seasons of various kinds must have borne very heavily on the primitive agricultural economies then existing, and it is clear that there were severe effects on the health of people and animals and in diseases affecting the crops. The most fearsome episodes of the times in these regards were the visitations of the great plague, the Black Death, as well as the occurrences of sheer starvation and of ergotism, a disease (claviceps purpurea) also known as St. Anthony's Fire, contracted from bread made with flour affected by the ergot blight of blackened grain in damp harvests. (The symptoms of ergotism are among the most horrifying produced by any disease: hallucinations and convulsions, feelings of icy chill followed by a burning sensation in the limbs which quickly blackened and became gangrenous, shrivelled and dropped off.) Many "murrains" or plagues among cattle were also recorded in the fourteenth and fifteenth centuries in Europe and caused great mortality among the animals. However, the relationship between epidemics and climatic fluctuations are not yet sufficiently understood (see Appleby, in Rotberg and Rapp 1981).

The warm climate of the twelth and thirteenth centuries in Europe can, to some extent, be traced back to the tenth century when there seem to have been great droughts, which have also left their mark in the long tree-ring series of oaks in central and western Germany (see the tabulated results of work by Huber and Giertz-Siebenliest (1969), and by Hollstein, in Lamb 1977, pp. 595-602). At that time, also, farm settlements were spreading up to 200 m higher than before on the hill country in Norway, and doubtless elsewhere in northern Europe.

Cultivation of the vine was also spreading farther north in many parts of Europe and ultimately spread to greater heights than today in Baden (southwest Germany) and even in England, where one reported medieval vineyard site in Herefordshire was 200 m above sea level. In the thirteenth century, and probably right from A.D. 1000, wheat was grown as far north as Trøndelag (63 to 64°N) in Norway. As early as 880 it is reported that "corn" (probably barley) was grown by Ottar, in the northern province of Troms as far north as Malangen (69 1/2°N 18 1/2°E), where no grain was grown between about 1550 and 1700 (Solvang 1942).

Cultivation spread by the 1280s to heights up to 320 m above sea level in northern England, near the Scottish border, and actually to 425 m on the Lammermuir Hills in southeast Scotland (Parry 1978). The same century has been described as the "golden age" in the history of Scotland, before the Highland troubles began, and it seems fair to deduce that farming was successfully carried on farther up the glens than in later times, when the

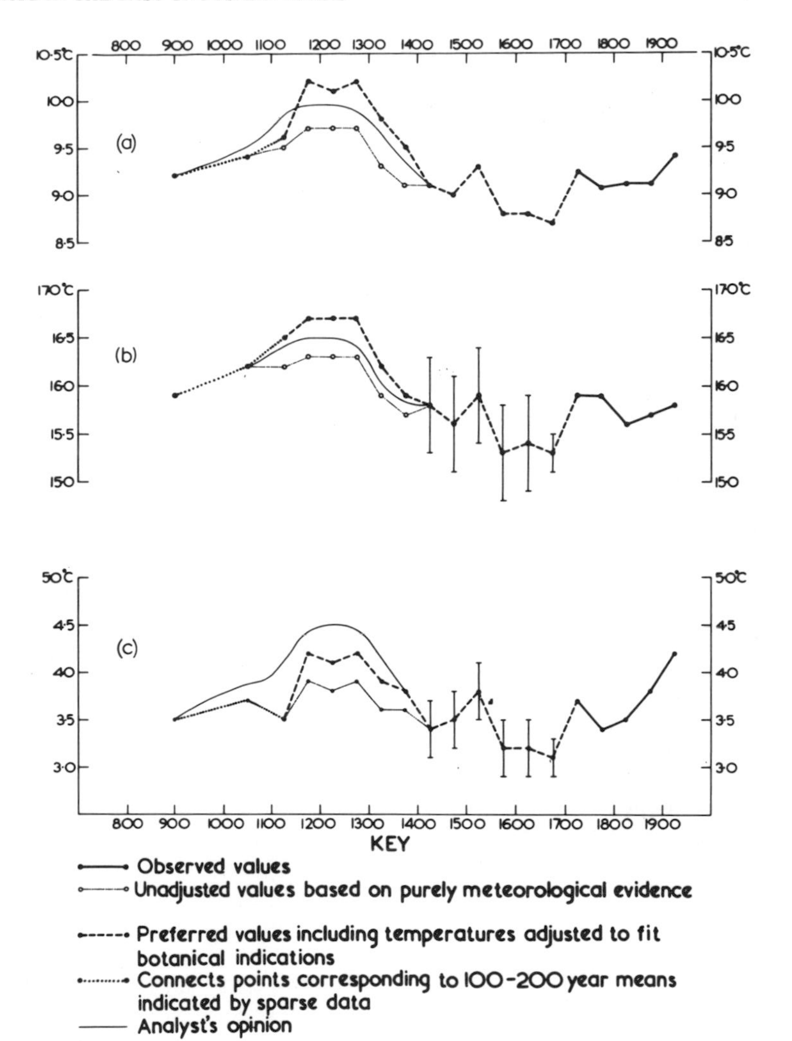

Fig. 2.5. Mean temperatures apparently prevailing in central England: successive period averages (each half-century since A.D. 1100 up to 1900-1949): (a) averages for the whole year; (b) averages for high summer (July and August); (c) averages for winter (December, January and February).

inhabitants of the remote valleys among the mountains took to raiding cattle from the lowlands.
The most accepted assessment of the overall average temperatures prevailing in central England between about 1150 and 1300, based on statistical assessment of the frequency of seasons of various unmistakable characters and on the limits of agriculture, etc., puts the level at 0.5 to 0.8°C above that of 1900 to 1950, and 1.2 to 1.4°C above that prevailing between 1550 and 1700.
Figure 2.5 displays the probable course of the prevailing temperatures in central England by a set of successive period-mean values (50-year means for the successive half-centuries from A.D. 1100 up to 1949) statistically derived from the summer and winter index values described earlier (p. 008). The various estimates of medieval temperatures and those which ensued in later centuries can be seen in the figure. The course of history can be studied in terms of the decade by decade dominance of wet or fine summers and mild or cold winters across Europe from west to east near latitude 50°N, derived from the series of summer and winter index values for Britain, central Germany and European Russia. Not too much weight must be put upon the index values for individual decades in case of incompleteness of reporting, but the general tendencies over periods of 50 years or longer seem (from comparisons with various other data) to be reliable. It appears that the course of climatic history over these centuries has been quite similar in the British Isles and central Europe; and the same seems to be true for Scandinavia, but some differences of timing become important in eastern Europe.

2.3.2. *The unsettled and deteriorating climate in the Late Middle Ages*

In Europe, the first signs of serious climatic disturbance came with a number of great wind-storms and sea floods over the low-lying coasts, for instance around the North Sea, in the thirteenth century (Fig. 2.6). The reported drowning of 100,000 to 400,000 people in some of these incidents places them among the worst ever recorded weather disasters. It was later reported that between the floods of 1240 and 1362 over 60 parishes accounting for approximately half the agricultural output of the diocese of Schleswig (Slesvig) had been "swallowed by the salt sea", off what is now the North Sea coast of Denmark and Germany. Climatologically, it seems probable that the phenomenon was compounded of (i) a rather high (and perhaps still rising) world sea level after some centuries of a warm climate over much of the globe and melting glaciers (particularly in Greenland and the Antarctic), and (ii) generation of increased storminess when cooling of the Arctic set in and strengthened the thermal contrast between middle latitudes and the Greenland-Iceland region.

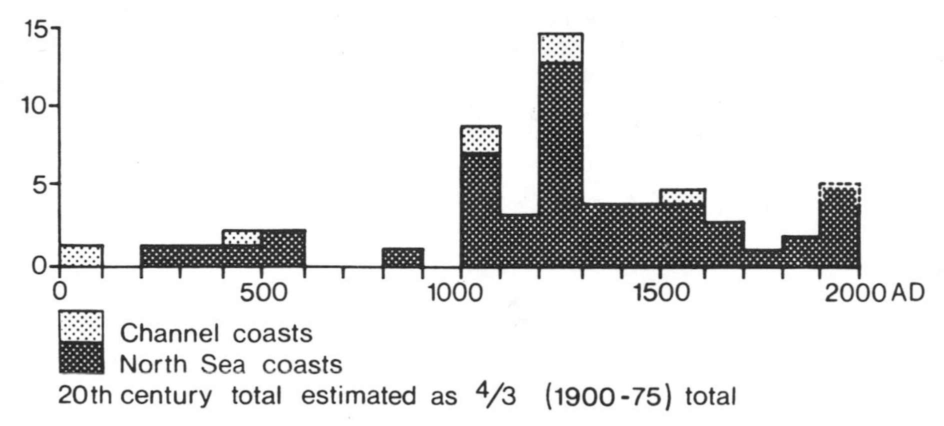

Fig. 2.6. Numbers of disastrous sea floods each century of which reports survive to our own times. (Restricted to cases which caused permanent losses of land to the sea and/or cost many human lives.) (For details see Lamb 1977, pp. 120-126).

After 1300, additionally, the tidal range was increasing and reached a maximum in the early 1400s: this doubtless increased the range of some of the storm tides (Lamb 1978).

The advance of agriculture and the northward and upward spread of vineyards on the hills of Europe seems to have ceased about 1300-1310. In England, the cessation was abrupt, coming as it did after a few decades in which the spread had been vigorous. And it is clear that, apart from the great storms at sea and floods of the low-lying coasts in the thirteenth century, the first bad climatic shocks came at this time. Particularly severe were the effects of the run of very wet (and presumably also colder) summers in the decade following 1310: the series was unbroken from 1313 to 1317, and in 1315 the harvest failures were more or less universal in Europe. Over the following three years, the population, which had doubtless expanded in the more genial times of the preceding two centuries, suffered one of the direst famines in the history of our continent. In some areas a substantial proportion of the population died, and there were outbreaks of cannibalism. In Britain and Germany, at least, it is known that some whole villages succumbed or were abandoned at this time, and in Denmark the beginnings of a similar movement are known to go back to 1330 or earlier. Detailed studies of some counties in the Midlands of England show that the villages which suffered the heaviest declines of population in these years were generally the ones which failed to survive the arrival of the plague 30 years later (see also p. 42).

Reference to the tentatively derived history of England based on the same summer and winter indices in Fig. 2.7 indicates that there was no persistent increase of rainfall at this time, although the autumns seem to have become wetter.

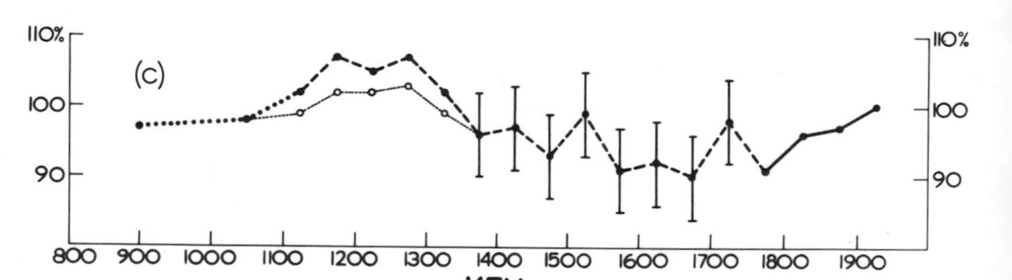

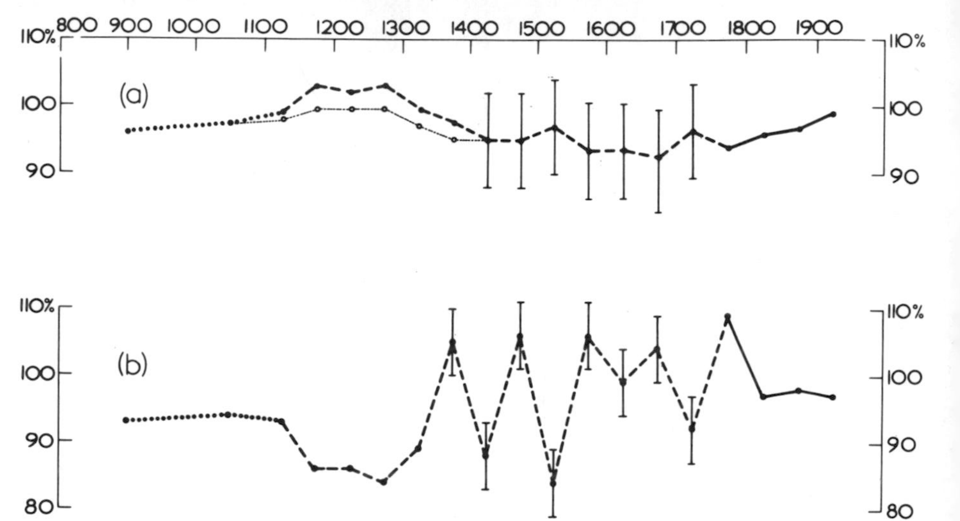

Fig. 2.7. Mean rainfall over England and Wales, expressed as percentages of the 1916-1950 average: successive

period averages (each half-century since A.D. 1100) up to 1900-1949: (a) averages for the whole year; (b) averages for high summer (July and August); (c) averages for the remainder of the year (September to June inclusive).

From 1750 onwards, the values (linked by the bold lines) are derived from carefully homogenized series of raingauge measurements (Nicholas and Glasspoole 1931). Before 1750 the values linked by the fine lines and broken lines rest on statistical derivation from the Lamb summer wetness and winter severity indices. Possible adjustments are indicated by other curves in the high Middle Ages. The error bars indicate $\pm$ 3 times the standard error of the statistical derivation.

It also seems certain that the ground became on the whole wetter because of the apparently sharp reduction of the temperatures prevailing and the consequent decrease of evaporation (see Section 3.2). Exceptions obviously occurred in the years with great summer droughts that were prominent in some decades later in the fourteenth and fifteenth centuries. (The latter seem to have been much more prominent in eastern Europe than in the west.) There appears to be confirmation of the general trend towards wetter ground in the rather widespread evidence of regrowth of the peat bogs investigated in western, central and northern Europe marked by a "recurrence surface" dated around A.D. 1200 or 1300, and in reports, notably from England, of high levels of the rivers in the fifteenth century.

By the beginning of the fifteenth century, wheat growing seems to have been rather generally abandoned in northern Europe. Even in Denmark, English visitors to a royal wedding in 1406 reported much uncultivated, sodden ground and remarked that wheat was grown nowhere -although this is more likely to have been true of for Jutland than for the islands farther east. In Norway, the change had been particularly catastrophic; the plague which arrived in 1349 spread even more rapidly and virulently than in many other parts of Europe, and it was reported that two-thirds of the population died (Holmsen 1961, p. 332): not only was corn-growing given up in the north of the country; the upland farms were generally deserted and not re-occupied for two hundred years. The tax returns from many districts in the 1430s had fallen to about one-quarter of what they had been around 1300 (Holmsen 1978).

The abandonment of villages and farms assumed the proportions of a characteristic phenomenon of the times over all northern, western and central Europe. It seems to have been most rife in the early 1300s and in the middle decades of the fifteenth century, between about 1420 and 1480; where precise dates are known, they are mostly the years and decades of severest weather, particularly those associated with the wettest summers or bitterest winters (though in some places the occasional runs of droughts in some decades of the fourteenth and fifteenth centuries may have proved insurmountable). Examples may be seen in the wave of abandonment of village sites in Britain and Germany between about 1315 and 1340 and again over the thirty to fifty years after about 1420 *. The 1430s, in particular, were a remarkable decade, in which most of the winters seem to have produced long spells of severe weather and a number of very dry, hot summers were interspersed with very wet spells in the preceding and following months and some whole summers (notably 1438) that were notably wet all through over an area stretching from England to Germany and the Alps. Most probably these various kinds of severe weather were associated with a highly abnormal frequency of presistent anticyclones over Scandinavia (see Fig. 2.3), while Iceland is believed to have enjoyed rather easier conditions than for a long time previously -presumably with southerly winds prevailing on the western side of the Scandinavian anticyclones.

Other symptoms of the stresses induced by the various kinds of prolonged severe weather in and around the 1430s may be seen in the fact that the Highlands of Scotland were in their worst turmoil at that time: savage battles between the clans had begun in the previous century, but it was between 1426 and 1433 that the economy of the ancient earldom of Mar in the central eastern Highlands finally broke down and was resigned to the king (Boece 1536; British Association, Aberdeen 1953). In the 1430s, after several of the severe seasons were reported to have affected Scotland, the Highland population was reduced to making bread from the bark of trees. The same is believed to have happened in Scandinavia, and it was circa 1435 that the marginal farm village at Hoset (63°24'N, 11°10'E, 350 m above sea level) in central Norway, east of Trondheim, was deserted after having been opened up in the early middle ages (Salvesen et al. 1977). The upset in the social and political life of Scotland culminated in the murder of the king (James I) while hunting near Perth in 1436, and it was then decided that none of the former royal residences near the Highlands -neither Perth, nor Scone, nor Stirling, nor even Dunfermline- could offer security for the king's person any more: so it was decided to move the capital to the fortress of Edinburgh castle (Henderson 1879).

* Of nearly 50 deserted village sites in Oxfordshire and 34 in Northamptonshire in the English Midlands, for which adequate population figures of tax-paying tenants at various dates have been published, only about 10% were attributable to the Black Death. All appear to have suffered serious decline in numbers in the years of disastrous summers and famine between 1313 and 1325.

Tree-ring series from England and from Lapland confirm the poverty of the growing seasons between about 1420 and 1460 or 1470 and in central Germany around 1440-70, although there was worse to follow in the late sixteenth and seventeenth centuries (Lamb 1977, pp. 458, 593, 601).

Parry (1978) finds that the retreat of cultivation from the heights on the hills of southern Scotland closely paralleled the stages in the change of prevailing temperatures indicated by Fig. 2.5.

Turmoil and depopulation of the countryside were widely reported in other parts of Europe also in the period between about 1420 and 1485, and in the severe winters in the 1430s wolves were unusually troublesome in various places, including England (see Lamb 1977, p. 459 and footnotes). The price of agricultural land in Denmark had already fallen by a half from its late thirteenth century (1259-99) level to the 1330s and '40s -before the arrival of the plague- and in the mid-fifteenth century vacant farms were commonplace: on the island of Falster the cultivated area is reckoned to have declined by 25% since the thirteenth century (Christensen 1938).

Old village sites continued to be abandoned in the sixteenth century and after, but the cases at this time in central and northern Europe (including Britain) were more usually part of the deliberate redesigning of the landscape by the new landholders after the Reformation. Exceptions which were attributable to the climate can be seen in the 1690s in Scotland and Norway (where Hoset was once again abandoned, this time for over 230 years).

We may envisage the commonest direct causes of abandonment of village sites and farmland, where this was climatically caused, as:

(i) Shortening of the average growing season, as the prevailing temperature level fell (see Fig. 5.3). This shortening probably amounted to about 3 weeks in England already by A.D. 1400, the season in the seventeenth century ultimately becoming about a month to 5 weeks shorter than it had been before 1300.

(ii) Increased wetness of the ground, spread of marshes and increased size of lakes and rivers. These tendencies seem to have reached their maximum in England in about the fiftheenth century, but farther north, in Scotland, and perhaps in northwest England and also in the west of Ireland, continued to develop further right up to the coldest period in the late seventeenth century.

There were, however, other sites abandoned -e.g. many in Norfolk in eastern England- to which these reasons can hardly apply and for which the causes often cannot be identified yet. Exposure to wind and snow may account for some of them. And perhaps some failed in the occasional drought years.

Some sites were lost to the advancing sea, as noted on p. 38. Others were destroyed in another way by the increased storminess and range of the tides: being overwhelmed by tremendous storms of blowing sand. Many settlements and much farmland were buried for ever by sand in this way in incidents on the coasts of northwest Europe between about A.D. 1300 and 1800. Examples are known on the coasts of France, southwest England, Wales, Scotland and the Hebrides, Denmark and the Netherlands. Sand-drift even became important at a few places inland in the Netherlands and East Anglia.

2.3.3 The climax of the Little Ice Age

Data from England, Norway and continental Europe indicate an easing of the climatic decline -or even some recovery- around

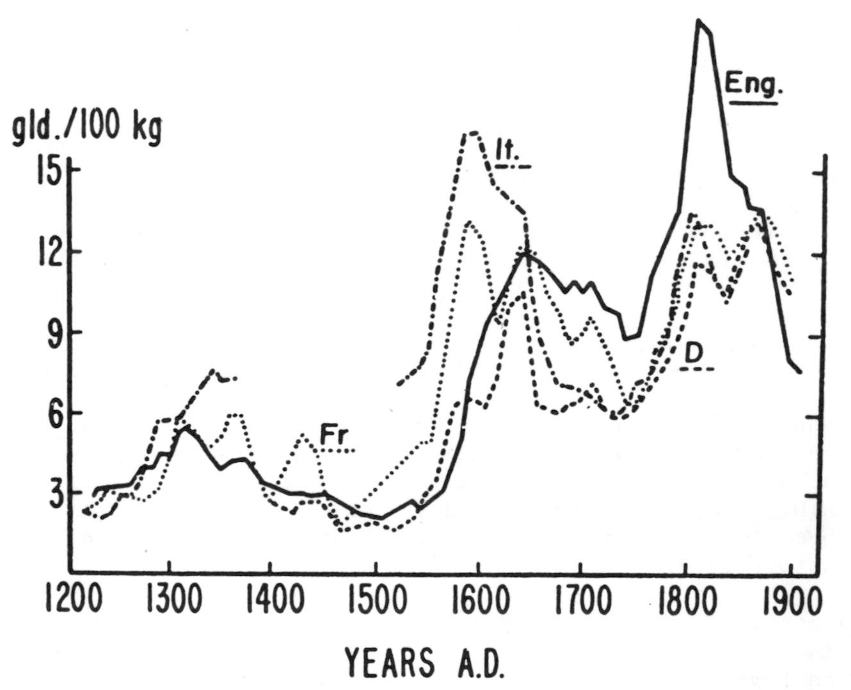

Fig. 2.8. Prices of wheat prevailing in England, France, the Netherlands (D) and northern Italy (25-year averages) from A.D. 1200-1900. Prices as given in Dutch Guilders per 100 kg wheat in De Landbouw in Brabants Westhoek in het midden van de achttiende eeuw, Wageningen, Netherlands (Veenman). (Figure kindly supplied by Professor L.M. Libby, University of California).

the early 1500s, as may be seen in Fig. 2.5., and in the records of wheat prices in various European countries (Fig. 2.8). In Norway, land values began to recover and abandoned farmland on the heights to be reoccupied.

The fisheries of the west and northwest coast of Norway had continued to yield well from 1300 to 1500 (with the warm Atlantic water current evidently continuing strong or even strengthening) and played an increasingly important part in the nourishment of the population in north Norway -as in Iceland during the long decline of agriculture. This situation seems to have continued with little change until about 1600, but soon after that there are signs of decline of the fisheries and a decline of the population of places such as the Lofoten Islands and Vesterålen; this became a steep decline in the 1660s and 1670s (Lindbekk 1978, e.g. p. 92). These were the places most dependent on the fishery in the Norwegian Sea, and this may be one of a number of symptoms of the spread of Arctic water during the later sixteenth and seventeenth centuries. At first, this watermass increased in the western part of the Greenland Sea, where it brought the ice south in large quantities to block the Denmark Strait between Iceland and Greenland in a number of years in the 1580s and after; the cold surface water seems to have reached the area of the Faeroe Islands (61°N) at times after 1615-1625 and persistently from 1675 to 1704; it apparently from 1675 to 1704; it apparently spread still farther south and east to near Shetland and the entire Norwegian coast in the 1690s, particularly in 1695.

2.3.3.1 Temperature considerations and snow-cover, glacier advance, etc.

The decline of prevailing temperature levels in continental Europe seems to have been equally abrupt and took place as early as in Iceland and the Denmark Strait. The period from about 1550 to 1700 seems to have been the climax of the Little Ice Age development so far as most of Europe is concerned, but there were further fluctuations and glacier advances after 1700 and the Little Ice Age can well be regarded as continuing to 1850 (when the Alpine glaciers were at their maximum), or even to the 1890s. By the late sixteenth century, there are already a few reliable weather diaries from which some details of the weather in Europe can be extracted. Daily observations at Zürich show that snow accounted for twice as large a proportion of the winter precipitation between 1563 and 1576 as in the immediately preceding years 1546-62: Flohn (1949) equated this to a 1.7° drop of the mean winter temperature between these two groups of years. Observations by the Danish astronomer Tycho Brahe from 1582 to 1597 on an island in the Sound, between Denmark and Sweden, show that the commonest wind direction was S.E. and that N.W. and W winds were as frequent as winds from the S.W. (which have been commonest over the last 100 years). Another calculation by Flohn (1967) indicated from this that the mean winter temperature at that time in the region of Denmark must have been 1.5°C lower than around 1900. In Hessen and Switzerland in the 1600s, the deficit may have been significantly greater than this.

Figure 2.5 indicates overall mean temperatures in England between 1550 and 1700 just 0.7 to 0.8°C below the average for 1900 to 1950, the departure being however greater in the case of the winters, 1.0 to 1.1.°C below the 1900-1950 average. It is certain that some of the severest winters of that time showed much greater departures from modern experience than this: in 1608, the ice was reported to be several feet thick on the River Thames in London, and in 1684 the ground was frozen to a depth of 4 feet (1.2 m) in Somerset, 3 feet (90 cm) in Kent and over 2 feet in northwest England. In 1684, also, ice formed on the sea in a belt 3 miles (5 km) wide on the Channel coast of Kent and 2 miles wide on the French side; there was ice on the North Sea up to 16 miles (25 km) out from the coast of Holland and Flanders. The winter of 1683-4 appears as the coldest in the central England record since 1659, though it was probably equalled by the winter of 1607-8 (which may have exceeded it in the Netherlands and Germany). Table 2.1 gives the mean temperature over the three winter months of the seven coldest and seven mildest winters in the long English series of thermometer measurements.

TABLE 2.1.: Mean temperatures over December, January and February of the seven coldest and seven mildest winters in central England between 1659 and 1979. (Average winter 1850-1950: 4.0°.)

Winter	1683-4	1739-40	1962-3	1813-4	1794-5	1694-5	1878-9
°C	-1.2	-0.4	-0.3	+0.4	+0.5	+0.7	+0.7
Winter	1868-9	1833-4	1974-5	1685-6	1795-6	1733-4	1934-5
°C	6.8	6.5	6.3	6.3	6.2	6.1	6.1

The severest individual months were all rather closely comparable: mean temperature (January in each case) 1684 and 1795 -3.0 or 3.1°, 1814 - 2.9°, 1740 - 2.8°, 1963 - 2.1°, 1716 - 2.0°. Differences in the effects of the winters concerned seem therefore to have had more to do with the length of the frost period than with the temperature of the coldest month. Differences in the amounts of snow covering the ground must also have been important.

As Table 2.1 indicates, there were also some very mild winters in the Little Ice Age period, and the close proximity of both extremes in 1795 and 1796 -and on other occasions within a few years- is remarkable.

Some of the winters during the harshest phases of the Little Ice Age were very long. The winter of 1564-5 brought a 2-month frost period in England from mid-December to mid-February, and on the continent the frost continued until the end of February, in Bohemia till early March. This was undoubtedly the longest event of the kind since the comparable winters in 1431-2 and 1434-5. The winter of 1607-8 was equally long, but was exceeded in Switzerland by 1613-14 when the snow lay for about 150 days in the region of Berne- as occurred again in 1784-5 (Pfister 1978). An unbroken frost period of 2 months' duration was reported in France (Paris) and the Netherlands in the winter of 1657-58, for which there is one report of 102 days snow-cover in England. Another case was reported in 1684-5 when Zürich had 112 days with snow lying, the previous winter having given a frost period more than two months long in England (London) and Denmark. The winter of 1694-5 also produced at least 50 days of snow-cover in northeast Hertfordshire, just north of London. (The duration in 1962-3 for comparison in southeast England was from 50 to 61 days and in Zürich 86 days.)

Pfister's studies of the period in Switzerland have produced the most detailed record so far available for anywhere, complete from 1525 to 1829. I am deeply indebted to him for enabling me to use, before publication, his data which show the course from decade to decade of his warmth and wetness indices, based on the frequencies of various types of weather and snow conditions reported mainly in the Berne and Zürich areas. Pfister found that one of the most remarkable features of the period is the coldness of the month of March. This month showed the greatest average departure from more recent experience. Reasonable estimates of the departure of the mean temperature of March in Switzerland from the twentieth century average seem to be -1.2°C for the whole period 1525 to 1829 and about -1.7°C for 1560-99 and 1620-1769. In two separate decades, the 1650s and 1690s, the mean departure seems to have been more than -3°C. In all the severest years, March in Switzerland was a full winter month, very cold and snowy (Fig. 2.2 illustrates one such year). And, as Pfister (1975, 1978) has pointed out, this led to very severe effects on agricultural production. Persistence of the snow-cover through March, and in a few cases through April, meant that the peasants ran out of hay and had to feed the cattle on straw and pine branches: many cows were slaughtered. Even more serious were the failures of the grain harvest in these years, apparently because of rot due to the attacks of a parasite Fusarium nivale which is active under prolonged snow-cover, particularly in spring (see also Section 5.2). Attacks of this parasite have been reported in recent times on crops in north Germany and Scandinavia, but have been unknown in Switzerland, presumably because there have been no such springs since the eighteenth century.

The more modest values derived for the 50- to 150-year mean departures from modern experience of the prevailing temperatures in England (as also in Switzerland) between 1550 and 1700 (p. 37 and Fig. 2.5), as compared with the departures in the individual most extreme years (Table 2.1), agree with evidence of a wider range of variability from year to year, and from one group of a few years to the next, in the Little Ice Age period. Signs of this can be seen in Table 2.1 in the occurrences of opposite extremes within a few years of each other. The standard deviation of winter temperature in England and the Netherlands seems to have been 40-50% greater in the severest part of the Little Ice Age period than in the early decades of the present century when the westerly winds from the ocean were extraordinarily dominant (Lamb 1977). The summer temperatures were also more variable from year to year in the Little Ice Age, the standard deviations again being 30% greater than in the equable decades of the first half to the present century.

Another symptom of this variability during the long period of colder climate is to be seen in the Baltic ice as indicated by the 450-year record of the dates of opening of the port of Riga: the earliest as well as the latest dates of opening occurred in the same decade -in 1652 and 1653 the port was usable from February 2 and 3, respectively, but in 1659 not until May 2. (A similar comparison seems likely to have been produced recently by the years 1975 and 1966.) Estimates of the maximum ice area each winter since 1701 have been given by Alenius and Makkonen (1981).

Despite the generally lower temperature level, there were also some heat waves and some hot summers during the Little Ice Age period. There were heat waves in mid to late June and July 1665 in England when the last great visitation of the plague was rampant in London, and the next summer in 1666 (with a mean temperature of 16.7°C) ranks as the twelfth hottest in the central England record since 1659. Both of these were drought summers and that of 1666 led up to the great fire of London. Much of Europe experienced prolonged heat and drought in the successive summers of 1717, 1718 and 1719 (though in the first of these the persistent warmth did not extend as far west as England). Table 2.2 lists the mean temperatures of the fourteen hottest and fifteen coldest summers in the 320-year English record.

TABLE 2.2.: Mean temperatures over June, July and August of the fourteen hottest and fifteen coldest summers in central England between 1659 and 1979. (Average summer 1850-1950: 15.2°C.)

Summer	1826	1976	1846	1781	1911	1933	1947	
°C	17.6	17.5	17.1	17.0	17.0	17.0	17.0	
Summer	1868	1899	1676	1975	1666	1719	1762	
°C	16.9	16.9	16.8	16.8	16.7	16.7	16.7	
Summer	1725	1695	1816	1860	1823	1674	1675	
°C	13.1	13.2	13.4	13.5	13.6	13.7	13.7	
Summer	1694	1888	1922	1812	1862	1698	1890	1920
°C	13.7	13.7	13.7	13.8	13.8	14.0	14.0	14.0

It is again a noticeable feature of Table 2.2 how many times opposite extremes occur within a few years of each other.

There was also a noteworthy geographical pattern of the temperature anomalies which characterized the Little Ice Age. From the earliest surveys of sea surface temperatures, between 1780 and 1820, it appears that there was a large area of the Atlantic Ocean between latitudes 20 and 40-45°N that was somewhat warmer than now, while the equatorial zone as well as all the higher latitudes were colder than now. North America as well as Europe was having a colder climate than now. The greatest departures so far determined of the long-term (30- to 50-year) average temperatures from twentieth century averages are found in northern Scotland and southern Norway and seem probable also in southern Iceland, all these places being close to the area of ocean where the intruding polar water current seems to have eliminated for a time the warm, saline North Atlantic Drift (of Gulf Stream) water from the ocean surface. A study by Matthews (1977) seems to establish that, averaged over all seasons of the year, prevailing temperatures in southern Norway from 1700 to 1725 were 1.7°C below the 1949-63 average, and by implication the departure in the preceding quarter century was probably about -2°C. In northern Scotland, the evidence of permanent snow on the tops of the highest mountains and of permanent ice on one or two high-level lakes among the mountains in the seventeenth century (Lamp 1977) suggests over-all average temperatures about 2 or 2.5°C below modern values. (In southern Iceland, the deficit may well have been almost 3°C, but this has not been determined. The longest continuous record of thermometer observations in Iceland, from Stykkisholmur on the west coast, shows that prevailing temperatures rose by about 1.5° from 1846-1870 to 1920-50.) It also seems possible that the deficit was greater on the higher levels of the Alps than elsewhere in central Europe.

It seems clear from the Swiss records studied in detail by Pfister, as can also be clearly seen from the temperatures in central England (Fig. 2.9), that the average conditions of the Little Ice Age time -especially around 1690- were colder than now in all seasons of the year.

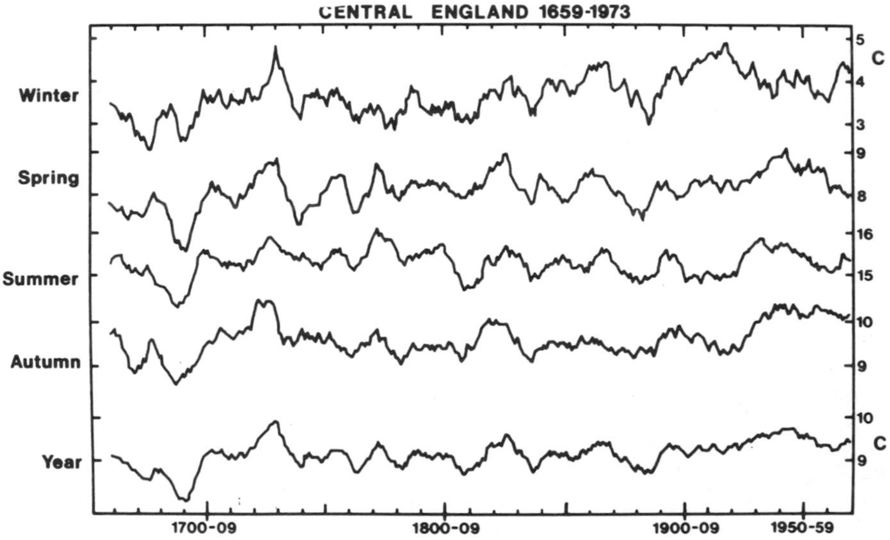

Fig. 2.9. Temperatures in central England: overlapping 10-year means for each season of the year and for the whole year, from 1659 to 1978. (Values from Manley (1974) updated by P.D. Jones, Climatic Research Unit, University of East Anglia, Norwich.)

The autumns of the late seventeenth century -so far as can be judged from the reports available -were also wetter than at most other times and in Scotland snowy. Of the period 1693-1700 in Scotland there are many dire accounts from parish records in the Old Statistical Account (Sinclair 1791-9). These were summarized by Graham (1899) in part as follows:

> "... During these disastrous times the crops were blighted by easterly 'haars' or mists, by sunless drenching summers, by storms, and by early bitter frosts and deep snow in autumn. For seven years the calamitous weather continued, the corn barely ripening, and the green, withered grain being shorn in December amidst pouring rain or pelting snowstorms. Even in the months of January and February, in some districts many of the starving people were still trying to reap the remains of their ruined crops of oats, blighted by the frosts, and perished from weakness, cold and hunger."

In the upland parishes of Scotland, even in Midlothian, it was reported that one third of the people died in those years (cf. also Section 6.3.7). Many were buried in mass graves. And whole villages disappeared (Graham 1899). These experiences, bitterly described at the time in the old Scottish parliament, probably made the Union of Scotland with England, which took effect in 1707, inevitable.

In Iceland, parish records report that whole farms disappeared under the increasing glaciers and ice-sheets at this same time (Lamb 1977, Chapter 12). And at various times between about 1600 and 1850 there was great alarm among the villages and farm communities in the Alps (e.g. around Chamonix and Grindelwald) and in Norway (here mainly between about 1670 and 1750) over the advancing glaciers (Ladurie 1967, 1971; Hoel and Werenskiøld 1962). Many pastures and some whole farms and villages were in fact overrun, and there was a general reduction of yields, which meant that taxes also had to be lowered.

2.3.3.2 Rainfall, wetness of ground, floods, landslips, etc.

Reduction of the total precipitation over England by 7-10% below the average downput in the first half of the present century is indicated by the values in Fig. 2.7 derived from the summer and winter indices, and a 15-20% reduction by the actual raingauge measurements in the very dry period of the 1740s and '50s. The record for London prepared (standardized as for Kew observatory) by Wales-Smith (1971) indicates that the downput averaged over the whole 53-year period 1697-1749 was only 90% of the 1916-50 average.

This reduction seems reasonable on the basis of a lower average moisture content in the atmosphere on account of the colder seas. It probably therefore applies also to other parts of the European flatlands in about the same latitude. Nevertheless, there were variations from year to year and from decade to decade, depending on the positions occupied and paths followed by the cyclonic activity and differences in the wind pattern.

There is evidence -e.g. the siting of houses in valley bottoms- that the ground in England from the late sixteenth to mid eighteenth centuries may have been mostly drier than in the present century and certainly in comparison to the fifteenth century. As mentioned on p. 43, this was not the case in Scotland, probably because the higher latitude and (more than proportionately) lower temperatures reduced the evaporation. There are many mentions in documents of the time and later of the wetness of the valley bottoms in lowlands and highlands alike in Scotland in the seventeenth century, of the greater size of the lochs, and of the existence of many small lochs where none now exist, as well as of the rumoured deterioration of the valleys from their condition some time earlier, in the middle ages.

The enhanced variability noted in the temperature history of the Little Ice Age also applied to the precipitation. Extremely wet periods seem to have occurred (a) when the prevailing tracks of cyclonic activity were transferred south, and (b) with stationary, or slow-moving, cyclonic situations, particularly in the areas affected by warm, moist southerly winds. Thus, the period was not only one of advancing glaciers and sea-ice, but also of avalanches, landslides and occasional bursts of peatbogs swollen by spells of prolonged rain. Records of such events are known from Norway, Scotland, Ireland, northern England and elsewhere. In the Alps, some valleys (e.g. Saastal) experienced a terrible succession of disasters with advancing glaciers sometimes crossing the valley bottoms where they dammed the rivers and formed lakes which later (and repeatedly) broke the ice-dams, producing a devastating torrent flood.

In southern Europe, in Spain, as in parts of Turkey, the sixteenth and seventeenth centuries seem to have produced successions of severe droughts and flood years alternating with each other. The effects on the economy were disruptive, and in Turkey a time of riots and uprisings in Anatolia in the late sixteenth and early seventeenth centuries has been attributed to the droughts. Travellers from Europe described an empty, parched landscape with deserted villages (W.J. Griswold, unpublished). And it was in the uplands of the Languedoc in the south of France that Ladurie (1971) found it necessary to attribute the many harvest failures and other agricultural troubles in the seventeenth century to the fluctuations of the climate.

Fragments of knowledge so far at our disposal indicate that at the height of the Little Ice Age development in the seventeenth century there were also climatic vagaries in other parts of the world which, if they occurred nowadays, would have a serious impact on the whole world economy. There seem to have been much more frequent failures and interruptions of the Indian monsoon than in our own century. And in Africa from about 1580 to 1650, and again from about 1680 to 1770, there were frequent famines and drought in places in the Sahel zone across Africa near the southern fringe of the Sahara, though at these same times the yearly floods of the Nile were particularly high.

2.3.3.3 *Storms and sea-floods, etc.*

Descriptions of a number of great storms at sea and on land in the times we are describing, and accounts of changes of the coast by blowing sand closing harbours and burying human settlements, suggest that during the cold climate period the storm activity in latitudes around 50 to 60°N sometimes reached an intensity not experienced in the earlier part of the present century. (The storm which crossed England on 3 January, 1976, and that which caused sea damage to the Channel coast near Portland in the south of England on 13 February, 1979, may, however, prove to be comparable with some of the Little Ice Age storms. Meteorological analysis of the weather reported day by day during the Spanish Armada expedition in the summer of 1588 indicates a speed of travel of the cyclone centres which on at least six occasions in August and September of that year appears to correspond to jetstreams somewhat stronger than the statistically probable extreme of twentieth century experience for those months of the year (Douglas et al. 1978). It seems likely that this development must be associated with the intense thermal gradient developed at times between latitudes near 50 and 60°N when the limit of the Arctic Ocean ice has advanced to Iceland.

The records of sea floods on the Dutch coast (Gottschalk 1971, 1975, 1977) also suggest a maximum -albeit probably a secondary maximum- of these occurrences in the late sixteenth and seventeenth centuries. This may also be attributed to an increase of storminess at that time and probably to a greater frequency of the major cyclone centres passing in latitudes close to this part of Europe. It is unlikely that the general sea level was as high as it had been in the middle ages around A.D. 1000 and between 1200 and 1400 or as in the twntieth century, though the differences may be estimated at only 10-50 cm. Cottschalk also reports a further peak of sea floods on the Netherlands coast in the early 1400s: this may have something to do with the increased tidal range at that time.

2.3.4 *The recovery from the Little Ice Age to the twentieth-century climate*

The long recovery from the cold climate which culminated in Europe in and around the 1690s to the climate we know can be traced in the records of the readings of meteorological instruments. Despite the existence of the instrument records, it is not widely known that the agricultural revolution begun in the eighteenth century took place against a background of improving climate -particularly a warming of the summer climate- and it is likely that some of the success should be attributed to the easier conditions provided by Nature.

Many details of the sequence of changes of the temperature level, rain- and snowfall, sunshine and cloudiness, winds and windflow patterns, in graphs, maps and tabulated values for each month of the year, are given in von Rudloff's (1967) history of the climatic variations and fluctuations in Europe from the beginning of the instrument records (cf. also Chapter 3). Other details and tabulated items will be found in Lamb (1977) and a wealth of local and particular detail of extreme events in Britain and their effects in Manley (1952).

The main course of the recovery of the temperatures in England can be followed in Fig. 2.9, a history which seems to parallel that indicated in most parts of western and northern Europe. In central Europe there are differences as regards which individual winters were most severe -e.g. in Switzerland 1613-14, 1684-5 and 1784-5 stand out, whereas in northwestern and northern Europe 1607-8, 1683-4 and 1783-4 produced more extreme conditions- and at least in south Germany and Austria the warmth of some later eighteenth century summers seems to have been a greater anomaly than in England.

These minor differences seem readily explained meteorologically by the great frequency of stationary blocking anticyclones in this period. Since these systems characteristically occupy somewhat different positions from one case to the next, the persistent warm southerly winds on their western flanks and cold Northerly winds on their eastern flanks affected different longitudes in Europe in different years. This also explains the enhanced variability from year to year, and from one group of a few years to the next, during the Little Ice Age period when "blocking of the westerlies" by stationary systems seems to have been much more prominent than in the present century, apart from the most recent years since about 1968.

Soon after 1700 some warming of the climate was apparent, and in the 1730s -only forty years after the coldest decade in the record- a remarkable peak of warmth, equalling the warmest part of the present century, was reached. The warmth of the 1730s seems to have affected the health and longevity of the people in northern countries, and can be traced in the population statistics from Iceland and Sweden. Lamb (1977, p. 12) has pointed to the probability that it also affected the social customs in Scotland, and it certainly led to the reintroduction of wheat growing in areas of the country where its former use had been forgotten. These variations were largely paralleled in Switzerland, where Pfister finds that Zürich had a mean duration of snowcover amouting to 70 days from 1680 to 1700, 75 days in the 1690s, whereas from 1705 to 1714 the average was only 42 days, equalling that of the 1920s, the mildest part of the present century (although urbanization has reduced the figure to about 37 days since about 1943). These figures illustrate an important feature common to the Little Ice Age and our own times, namely the clustering of particularly warm and cold years in runs and the abruptness of climatic changes, which can be costly for the economy.

As Fig. 2.9 shows, there was an abrupt reversion of the prevailing temperature level after the 1730s to conditions only a little less severe than in the seventeenth century. This was ushered in by the historic, long winter of 1739-40 (cf. Table 2.1), which affected the whole of Europe from Spain and Portugal and the British Isles in the west to the Ukraine in the east: the French harbour of Dieppe was frozen, the ice did not clear from the port of Riga until 25th April and it was reported that the frost did not entirely clear from the subsoil in eastern England (Suffolk) until June. The frost was, however, more moderate than usual in central Sweden; and Iceland (as well as perhaps northern Scandinavia) had a mild winter. The summer of 1740, with a mean temperature in central England of 14.3°C (cf. the cases in Table 2.2), also ranked as very cool, and there is no doubt that that year (overall mean temperature of 6.8°C) -the coldest in the record- must have imposed a great fuel demand. Reasonable estimates (e.g. by Manley 1957) suggest that the fuel demand in such a year would be double that in the warmest years of the present century (e.g. 1949 and 1959, with mean temperatures of 10.6 and 10.5°C respectively).

The later eighteenth century summers, unlike the other seasons of the year, became warm all over Europe, in general probably the longest sustained period of summer warmth in the instrument record. This phase lasted broadly from the later 1740s to 1800.

After great dryness in the 1740s (mean yearly rainfall in England and Wales 83% of the 1916-50 average, in the summers 91%) there was, however, an important change to wetter years, particularly to the wet summers of the 1750s (mean summer rainfall in England and Wales 122% of the 1916-50 average), 1760s (119%) and 1775-84 (116%): 1763 ranks as the wettest summer in the England and Wales record (181%), and 1768 (164%) was surpassed only by 1829 (168%), 1852 (166%) and 1860 (169%).

The nineteenth century was, in general, a century of cool summers, including some of the coldest in the record, apart from a few exceptional warm years such as 1826 (see Table 2.1), 1846, 1868 and 1899. The coolest summers were in the decades 1810-19 (central England average 14.7°C), 1840-49 (15.0°) and 1880-89 (14.9°). It was, however, a combination of warmth and wetness which produced the potato blight famine disaster in Ireland in 1846-48.

It was not until 1896 and after that a persistent trend towards greater warmth in all seasons of the year set in. Growing seasons lengthened. And recession of the glaciers, which had begun a little in a slow and fluctuating manner after 1860, became increasingly rapid from about 1925 to 1960. After 1920 the Arctic sea ice was rarely seen near Iceland until its return in the 1960s, and the open season for shipping in Spitsbergen increased from three to seven months of the year.

For most of Europe the climax of the twentieth century warmth was reached in the years 1933 to 1952 (von Rudloff 1967), in spite of the consecutive severe winters 1939-42.

The change of climate from the eighteenth century is registered clearly in changes of the prevailing wind patterns, particularly in the frequency of mobile Westerly wind situations which is known for the British Isles from counts of the frequency of these and the other, more blocked, situations (Lamb 1972) on daily weather maps for the years 1781-86 and 1861 to the present date (Fig. 2.10). This is a variable which is intimately related to the mechanism of climatic changes, being significantly correlated with a wide range of other variables in world climate including probably both world temperature and the temperatures prevailing in Europe.

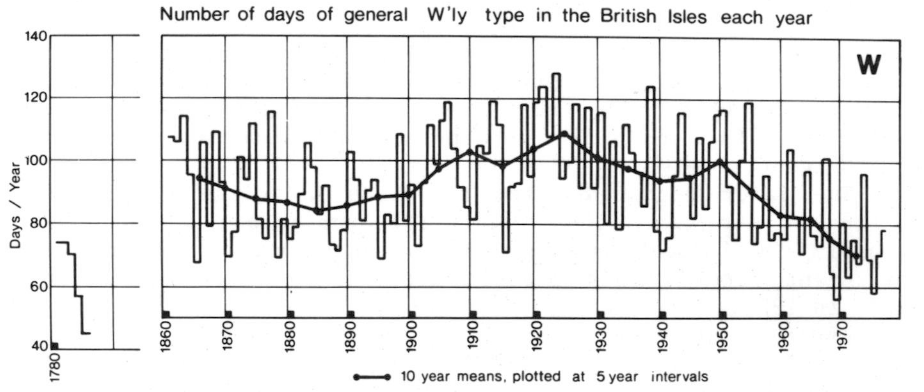

Fig. 2.10. Number of days each year with general Westerly winds over the British Isles (and 10-year averages plotted at 5-year intervals at the middle of the period concerned): 1781-85, 1861-1978.

(10-year totals of the number of days with South-Westerly surface wind in London from 1730 to the 1970s gave a correlation coefficient of about + 0.4 with the central England temperatures, significant at the 5% level.) Thus, the recent decline of the frequency of Westerly situations, and renewed increase of stationary, "blocked" situations, must attract interest. It is undoubtedly connected with the recurrence since about 1960 of diverse extremes somewhat as in the Little Ice Age period. There has also been a renewed frequency of occurrence of gale, severe gale and storm situations over the seas between latitudes 50 and 60°N near the British Isles and in the North Sea (Fig. 2.11).

GALE INDEX SURVEY BASED ON PRESSURE DISTRIBUTION

(a) Over the North Sea

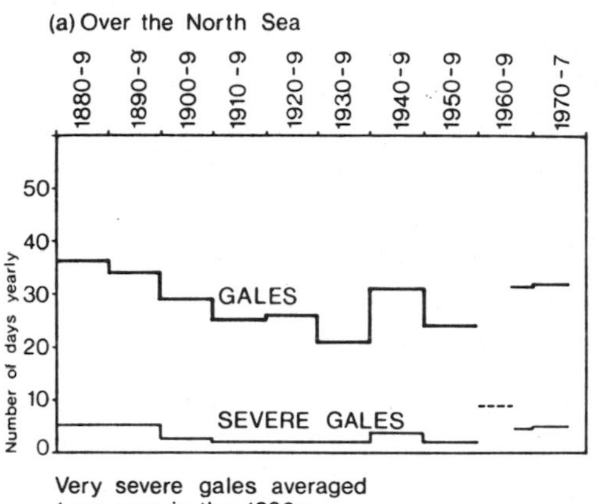

Very severe gales averaged
1 per year in the 1890s
1 in 2 years in the 1880s and 1970s
1 per decade between 1910 and 1939

(b) Over the British Isles/ East Atlantic

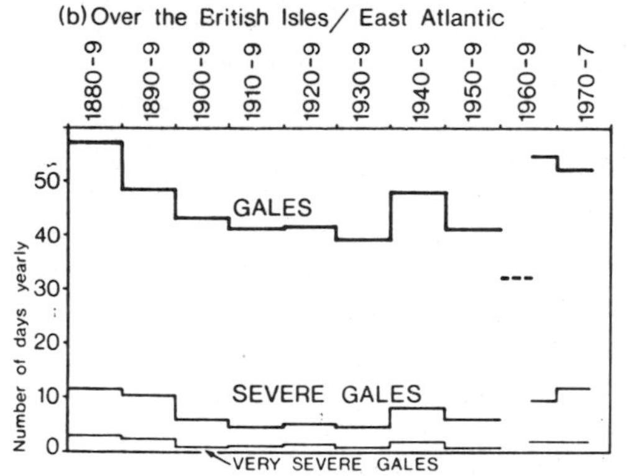

Fig. 2.11. Frequency decade by decade since 1880 of gale situations over the North Sea, British Isles and nearest parts of the Atlantic Ocean between latitudes 50 and 60°N. Derived from measurements of atmospheric pressure gradients on daily synoptic weather maps.

The variations in frequency of winds from the ocean implied by the history of the Westerly wind situations in Fig. 2.10 are also broadly paralleled by the record from decade to decade of the rainfall over England and Wales (Fig. 2.12), which shows some decline in the 1970s (93% of the 1916-50 average) after the climax in the previous half century.

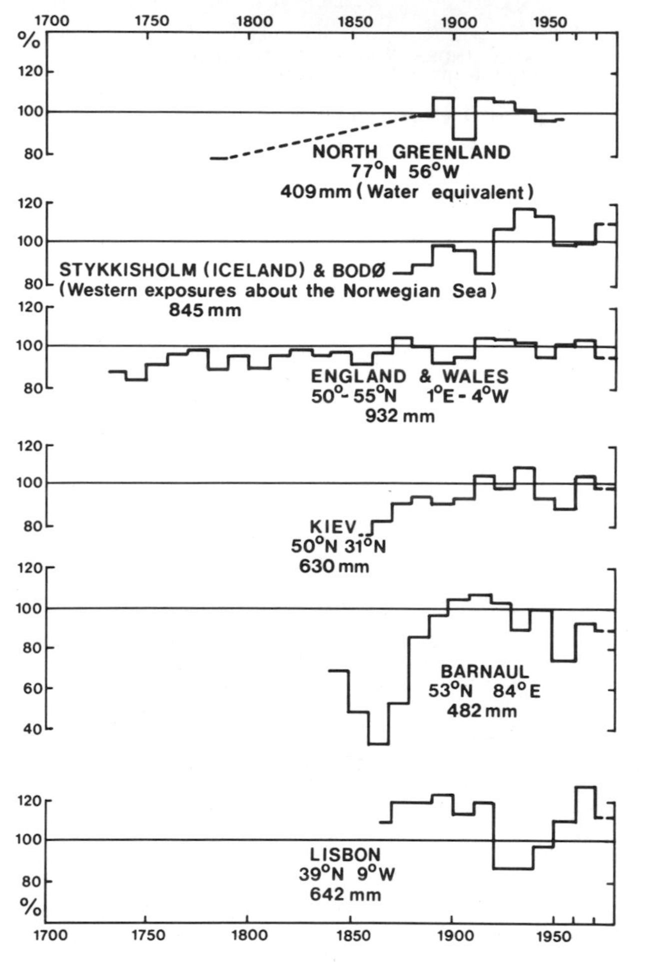

Fig. 2.12. Rainfall (or total downput of rain and snow measured as equivalent rainfall) at various places decade by decade, as percentages of the 1900-1939 averages (which are given).

(Fig. 2.12 also shows the rainfall changes in some other parts of Europe and as far east as Soviet central Asia: all these records apart from Lisbon in the south, which comes within the range of the subtropical regime, show a history broadly in line with that of England and Wales up to the latest decades for which figures were available.)

Further points of interest emerge from studies of the long history of the winds. In the case of surface winds prevailing over the lowlands of southern, central and eastern England an outline history covering the last 600 years can be presented (Fig. 2.13).

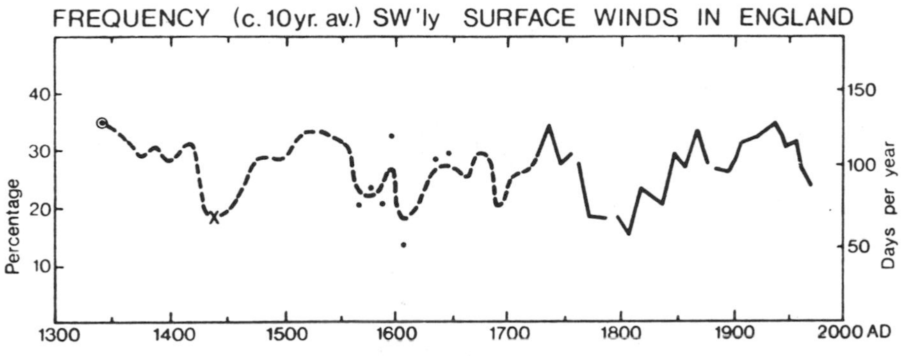

Fig. 2.13. Frequency of South-Westerly (SW) surface winds in eastern England. Curve represents 10-year means. (Based on complete daily observations in the London area from 1668, more indirect data earlier including various weather diaries -notably near Lincoln about 1340 and in Denmark, Germany and Switzerland in various years between 1546 and 1650).

This is derived from actual daily observations in, or close to, London from 1668 and rests on more indirect evidence for earlier years. The first point on the graph is based on a record of about five years' daily observations in eastern England (mainly near Lincoln) around 1340. Other points are derived from preliminary seasonal weather mapping stages of the Historical Weather Mapping Project (funded from 1974 to 1980 by the Rockefeller Foundation) in the Climatic Research Unit, University of East Anglia, Norwich, and from weather diaries covering most of the years 1546 to 1650 at various points in western and central Europe. The frequency variations of surface winds from the S.W. have been chosen for most study and for presentation here, because this is the most frequent individual wind direction and its frequency corresponds most closely to that of the mobile Westerly type (Fig. 2.10) defined in terms of the direction of steering of the travelling weather disturbances (cyclones and their fronts). Nevertheless the frequency of this pattern is seen to undergo substantial long-term variations. From Fig. 2.12 it appears that a recurring fluctuation of about 200 years' length may be an important element

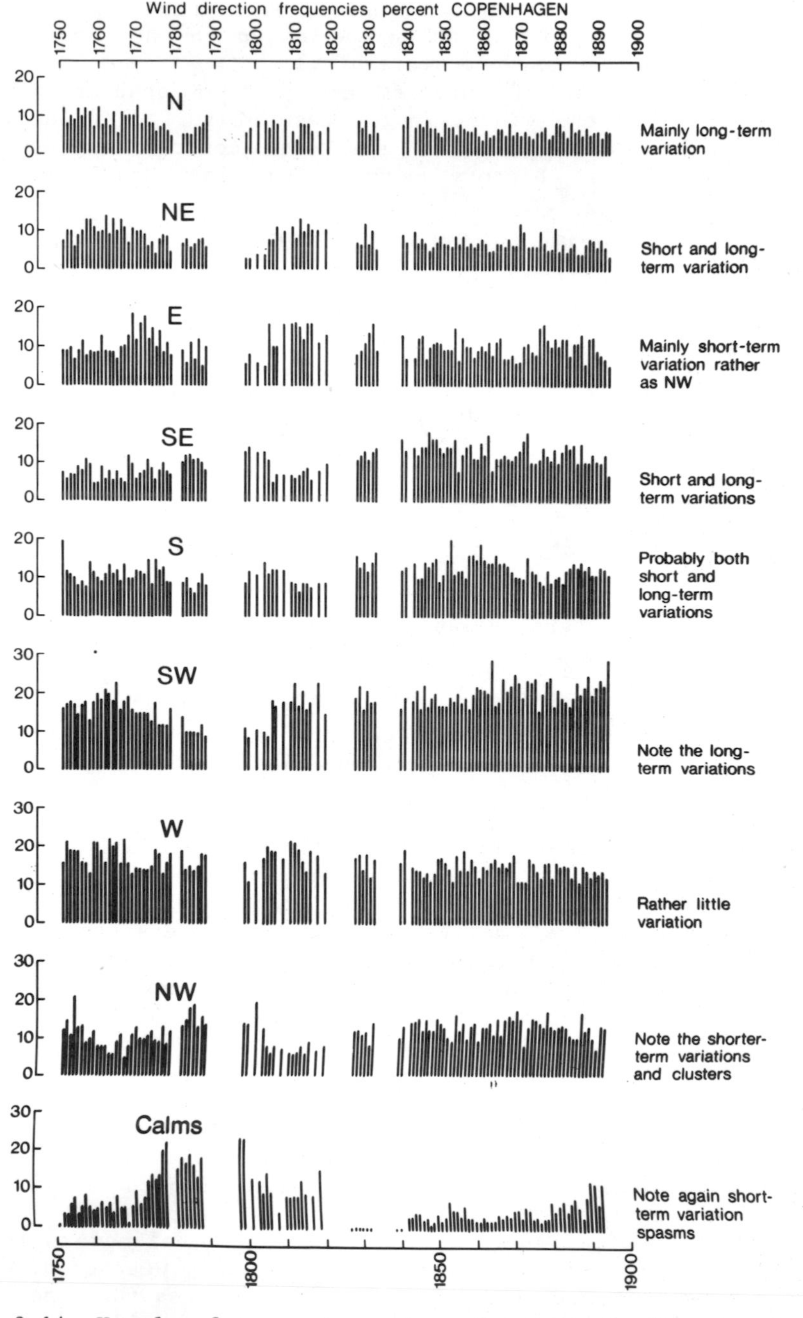

Fig. 2.14 Yearly frequencies of surface winds from various directons at Copenhagen, 1751-1893.

of the history. If this be assumed likely to continue, an outline long-term forecast is implied. Such a forecast would, however, carry the additional uncertainties that the regime would be changed if either a rise of world temperature (e.g. as a side-effect of Man's activities) or a world cooling (e.g. due to an outburst of explosive volcanic activity polluting the stratosphere) should supervene.

A history of the wind direction observations at Copenhagen since 1751 is also available, and this is presented here in diagram form (Fig. 2.14). Over the 14 decades available for study at both London and Copenhagen the two series are satisfactorily correlated. The correlation is closest in the case of the total frequency of S.W. and S winds at Copenhagen compared with that of the S.W. winds alone at London ($r = +0.72$). A very long record of wind observations, from about 1700, is also available for Amsterdam (Labrijn 1945). There as at Copenhagen, the frequency of S winds appears to be correlated with the frequency of S.W. surface winds in London.

Figure 2.14 also shows that the frequency of S.W. winds is dominated by a long-term variation. By contrast, the frequency of winds from some other directions, notably N.W. and E, gives much more prominence to variations affecting shorter runs of just 10-25 years. These are presumably the wind directions at Copenhagen most associated with Greenland-Iceland-British Isles or Scandinavian anticyclones, respectively. The same feature is very strongly seen in the occurrence of calms at Copenhagen, which are presumably most often associated with anticyclones over southern Scandinavia or Denmark itself. The phenomenon may be related to the frequently observed "clustering" (or runs) of years of anomalous but somewhat like character.

Study of the variations in the frequency of the various types of British isles wind and weather pattern also shows that the variability is greatest in the case of the North-Westerly and Easterly wind types. The ratio of the standard deviation of the yearly frequencies to the mean from 1861 to 1973 was 37.1% for N.W. and 29.4% for E, compared with 18.0% for W, 18.9% for Cyclonic central patterns and 19.1% for Anticyclonic central patterns.

2.4 Conclusions

The sequence of climatic regimes surveyed in this chapter reveals that the climate continually varies, fluctuates and changes, sometimes abruptly, and that the time-scales of fluctuation and the duration of the various regimes range widely. All these things have occurred and continue to occur at times and in circumstances in which any effects of Man's activities cannot be held to have played any significant part in the phenomenon. Hence, any variations which now or in the future may occur as a result of some activities of Man will be additional to the ongoing natural variability of climate, though they may also interact with the natural variations.

This means that the most precise knowledge of the natural variations of climate which scientific research may be able to produce is needed, if we are to be able to detect what are the additional effects due to Man. The natural variability may, in this connexion, be considered as "noise": the range of any variability which, for lack of physical understanding and predictability, has to be treated as random must tend to obscure any incipient trend attributable to human disturbance of the regime or of the environment, and doubtless means that it will be impossible to establish clearly any anthropogenic effects until they have already attained some magnitude. Global trends may also be obscured, or put in doubt, by a tendency for opposite variations south of about latitude 35 to 50°S as compared with the rest of the world.

The last thousand years are seen to have produced a large--amplitude swing of the prevailing climatic regime from a time of greater warmth than at any other period in the last several millennia to the coldest global regime since the last major ice age and then a recovery to the present century.

Not only a change of prevailing temperature levels was involved. We can discern:

(i) changes of prevailing temperature affecting all latitudes but strongest in the higher latitudes of the northern hemisphere;

(ii) significant consequential changes of

(a) length of the growing season,

(b) frequency and duration of snow-cover,

(c) length of the winters,

(d) frequency and severity of frosts and depth of penetration of frost into the ground,

(e) rainfall and its seasonal distribution,

(f) evaporation, wetness of the ground, size and levels of lakes and rivers,

(g) frequency of droughts and floods,

(h) variability of temperature, rainfall, duration of winter, etc., from year to year and from one group of a few years to the next,

(i) frequency and severity of gales,

(j) sea level and frequency of flooding of low-lying coastlands, including permanent erosion of the coasts,

(k) storm damage to coasts, harbours and farmland by blowing sand and alteration of coasts and landscapes thereby,

(l) growth and retreat of glaciers

(m) increase and decrease, and variable distribution, of sea ice,

(n) variations of sea surface temperature and of the fisheries.

Much further information on the historical period reviewed in this chapter, and on the data available and methods of handling it, has lately been published in Lamb (1982) and Wigley et al. (1981).

2.5 Summary

Examination of the longest past records of climate that can be constructed, using not only the longest (standardized) runs of meteorological instrument measurements but also earlier documented descriptive reports and many kinds of fossil data, indicates that the natural climate continually fluctuates and changes. Moreover, all these changes have some geographical pattern so that the changes of prevailing temperature from one epoch to another are seldom (and the changes of precipitation almost never) in the same direction everywhere in the world; their magnitudes always differ in different wind regimes. The associated changes of the general wind regime and shifts of storm tracks do tend to be simultaneous over vast areas, possibly over the whole globe, but the effects on prevailing temperatures (and hence on the vegetation and crops, etc.) commonly take longer to mature and may follow a different course in some regions compared to others.

The various kinds of data available for reconstruction of the past record of climate are briefly surveyed. Samples of the earliest daily and seasonal weather maps produced are illustrated.

The chapter then proceeds to outline what has been learnt regarding the sequence of climatic developments, starting with a brief look at the first millennium A.D. The sections which follow describe in turn:

(i) the medieval warm epoch, which probably was at its height in western, central and northern Europe between about A.D. 1150 and 1300.

(ii) the unsettled and deteriorating climate in the Late Middle Ages;

(iii) the cold period, often known as the Little Ice Age, which followed and reached its climax in various parts of Europe at different times between about A.D. 1550 and 1850. As a good deal of detail is known about this period from reports made at the time, subsections deal with:

(a) temperature and snow-cover, glacier advances, etc., including tabulations of the warmest and coldest winters and summers in the long temperature record for central England, which starts in 1659. Comparisons are made with the known temperature changes over at least some large part of this period in other places, from Iceland to central Europe. The length of the winters is also discussed in relation to records of snow-cover in Switzerland since the 1600s;

(b) rainfall, wetness of the ground, floods, landslips, etc;

(c) storms and sea floods;

(iv) The recovery from the Little Ice Age climate to the twentieth century.

At appropriate places in every section of the chapter mention is made of various impacts of the climatic development on the human economy, particularly agriculture, also the changes affecting sea temperature and the fisheries.

Chapter 3

CLIMATE VARIABILITY AND ITS TIME CHANGES IN EUROPEAN COUNTRIES, BASED ON INSTRUMENTAL OBSERVATIONS

3.1. Temperature

3.1.1. *Changes of temperature level*

3.1.1.1. *"Normal", changes and fluctuations*

Maps of monthly and annual mean temperatures in Europe for the so-called "normal" climatic period 1931-60 have been published by WMO/Unesco (1970). Climate data for the observational period 1931-60 are also available in other forms (e.g. as climatic tables in the World Survey of Climatology. Volumes 5 and 6 covering Europe (Wallén 1970 and 1977). These data are generally recommended for use for alle purposes. And yet, we know that they do not constitute just one arbitrary set of 30-year data, but a very special one, not at all representative of climate over the last three centuries. Climate has changed since instrument records began. Descriptions of these changes are given in many publications, but as far as Europe is concerned most extensively by Hans von Rudloff (1967) in his book entitled "Die Schwankungen und Pendelungen des Klimas in Europa seit dem Beginn der regelmässigen Instrumenten-Beobachtungen (1670)" and by Hubert Lamb (1972 and 1977) in his "Climate: Present, Past and Future".

To give a preliminary indication of the climate changes that took place we may repeat some of the main conclusions of the above-mentioned authors.

Von Rudloff distinguishes "Klimaschwankungen" and "Klimapendelungen", Klimaschwankungen being changes of a climate element with a period of more than 30 years and an amplitude of more than half the standard deviation, computed with reference to the 1851-1950 mean, or to the mean of the whole record if shorter; Klimapendelungen being all other variations. On the basis of this definition he concludes that during the last 300 years the most marked real change of climate started in the year 1897 and included all climate elements, all seasons and the whole European area. At the time of writing (1967) he considers that this major change has still not ended in a sense that a longer period of opposite deviations has not started as yet.

According to von Rudloff, the 20th-century climate change can be subdivided as follows: from 1897 onward, enhancement of the maritime influence over Europe; 1909-1939 maximum of maritime influence; from 1940 onwards weakening of it and shift of the general warming from the winter to the summer half-year; after 1951, first signs of cooling of the summer half-year. Besides this major climate change von Rudloff discovers a large number of short term "Pendelungen" or climate fluctuations, in all seasons of the year. Detailed descriptions of these fluctuations will not be repeated here.

As far as temperature is concerned the main basis for von Rudloff's conclusions is seasonal means of temperature for a large number of European stations. In his book, the data are presented as 10-year running means of the deviations of the seasonal temperature form the 1851-1950 average.

Lamb centres his discussion on temperature variations since instrument records began around the long record for Central England. Decadal averages of this so-called Manley series are reproduced in Fig. 2.9. Lamb first takes a broad overall look at the series, describing the behaviour of temperature from about 1700 to 1945 as a one-way story: a trend towards greater warmth that was interrupted only by various shorter-term fluctuations. This does not contradict von Rudloff's statement about a warming period starting about 1900. It merely places the 20th-century warming in a broader perspective. This recovery from Little Ice Age temperatures reached its highest level between 1933 and 1952, approaching temperatures then which had not occurred over any prolonged period since the early Middle Ages.

Talking about the shorter-term fluctuations, which he considers to be representative for most of Europe, Lamb mentions the very rapid warming from the late 1600's to the 1730's followed by a sharp setback at 1740, initiated by the great winter of 1739-40. From 1740 to 1900, little real change occurred, apart from some individual colder decades in the late eighteenth and nineteenth centuries and a 20-year period of rather sustained warmth in the 1820's and 1830's.

Another important point treated by Lamb is the apparent parallelism of temperature variations in England with those of temperature averaged over the whole Northern Hemisphere or even the whole globe. He considers that this may be due to the fact that the British Isles are situated in the middle latitude zone of prevailing westerly winds, at the eastern limit of an ocean where winds and ocean currents freely and continually effect heat exchanges between all latitudes.

In order to illustrate some of the features mentioned above, 30-year running averages of seasonal mean temperatures have been computed for 8 European stations, two in each of the latitude belts 55-60°N, 50-55°N, 45-50°N and 40-45°N. The resulting curves for winter, spring, summer and autumn are shown in Fig. 3.1 (a)-(d). The curves display the seasonal mean temperature for the reference period 1931-60, which for each station and season is indicated by a dashed horizontal line. At the right-hand side of these dashed lines the particular mean temperature value for the period 1931-60 is given. Numbers of the ordinate define only a linear temperature scale and have no absolute meaning. Each 30-year mean value is plotted at the fifteenth year of the 30-year period to which it belongs. Plotted in this way the curves clearly show that, at all stations, the present century has been warmer than the foregoing century, with only a few exceptions in some of the seasons. Apart from this, a number of detailed observations can be made, which do not agree in every aspect with the details quoted above from the works of von Rudloff and Lamb. However, some of these differences may be simply due to the use of different time-averaging methods. That the warming trend cannot be interpreted only as an effect of urbanization is shown by the recent decline; moreover the station Valentia, on the SW coast of Ireland, represents truly undisturbed maritime conditions (see also Section 2.1.3.4).

3.1.1.2. Range of 30-year mean temperatures

From the foregoing it is clear that the 1931-60 period, as far as temperature in Europe (and many other parts of the globe) is concerned, may be considered exceptional rather than normal. This leads us to the question of what is really normal. Many climatologists think that in view of the variability of climate on so many time scales the word "normal" should be avoided. This is certainly wise when the word is used to indicate a certain reference period, such as is the case for instance with the period 1931-60. However, in order to indicate the most probable 30-year mean temperature at a certain location (based on a statistical analysis of the available instrumental record) the word normal might very well be used. Unfortunately, such normals cannot easily be computed from the frequency distribution of 30 year (or more generally n-year) mean temperatures, because of the shortness of most records and more importantly the occurrence of long-term climate changes. As a first approximation, which has been checked using different records, one can estimate the range of an n-year average with the formula $\Delta/\sqrt{n}$, Δ being the difference between the maximum and minimum temperature averages of the whole record. On the basis of this procedure which takes into account the climatic changes observed during the period of instrument measurements, one arrives at the following estimates of the range of variability of 30-year average temperatures in Europe:

For winter temperatures: 1.5 - 2.5°C for all stations between 45° and 60°N except for the western margines where values are

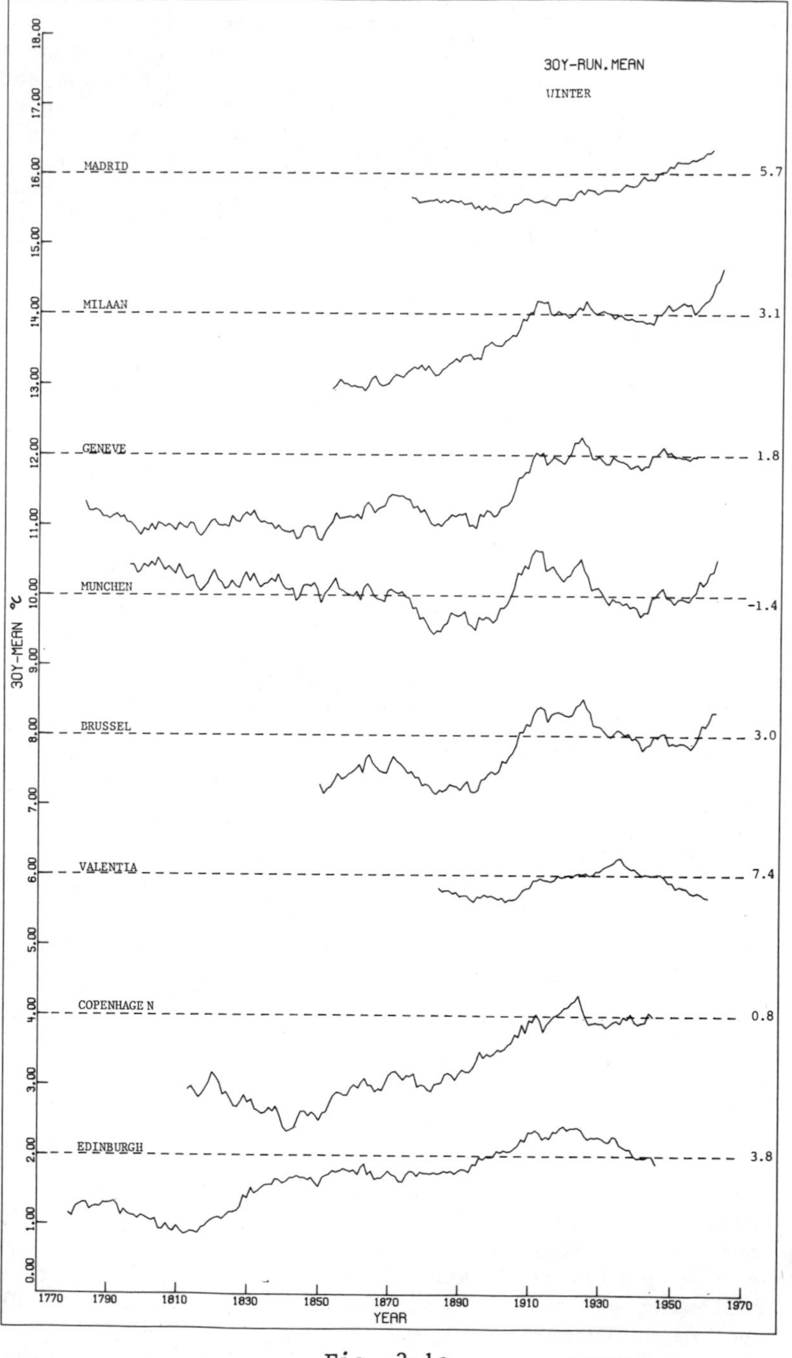
30Y-RUN.MEAN
WINTER
MADRID
5.7
MILAAN
3.1
GENEVE
1.8
MUNCHEN
-1.4
BRUSSEL
3.0
VALENTIA
7.4
COPENHAGE N
0.8
EDINBURGH
3.8
30Y-MEAN °C
0.00
1.00
2.00
3.00
4.00
5.00
6.00
7.00
8.00
9.00
10.00
11.00
12.00
13.00
14.00
15.00
16.00
17.00
18.00
1770
1790
1810
1830
1850
1870
1890
1910
1930
1950
1970
YEAR

Fig. 3.1a

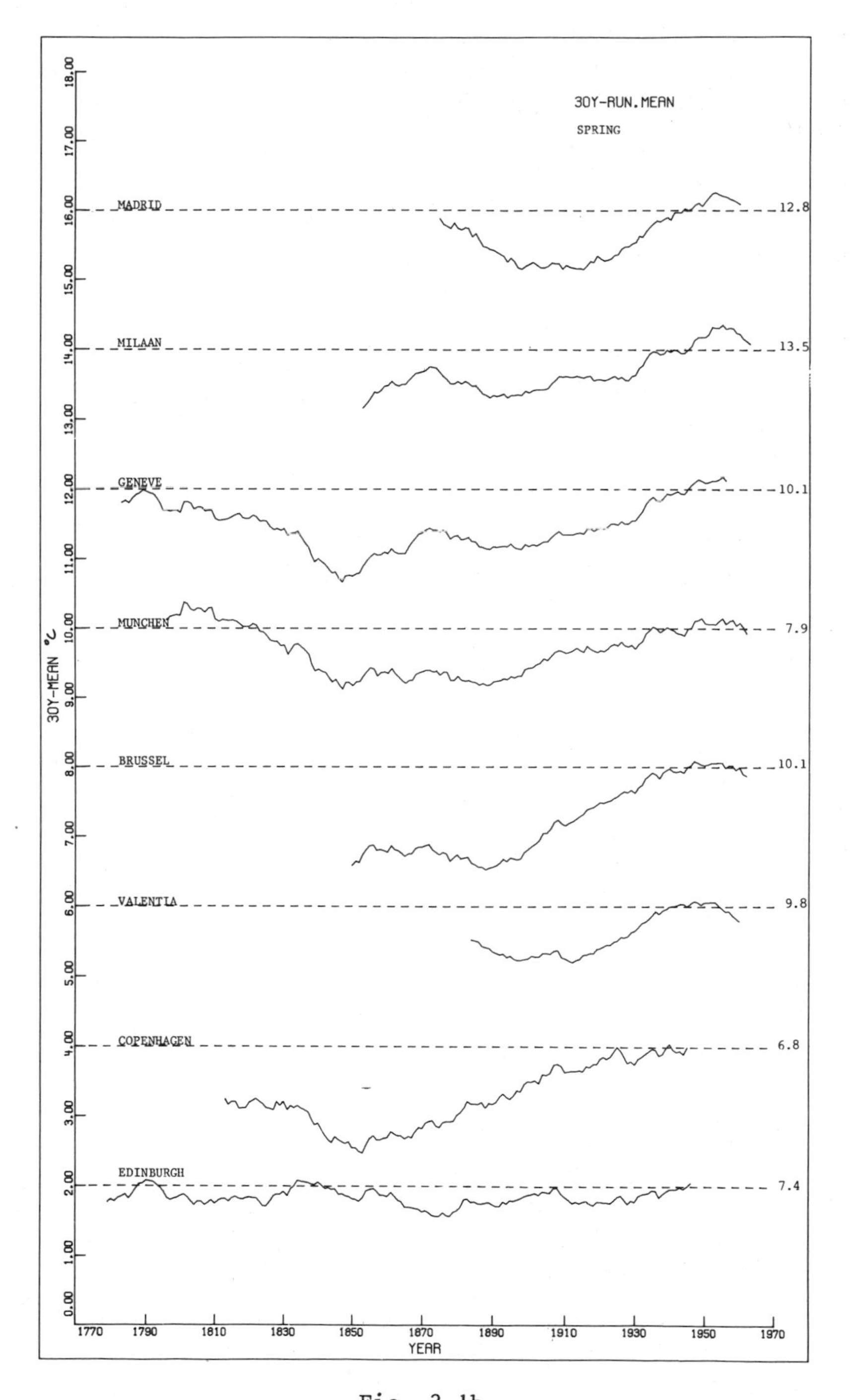
30Y-RUN.MEAN
SPRING
MADRID
12.8
MILAAN
13.5
GENEVE
10.1
MUNCHEN
7.9
BRUSSEL
10.1
VALENTIA
9.8
COPENHAGEN
6.8
EDINBURGH
7.4
30Y-MEAN °C
0.00
1.00
2.00
3.00
4.00
5.00
6.00
7.00
8.00
9.00
10.00
11.00
12.00
13.00
14.00
15.00
16.00
17.00
18.00
1770
1790
1810
1830
1850
1870
1890
1910
1930
1950
1970
YEAR

Fig. 3.1b

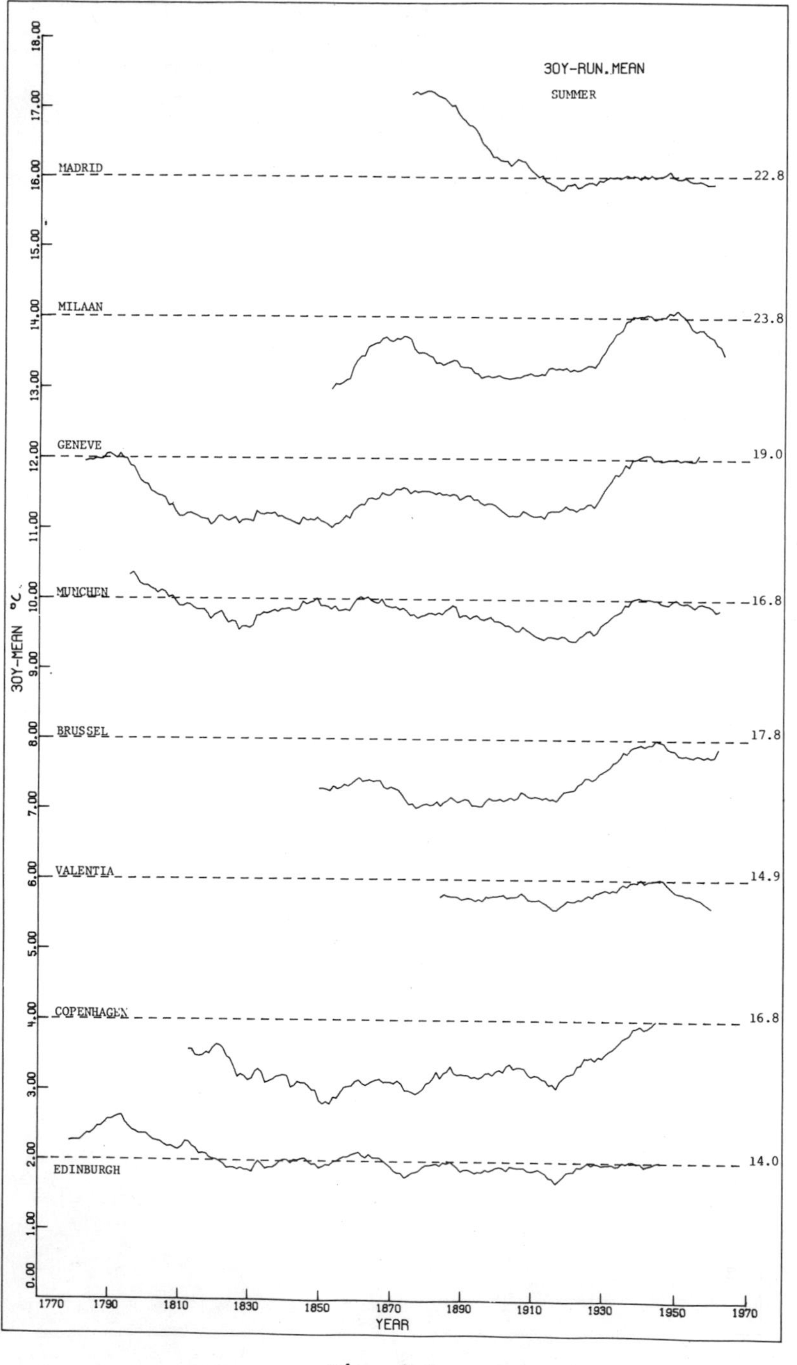
30Y-RUN.MEAN
SUMMER
MADRID
MILAAN
GENEVE
MUNCHEN
BRUSSEL
VALENTIA
COPENHAGEN
EDINBURGH
22.8
23.8
19.0
16.8
17.8
14.9
16.8
14.0
30Y-MEAN °C
0.00
1.00
2.00
3.00
4.00
5.00
6.00
7.00
8.00
9.00
10.00
11.00
12.00
13.00
14.00
15.00
16.00
17.00
18.00
YEAR
1770
1790
1810
1830
1850
1870
1890
1910
1930
1950
1970

Fig. 3.1c

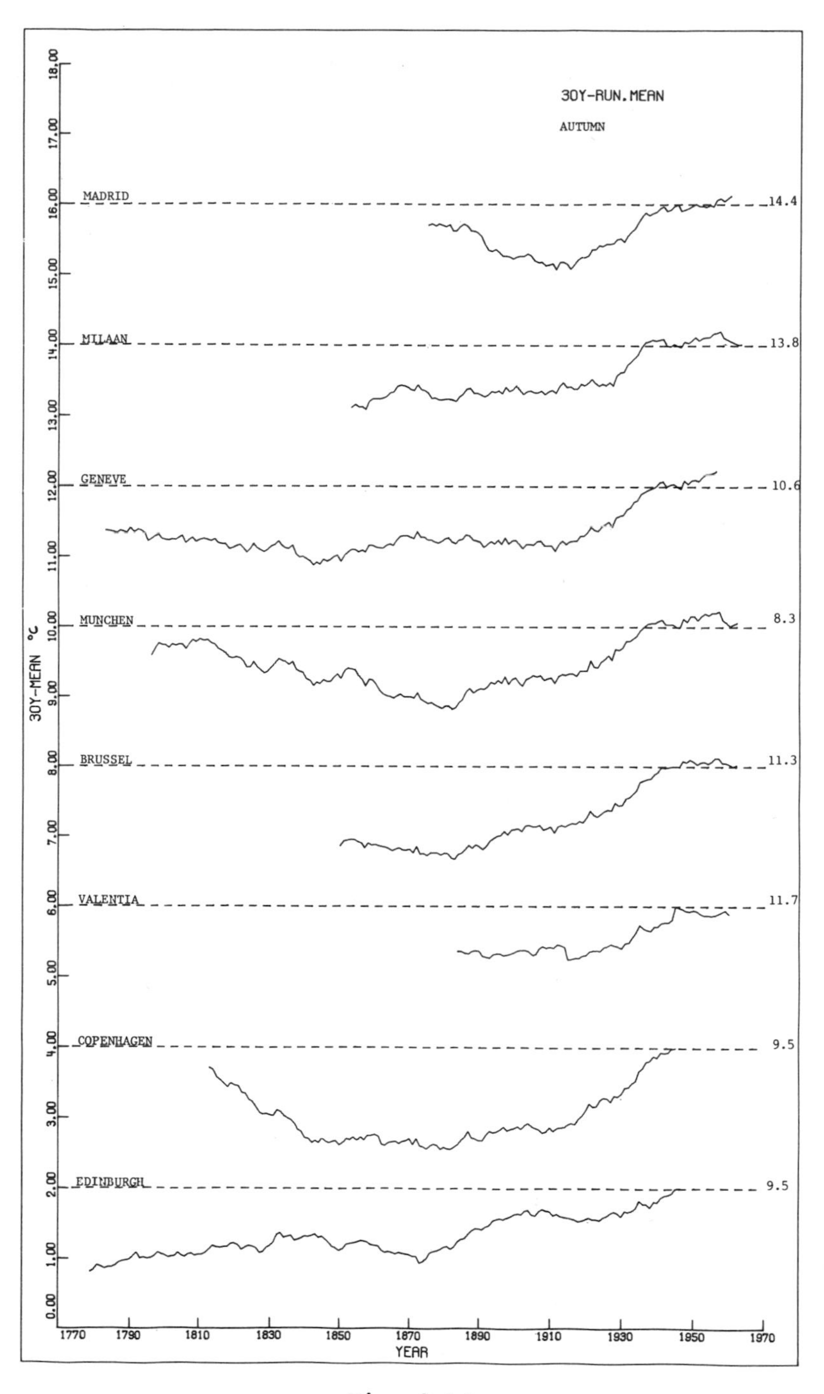
30Y-RUN.MEAN
AUTUMN
MADRID
14.4
MILAAN
13.8
GENEVE
10.6
MUNCHEN
8.3
BRUSSEL
11.3
VALENTIA
11.7
COPENHAGEN
9.5
EDINBURGH
9.5
30Y-MEAN °C
0.00
1.00
2.00
3.00
4.00
5.00
6.00
7.00
8.00
9.00
10.00
11.00
12.00
13.00
14.00
15.00
16.00
17.00
18.00
1770
1790
1810
1830
1850
1870
1890
1910
1930
1950
1970
YEAR

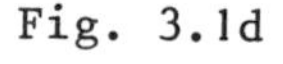
Fig. 3.1d

closer to 1°C for stations in the Mediterranean countries 0.5°-1.5°C.
For summer temperatures: at all stations 0.5°-1.5°C, with the lowest values in the maritime areas and higher ones over the continent.

Ranges of 30-year average spring and autumn temperatures are generally in between those for winter and summer, though in maritime areas they may be higher.

3.1.1.3 Coherence between seasons

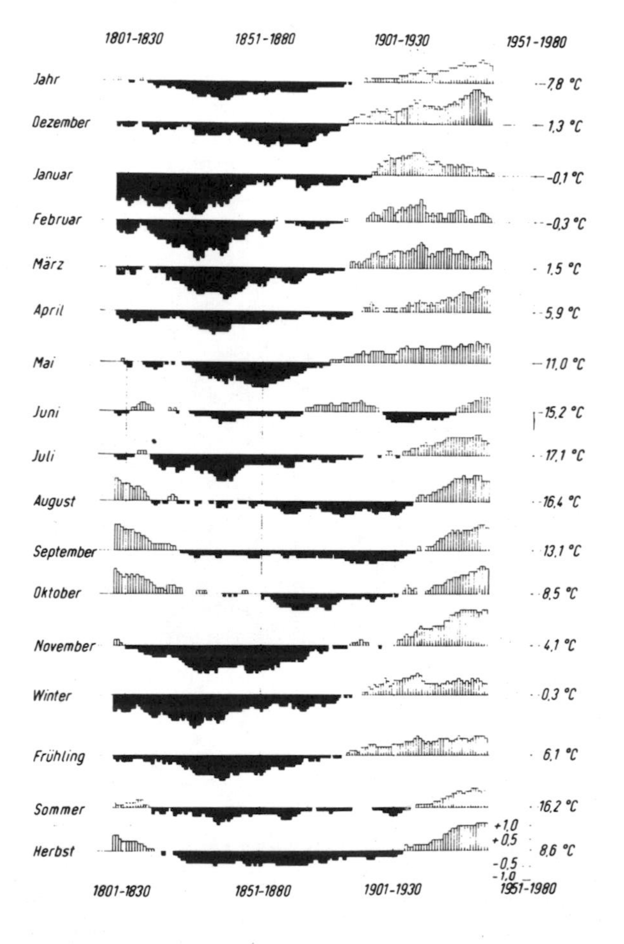

Fig. 3.2. Thirty-year running averages temperatures at Copenhagen. After von Rudloff (1967).

It does occur that changes are coherent in two adjacent seasons. In general, however, temperature changes in different seasons are not coherent or simultaneous. Especially the large warming and subsequent cooling in the present century did not run in parallel in all seasons. This again is illustrated by the 10-year running seasonal averages for Central England (Fig. 2.9) and in more detail on a monthly basis by the 30-year averages for Copenhagen (Fig. 3.2). A seasonal shift in the starting time and the maximum of the warming however does not seem to occur everywhere outside the European area. An example of strong parallelism is given in Fig. 3.3.

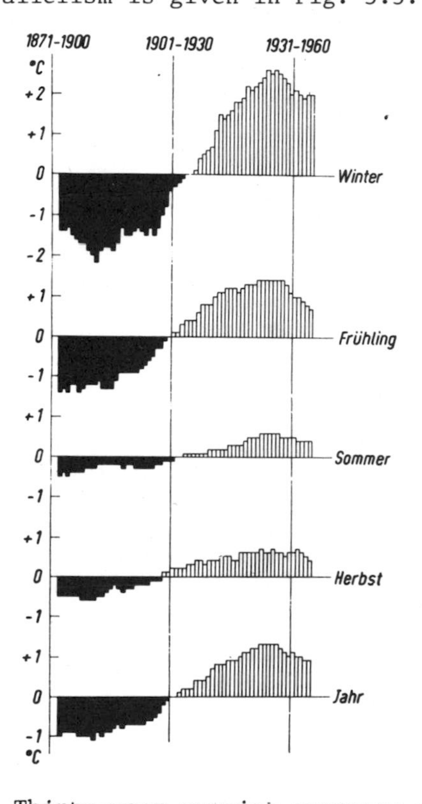

Fig. 3.3. Thirty-year running averages of temperature deviations from the 1875-1950 mean at western Greenland. After von Rudloff (1967).

3.1.1.4. Spatial coherence

Besides coherence between seasons at a particular location, spatial coherence of climate changes at different locations may also be considered. In general, climate change at one location will not be independent of changes at another location, because all places are connected by the atmospheric circulation. Moreover, changes in atmospheric circulation patterns are practically always the immediate cause of climate change. The interrelation, however, will very much depend on how far apart are the stations considered. For the area of the European Community they are so close together and under the influence of the same circulation types that large differences in timing, as well as in sign, of the temperature changes, may not be expected. Certainly not on a 10- to 30-year time basis. As may be judged from the curves in Fig. 3.1. (a)-(d) temperature changes over Europe are quite coherent over the whole area considered, though some remarkable incoherences exist also both in the phase and the amplitude of certain events.

In comparing temperature changes in the maritime and continental areas of Europe over the last century Meyer zu Düttingdorf (1978) has come to the same conclusion. She emphasises, however, that for the annual mean temperatures, as well as for winter and summer temperatures the patterns of change in the two areas are more identical than for spring and autumn values.

For places further apart trends in temperature may occur simultaneously, but be of opposite sign. An example of the latter is given in Fig. 3.4.

Furthermore, Schönwiese (1978) has found indications that a regional coherence of temperature changes exists only for the long-period part of the variations (i.e. for variability on a time scale of more than 10 years); shorter-period (1-10 years) variations are present at all stations but they do not show any regional pattern. He also shows a clear example of a time or phase shift of nearly 20 years between the start and maximum of the 20th century warming in Scandinavia and Central Europe based on 10-year running yearly mean temperatures.

The mere fact that for winter temperatures an opposite phase shift between North and South/Central Europe is observed (see von Rudloff (1967), table 39) illustrates the complexity of the matter.

3.1.1.5. Present temperature trends

Finally, something must be said about the present situation. As has been stressed before, the period 1931-60 was exceptionally warm, comprising the maximum of the great climate fluctuation of the present century. This should mean that temperatures from 1961 until now (1983) have been more often below than above the values of the 1931-60 reference period. In Fig. 3.5, this is verified for a small selection of stations: Valentia (Ireland) being under the direct influence of the North Atlantic Ocean, De Bilt, experiencing Atlantic and North Sea influence but strongly influenced by continental air masses too, Munich as a central European station, Madrid at the lower and Stockholm at the higher European latitudes.

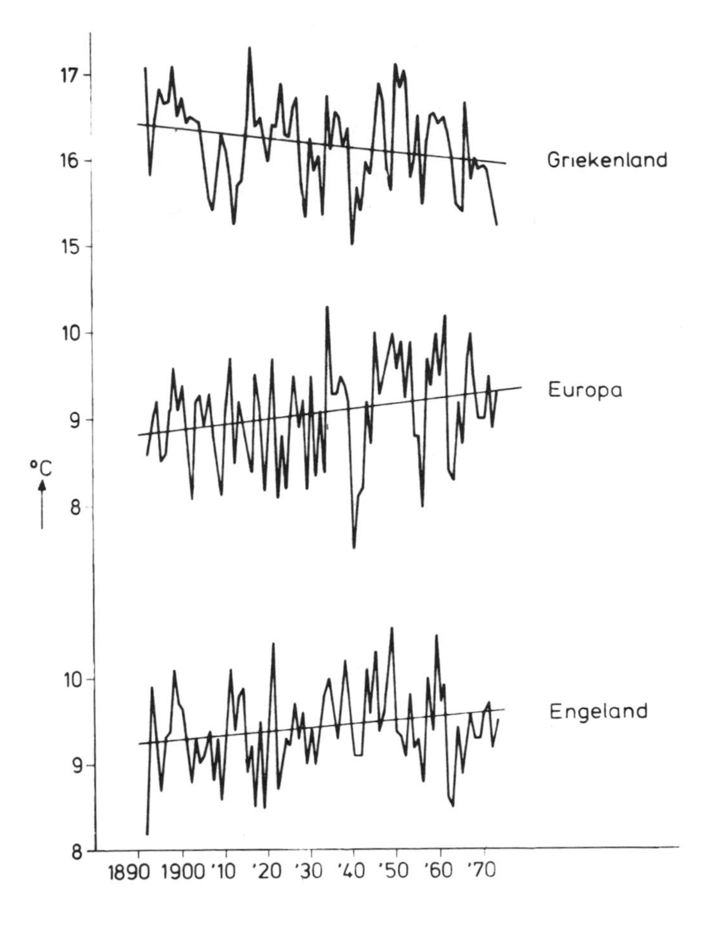

Fig. 3.4. Trends of yearly mean temperatures in Greece, Central Europe and Central England. After P.J. Rijkoort, Internal Report, KNMI, 1975.

		1961	62	63	64	65	66	67	68	69	70	71	72	73	74	75	76	77	-78	79	80
winter	Stockholm	++	-	--	0	0	--	0	--	-	--	+	++	++	++	++	+	-	0	--	--
	Valentia	0	0	-	0	-	0	0	0	--	0	0	0	+	0	0	0	--	-	-	0
	De Bilt	++	0	--	-	0	+	++	0	-	--	+	+	+	++	++	+	+	+	--	+
	Munich	++	+	--	--	0	++	++	+	0	-	-	++	++	++	++	+	+	0	0	++
	Madrid	+	+	-	+	+	+	0	0	0	+	-	+	+	+	++	+	+	++	+	+
spring	Stockholm	++	--	0	+	-	-	+	+	-	-	0	0	++	+	+	0	0	0	0	0
	Valentia	+	-	-	0	0	0	-	-	-	-	-	-	0	0	-	0	-	-	--	-
	De Bilt	+	--	0	0	-	0	+	0	-	-	-	0	-	0	-	-	0	0	-	0
	Munich	+	--	0	0	-	+	0	+	0	--	-	0	-	0	0	-	0	-	0	--
	Madrid	++	0	0	+	+	0	0	-	-	0	--	-	0	-	--	0	0	-	-	0
summer	Stockholm	-	--	0	-	--	+	0	+	++	0	0	+	++	-	++	0	--	-	0	0
	Valentia	-	-	-	-	-	0	-	0	0	0	0	-	-	-	+	+	0	-	-	-
	De Bilt	-	--	-	0	-	-	0	0	+	0	0	-	+	-	+	++	-	-	-	-
	Munich	0	0	0	+	-	-	0	0	-	0	0	-	0	-	-	0	-	--	0	-
	Madrid	+	+	-	0	0	0	0	0	-	0	--	-	0	0	0	0	--	-	0	0
autumn	Stockholm	++	0	+	0	-	0	++	-	0	-	-	0	--	0	+	-	0	0	-	--
	Valentia	-	-	0	0	-	-	-	+	0	0	+	-	+	+	0	-	0	+	0	-
	De Bilt	+	-	+	-	--	0	0	0	+	+	0	--	-	-	-	+	+	0	0	0
	Munich	++	-	++	0	-	+	0	+	+	+	-	--	-	-	0	0	0	-	0	0
	Madrid	0	+	+	+	-	-	0	+	-	++	+	-	0	-	0	--	+	++	0	0
year	Stockholm	+	-	-	0	-	-	+	0	0	-	0	+	+	+	++	0	0	-	-	-
	Valentia	0	-	-	0	-	0	-	0	-	0	0	-	0	-	0	-	-	0	-	-
	De Bilt	+	-	--	0	-	0	+	0	0	0	0	-	0	0	+	+	+	0	-	0
	Munich	+	-	-	0	0	+	+	0	0	0	0	-	0	+	0	0	0	-	0	-
	Madrid	+	0	0	+	0	0	0	+	-	+	-	-	--	0	0	0	0	0	0	0

Fig. 3.5. Seasonal and yearly mean temperatures for the period 1961-80 expressed as a deviation from the corresponding 1931-60 value. Key: ++ $\Delta T > +1.5°C$; + + $1.5°C > \Delta T > +0.5°C$; 0 + $0.5°C > \Delta T > -0.5°C$; - $|-1.5|°C > \Delta T > |-0.5|°C$; -- $\Delta T > |-1.5|°C$.

In general, a tendency for temperatures lower than the 1931-60 mean is apparent only in spring and summer (except at Stockholm). Winter temperatures at the selected stations show a warming trend, as is to some extent also the case for autumn. The yearly mean has shown some large and persistent negative deviations in the beginning of the period with zero to positive deviations in the 1970's. Madrid, however, shows the opposite trend.

These results are in agreement with those published by Cehak (1977) for Austria. Here, winter and annual mean temperatures increased significantly since the early sixties, whereas the summer temperature mostly showed a more or less significant decrease.

Temperature trends for the free atmosphere have also been analysed. See e.g. Dronia (1974), from which Fig. 3.6 (updated to 1978) is taken.

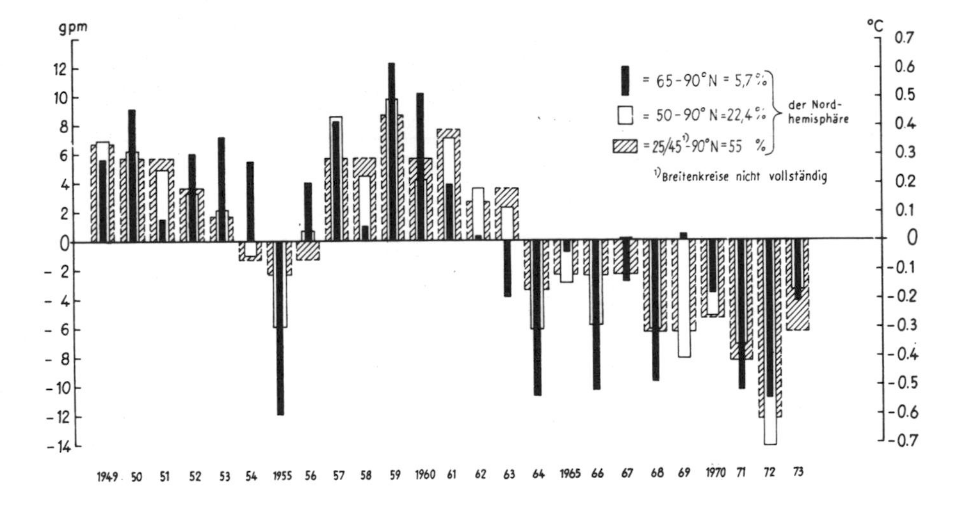

Fig. 3.6. Annual deviations of 500-1000 mb thickness from the mean 1949-1973. After Dronia (1974, updated).

The analysis however applies to the mean hemispheric situation and does not indicate whether the cooler conditions since 1962 also prevailed in the upper air of the European area nor how the different seasons behaved.

3.1.2. *Changes in variability*

3.1.2.1. *Temperature variability*

Here, we consider variability of time series consisting of seasonal or yearly mean temperature values. Variability occurs in many forms and various quantities may be employed to quantify it. Variance (σ^2) or standard deviation (σ) with respect to a certain mean value may be used.

In view of the relatively large contribution of extreme values to the value of σ, variability V or interannual variability I are sometimes preferred. Here, emphasis will be on the interannual variability; but in some instances standard deviations will also be used.

The magnitude of the interannual variability of seasonal mean temperatures in Europe is generally of the order of one degree Celsius. As is well known, variability is largest in winter and smallest in summer, except in the southern European countries, where I is nearly the same for all seasons. Further-

Table 3.1. Ratio of σ of monthly mean temperature to that of Central England (period 1900-1950 approx.). After Lamb (1977), slightly modified.

	Madrid, Spain	Ireland (eastern and inland	Netherland, Rhineland	Bergen, Norway	Western Iceland	Western Greenland near 65°N	Novaya Zemlya (south Island)
January	0.7	0.8	1.5	1.1	1.2	2.1	2.4
April	1.4	0.9	1.2	1.0	1.4	1.5	2.2
July	1.0	0.7	1.1	1.2	0.8	1.2	1.8
October	1.3	0.8	1.2	0.9	1.2	0.8	2.0

Table 3.2 Extreme values of non-overlapping 10-year standard deviations of monthly mean temperatures at De Bilt, over the period 1741-1940.

Month	J	F	M	A	M	J	J	A	S	O	N	D
Max σ	3.71	3.22	2.64	1.87	2.12	1.74	1.83	1.75	1.47	1.81	2.12	3.61
Min σ	1.45	1.78	1.25	0.94	0.93	0.59	0.77	0.61	0.66	0.77	0.70	1.72

After Van der Bijl (1952).

more, continental and inland stations show a larger variability than stations strongly influenced by the ocean or seas.

This may be further ilustrated by Tables 3.1 and 3.2.

3.1.2.2. Time changes of temperature variability

During the last ten years, climate change has become a problem of much worldwide concern. One reason has been the occurrence of disastrous climatic events in various parts of the world. The frequency of occurrence of climatic extremes seemed to have become larger everywhere and climatologists told the public that we had entered an epoch of a much more variable climate than in the recent past. This special kind of climate change -change of climate variability- started to receive considerable attention in publications and scientific meetings. Yet the picture is till far from clear. The claim that climate recently had become more variable could hardly be substantiated, except for some regions outside Europe. This does not mean however that time changes of climate variability have not occurred in Europe, but they have attracted relatively little attention compared to time changes of the temperature averages. Van der Bijl (1952) made an extensive study of the standard deviation of monthly mean temperature for non-overlapping ten-year periods for two centuries (1740-1940) as shown by the long De Bilt record. His purpose, however, was merely to analyse the frequency distribution of standard deviations. It turned out that the 20 x 12 values of $(n-1)s^2/\sigma^2$ obeyed a χ^2-distribution. (s = standard deviation of monthly mean temperatures in an n-year period, σ = standard deviation of monthly mean temperatures over the period 1741-1940, n = 10 in this case.) In the same study he found an apparent dependence between the 12 months, which could not fully be explained by month-to-month persistence of temperature. Most probably long-term climatic changes in temperature variability account for the rest.

The practical importance of temperature variability and its time changes was completely ignored in the study by Van der Bijl, as is also the case in other studies of climate; only recently has this aspect received more attention. This point has also been stressed by Lamb (1977, p. 485); his data show that the standard deviations over 20-year periods have varied by a factor of 1.25-1.50.

This has been verified for a longer period by computing the interannual variability of winter temperatures at De Bilt for 10, 30, 60 and 90-year periods for the period 1640-1979. The results are given in Table 3.3.

Table 3.3. Interannual variability of winter temperatures at De Bilt (in °C). Note the decadal values 1660-1699 and 1790-1849.

	over 10 years	over 30 years	over 60 years	over 90 years
1640-49	1.94			
1650-59	1.68			
1660-69	3.01	2.51		
1670-79	2.84			
1680-89	2.99			
1690-99	3.12	2.36		
1700-09	1.01		2.07	
1710-19	2.19			
1720-29	1.40	1.78		
1730-39	1.75			
1740-49	2.13			
1750-59	2.13	2.30		2.07
1760-69	2.61		2.22	
1770-79	1.07			
1780-89	1.97	2.14		
1790-99	3.39			
1800-09	2.42			
1810-19	1.58	2.30		
1820-29	2.91		2.27	
1830-39	2.55			
1840-49	2.78	2.25		2.09
1850-59	1.41			
1860-69	2.01			
1870-79	2.04	1.73		
1880-89	1.14		1.57	
1890-99	1.82			
1900-09	1.23	1.42		
1910-19	1.22			
1920-29	1.90			
1930-39	1.60	2.07		1.66
1940-49	2.72		1.77	
1950-59	1.29			
1960-69	1.80	1.48		
1970-79	1.36			

Interannual variability is somewhat larger than standard deviation but its variability is nearly the same as for standard deviation. The long record shows that estimates of variability as mentioned by Lamb are rather conservative. The De Bilt winter temperatures show the following ratio of maximum to minimum interannual variability:

For non-overlapping 10-year periods: 3.36,
For non-overlapping 30-year periods: 1.77 (1.94),
For non-overlapping 60-year periods: 1.45 (1.37),
For non-overlapping 90-year periods: 1.26 (1.12).
Between brackets the values of $3.36/\sqrt{n}$ (n = 3, 6, 9, respectively) are given, which give a good first-order approximation.

3.1.2.3. Coherence over Europe and between seasons

Curves of 30-year running interannual variability values of seasonal mean temperatures have been prepared for a number of European stations and are shown in Fig. 3.7 (a)-(d) (scale and plotting is the same as in Fig. 3.1). Inferences about the spatial coherence of time changes of interannual variability can be made, as well as conclusions about a possible parallel behaviour in different seasons. However, visual inspection of the various curves is enough to reveal that the pattern of time changes of interannual variability is very irregular, both in space and over the year.

For stations influenced by the same climatic regime (continental or maritime), and not further apart than 500-1000 km, interannual variability may run in parallel for some time; but parallelism in one season does not necessarily mean parallelism in other seasons. In some instances, phase changes between stations appear to occur in the same way as for changes in mean temperature level. An interesting case is the minimum of interannual variability in the first half of this century, often referred to in the literature. Staring with the winter season, we see that this minimum is most pronounced in the maritime areas (stations Edinburgh, Copenhagen, etc.) reaching its lowest 30-year value as early as 1890-1920. In Central Europe, this minimum was far less pronounced and did not occur before 1930-1960. In Southern Europe this minimum does not seem to be present. Spring shows about the same changes as winter in the maritime areas, but at the inland stations the situation is much more complex. In general, at the latter stations interannual variability of spring temperatures has shown an increasing trend from the beginning of this century on. The present century minimum of interannual variability of summer temperatures occurred around 1920-1950 at practically all stations except in Southern Europe. Summer is the only season with rather coherent changes over a large part of Europe. Not only does minimum interannual variability in the 1920-1950 coincide at many stations, but also the preceding maximum at about 1890-1920 and minima and maxima at about 1860-1890 and 1820-1850, respectively. Autumn is interesting in that it shows a close parallelism with summer for an extended period, which is then followed by a period of stronger coherence with winter. A clear-cut minimum in the present century is lacking at a number of stations. At the most northern stations most of the time interannual variability of autumn temperatures has been below the 1931-60 value, while at the other stations just the opposite occurred.

In general, 30-years interannual variability of seasonal temperatures may change nearly as abruptly as the 30-year seasonal mean temperature level. This appears to be the case all over Europe except in the Mediteranean area.

A quantitative measure of the coherence of interannual variability between seasons is given by the following example for De Bilt. For 27 non-overlapping 10-year periods (1710-1979) the interannual variability of January and July-temperatures yields a correlation coefficient $r = +0.32$, significant only at the 10% level. If this indicates a causal relationship, most probably this common cause is the frequency of zonal circulation patterns in both seasons alternating with periods of more meridional circulation (see Chapter 1).

Wallén (1953) had already found that the variance of the summer temperature in Sweden in the past 200 years tended to be greatest in periods with prevailing meridional circulation. Similarly Lamb (1977) found that the variance of winter temperatures in western Europe was greatest in periods with a minimum of westerly winds. Van Loon and Williams (1978) did not find a simple connection between variability and temperature trends; this was confirmed by extensive studies on data for European stations, where statistically significant (negative) correlations between mean temperature and interannual variability only were found for the winter season (Schuurmans and Coops, to be published in Monthly Weather Review, 1984).

3.1.2.4. Present trends in temperature variability

From Figs. 3.7 (a)-(d) and other available curves, Table 3.4 has been derived, indicating for each station and season whether the 30-year running interannual variability at present shows a rising (r), declining (d) or neutral (o) trend. The indications are quite rough and determined by visual inspection only.

Apparently, for summer temperatures the interannual variability in increasing nearly everywhere. For the other seasons the situation is rather mixed, increasing trends prevailing in autumn, while winter and spring show a decreasing temperature variability at most stations.

A more quantitative look at the situation for recent years (since 1961), also does not confirm the idea of a general rise of inter-annual temperature variability in Europe. For the same 5 stations treated in Fig. 3.5 the interannual variability of seasonal mean temperatures since 1961 has been compared with I-values for the reference period 1931-60 (see Table 3.5).

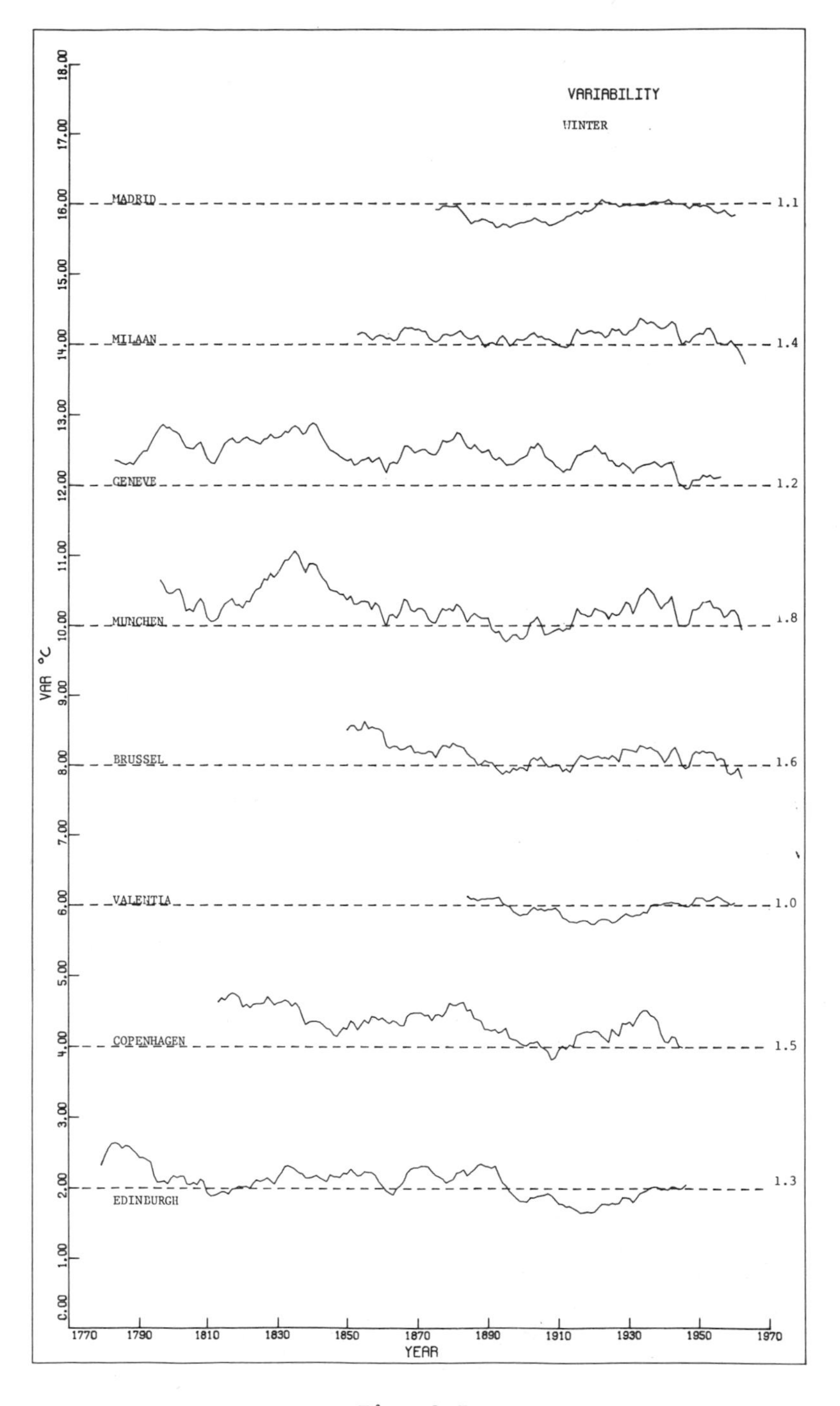
VARIABILITY
WINTER
MADRID
MILAAN
GENEVE
MUNCHEN
BRUSSEL
VALENTIA
COPENHAGEN
EDINBURGH
1.1
1.4
1.2
1.8
1.6
1.0
1.5
1.3
VAR °C
C.00
1.00
2.00
3.00
4.00
5.00
6.00
7.00
8.00
9.00
10.00
11.00
12.00
13.00
14.00
15.00
16.00
17.00
18.00
1770
1790
1810
1830
1850
1870
1890
1910
1930
1950
1970
YEAR

Fig. 3.7a

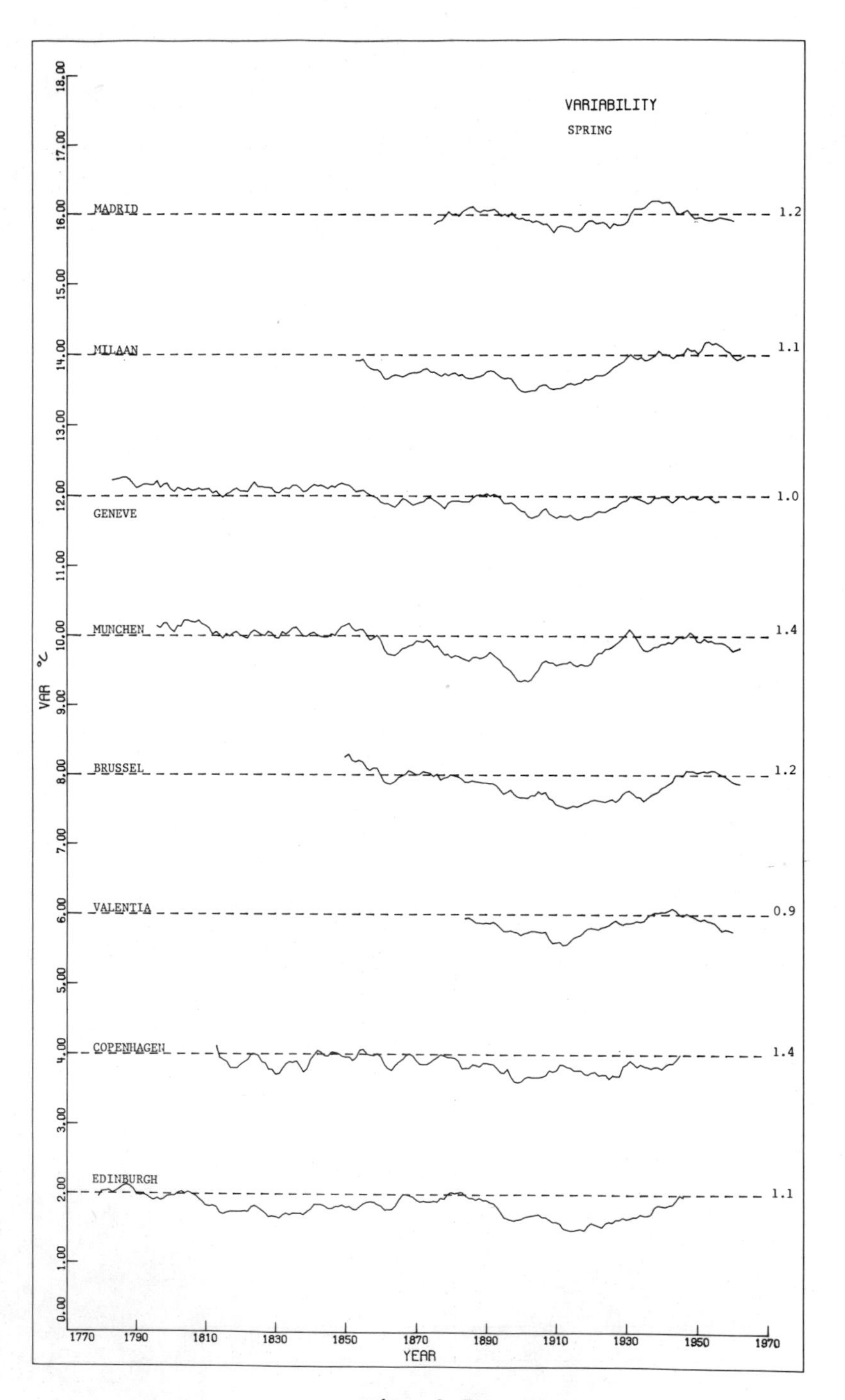
VARIABILITY
SPRING
MADRID
MILAAN
GENEVE
MUNCHEN
BRUSSEL
VALENTIA
COPENHAGEN
EDINBURGH
1.2
1.1
1.0
1.4
1.2
0.9
1.4
1.1
VAR °C
0.00
1.00
2.00
3.00
4.00
5.00
6.00
7.00
8.00
9.00
10.00
11.00
12.00
13.00
14.00
15.00
16.00
17.00
18.00
YEAR
1770
1790
1810
1830
1850
1870
1890
1910
1930
1950
1970

Fig. 3.7b

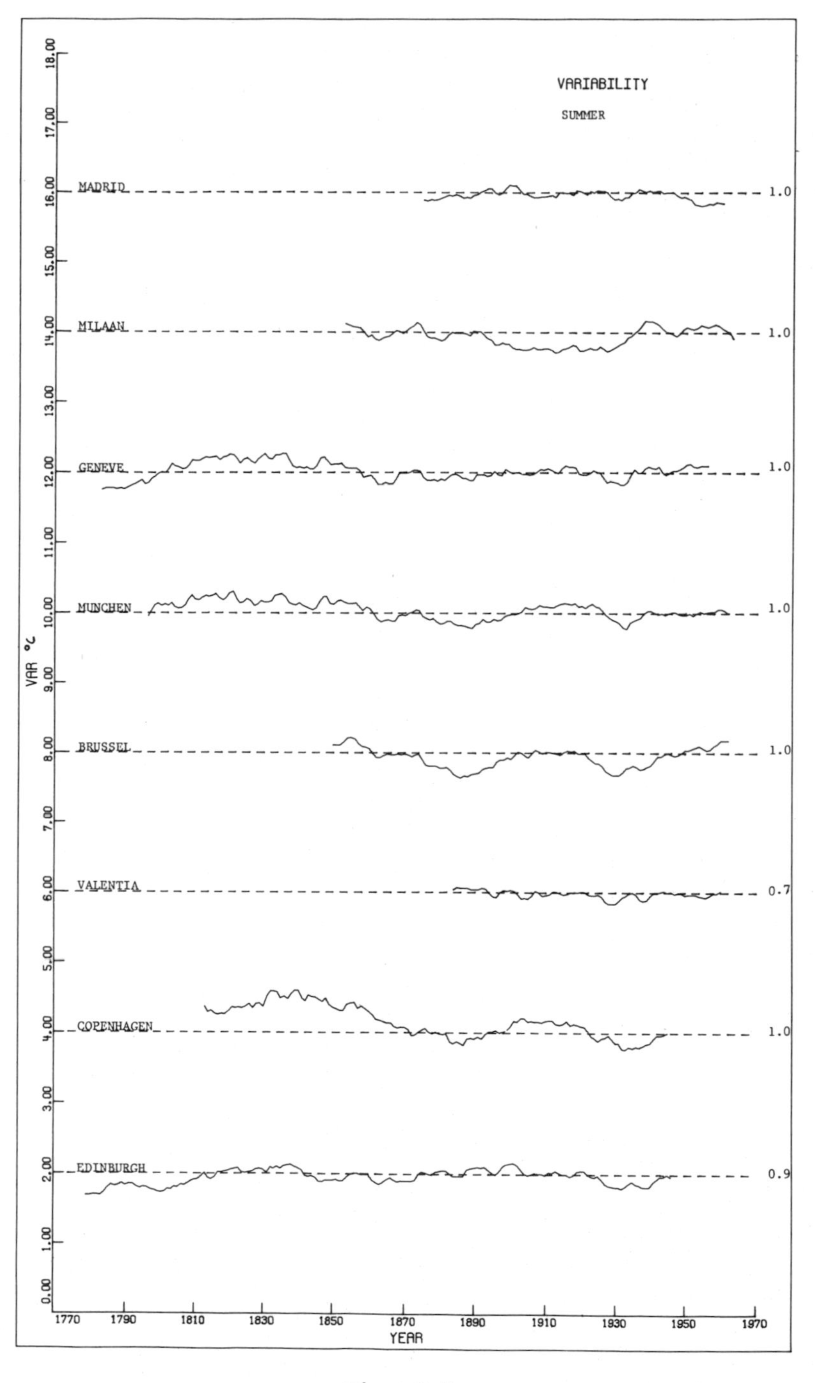
VARIABILITY
SUMMER
MADRID
MILAAN
GENEVE
MUNCHEN
BRUSSEL
VALENTIA
COPENHAGEN
EDINBURGH
1.0
1.0
1.0
1.0
1.0
0.7
1.0
0.9
VAR °C
18.00
17.00
16.00
15.00
14.00
13.00
12.00
11.00
10.00
9.00
8.00
7.00
6.00
5.00
4.00
3.00
2.00
1.00
0.00
1770
1790
1810
1830
1850
1870
1890
1910
1930
1950
1970
YEAR

Fig. 3.7c

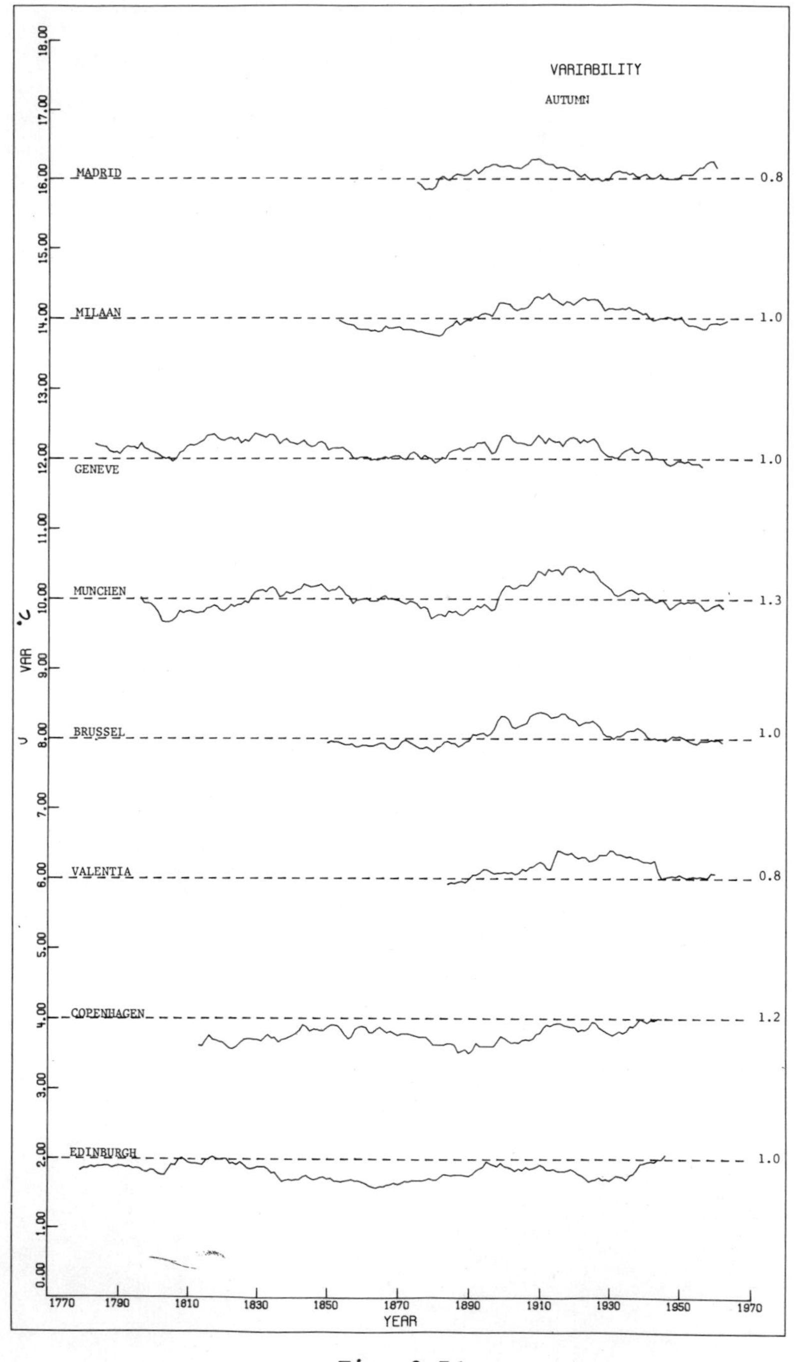
VARIABILITY
AUTUMN
MADRID
MILAAN
GENEVE
MUNCHEN
BRUSSEL
VALENTIA
COPENHAGEN
EDINBURGH
0.8
1.0
1.0
1.3
1.0
0.8
1.2
1.0
VAR °C
0.00
1.00
2.00
3.00
4.00
5.00
6.00
7.00
8.00
9.00
10.00
11.00
12.00
13.00
14.00
15.00
16.00
17.00
18.00
1770
1790
1810
1830
1850
1870
1890
1910
1930
1950
1970
YEAR

Fig. 3.7d

Table ,3.4. 30-year running interannual variability trends.

	Winter	Spring	Summer	Autumn
Geneva	r	o	r	d
Munich	d	d	o	d
Copenhagen	d	r	r	r
Edinburgh	r	r	r	r
Sonnblick	r	d	r	r
Säntis	r	d	r	r
Madrid	d	d	o	o
Milano	d	d	d	r
Valentia (Irl.)	d	d	r	o
Brussels	d	d	r	o

r- rising; d- declining; o- neutral.

Table 3.5. Comparison of I-values 1931-60 and 1961-79.

	winter		spring		summer		autumn	
	1931-60	1961-79	1931-60	1961-79	1931-60	1961-79	1931-60	1961-78
Stockholm	1.99	2.60	1.62	0.98	1.03	1.28	1.41	1.18
Valentia	1.00	0.71	0.92	0.45	0.69	0.42	0.77	0.84
De Bilt	1.73	1.69	1.08	0.78	0.91	0.92	0.92	0.81
Munich	1.77	1.56	1.36	1.03	1.00	0.73	1.28	0.99
Madrid	1.11	0.62	1.20	0.90	0.97	0.82	0.84	1.20

Taking into account that the period since 1961 is shorter than 30 years, the conclusion from Table 3.5 must be that at least for this selection of stations the seasonal mean temperature variability has not increased as compared to the variability in the 1931-60 period.

In general, one tends to conclude that the problem of changes and trends in temperature variability is rather complex, at least on time scales of the order of decades. While results of Ratcliffe et al. (1978) and Chico and Sellers (1979) agree with the conclusions presented here, results by Tavokol (1979) indicate, quite convincingly, that variability of climate, on a global scale, has increased in the 1960's and 1970's as compared with previous decades. Year-to-year variability of surface air temperatures in the Northern Hemisphere were particularly pronounced in the 1970's (Jones et al. 1982).

3.1.3. *Other aspects of temperature changes*

3.1.3.1. *Temperature extremes*

An important aspect of changes in temperature level and/or variability is the time change in the frequency of occurrence of temperature extremes. For the present century, Fig. 3.8. shows changes in the decadal frequence of occurrence of monthly mean temperatures which deviated more than 2 σ from the overall mean.

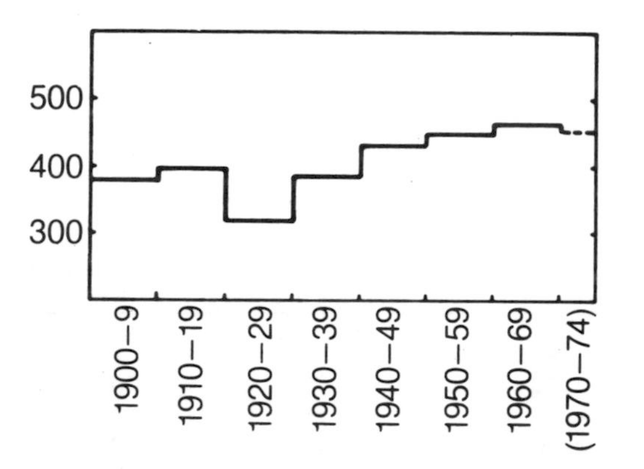

Fig. 3.8. Changes from decade to decade in the occurrence of "extreme" (see text) monthly mean temperatures. After Lamb (1982).

The picture is based on a global network of stations. The 1920-29 period shows a clear-cut minimum, while values remain at a relatively high level from 1940 until the 1970s. This picture is more or less confirmed by the result of the following analysis of extremes at a selection of 11 European stations. For each station and each individual month the year of the lowest and highest monthly mean temperature of the whole record are ranked according to 40-year periods starting with 1740-1779 and ending with 1940-1979.

The resulting distribution of extreme months, when properly weighted, shows the following values of "density" of extremes:

1740-79	1780-1819	1820-59	1860-99	1900-39	1940-79
0.07	0.16	0.16	0.14	0.15	0.18

The value for the 1740-79 period appears to be the lowest by far, but, in view of the fact that only a few very long records contributed to this period, its reliability may be small. The other values do not differ very much, but the highest value is clearly attained in the most recent 40-year period. The analysis however has its shortcomings; for instance, trends in the individual series have not been removed.

The mere fact that the frequency of extremes does vary in time makes it very difficult, if not impossible, to say anything reliable about the probability of occurrence of certain extremes. Answers will very much depend on the length of the series and the time period chosen for analysis. In general, it will be advisable to use the longest record available. The most important extremes as far as temperature is concerned are the extreme (cold or warm) winter and summer seasons or months. The areal extent of such extremes is usually quite large, comprising a large part of the European continent. On the basis of 13 long European temperature records, Von Rudloff (1967) compiled a list of the 10 coldest and 10 mildest European winters and the 10 hottest and 10 coolest European summers of the last two and a half centuries (up till 1967).

The top two years of the extreme seasons are given as follows:

	Winter		Summer	
	Coldest	Mildest	Hottest	Coolest
1	1829/30	1924/25	1947	1816
2	1962/63	1795/96	1826	1902

Unfortunately, temperature anomaly maps for these most extreme European winters and summers are not readily available, except

Table 3.6. Magnitude and areal extent of extremes in winters and summers.

	Winter		Summer	
	coldest	mildest	hottest	coolest
ΔT	-8°C	+5°C	+4°C	-3°C
A	whole of Europe		large part of Europe	

ΔT= maximum deviation from normal temperature attained over a certain area within the European Community; A= area where the temperature deviation is of the same sign.

for the more recent years. Table 3.6 however, gives an idea of the magnitude and areal extent of such extremes.

Not only extreme seasons, however, are of importance. Also extreme months (included in a season which on average is not extreme) may have a serious impact on society, especially in winter and summer. Examples of such extreme months for recent years are the very cold February 1956 (Fig. 3.9 (a)) and the very hot August 1975 (Fig. 3.9 (b)). Maximum deviations here are, of course, larger than for extreme seasons but the areal extent of the deviations is quite representative for the situation in extreme winter and summer seasons respectively.

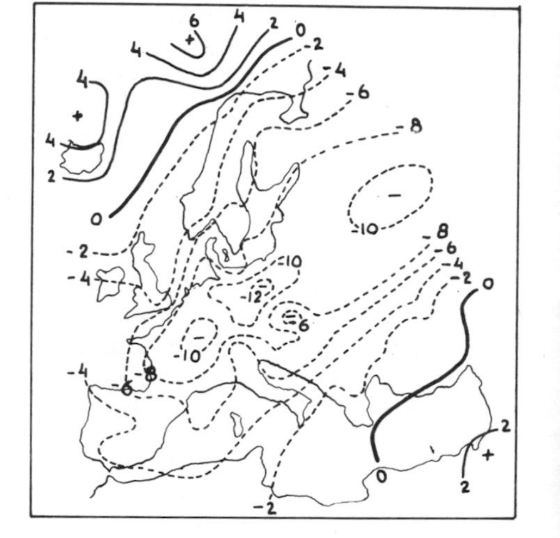

Fig. 3.9. Deviations of monthly mean temperature (in °C) from its reference value: (a) for February 1956 (reference period 1901-1930 and (b) for August 1975 (reference period 1931-60). After Deutscher Wetterdienst (1956, 1975). Note the spatial variability of the deviations.

3.1.3.2. Spectral distribution of temperature variance

Despite considerable effort by various climatologists, applying a range of different methods, real periodicities in the variation of temperature (except for the annual variation) have been difficult to detect. von Rudloff (1967) even concludes that "the non-periodicity of the variations of all climatic elements over the whole of Europe is so strong that periodic changes most often are merely to be considered an unimportant computational result".

From a practical point of view, von Rudloff is probably right, maybe with the exception of the quasi-biennial oscillation (QBO). This 2-3 year cycle shows up in time series of various climatological elements all over the world, see e.g. Landsberg (1962). Power spectra for temperature at Hohenpeissenberg (Germany, 1781-1966) and for Central England (1659-1973) have been published by Schönwiese (1969) and Shapiro (1975), respectively. In both spectra, the QBO is in evidence: Hohenpeissenberg with maxima at 2.2 and 3.3 years and Central England at 2.1 years. A secondary maximum at 11 years (the solar cycle) in the Hohenpeissenberg spectrum fails to show up in the Central England spectrum and in many other long records.

The long record of winter temperatures at De Bilt (1634-1977) does not reveal significant peaks in the power spectrum other than the QBO and a significant depression at 4-6 years (see Fig. 3.10).

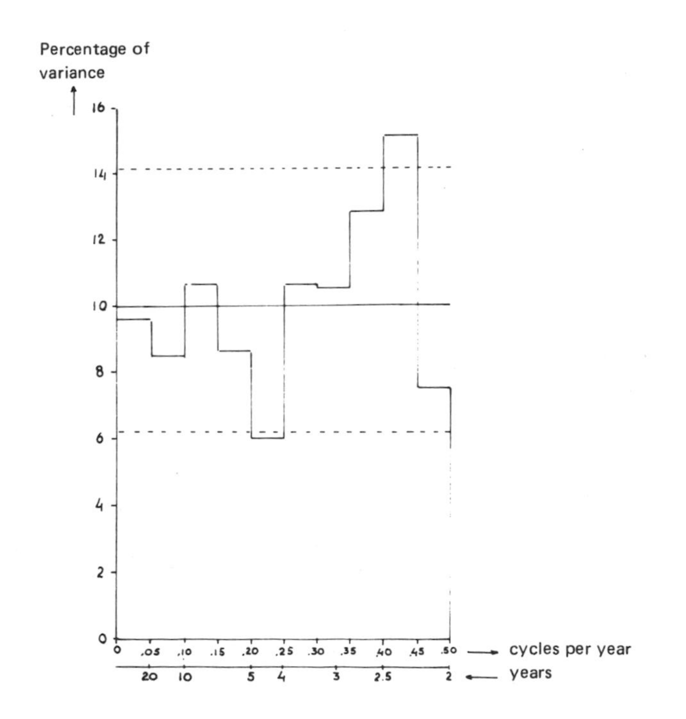

Fig. 3.10. Variance spectrum of winter temperatures at De Bilt for the period 1634-1977. The expected contribution in each of the 10 frequency intervals of equal width is indicated by the solid line; the dashed lines indicate the 5 and 95% confidence limits. After van den Dool, Krijnen and Schuurmans (1978).

Apparently power spectra of monthly or seasonal mean temperature at European stations have only the QBO peak in common.

In spite of the great theoretical interest of such hidden, but wide spread, periodicities, it should be recognized that their contribution to the total variance is in many cases rather small and their practical applicability limited.

3.1.3.3. *Duration and shift of "seasons"*

In the application of climatological data, the duration and timing of certain climatological periods of the year are of more importance and relevance than the mean temperature of the meteorological seasons. This is certainly the case with the vegetation period (or growing season) and the frost-free period of the year.

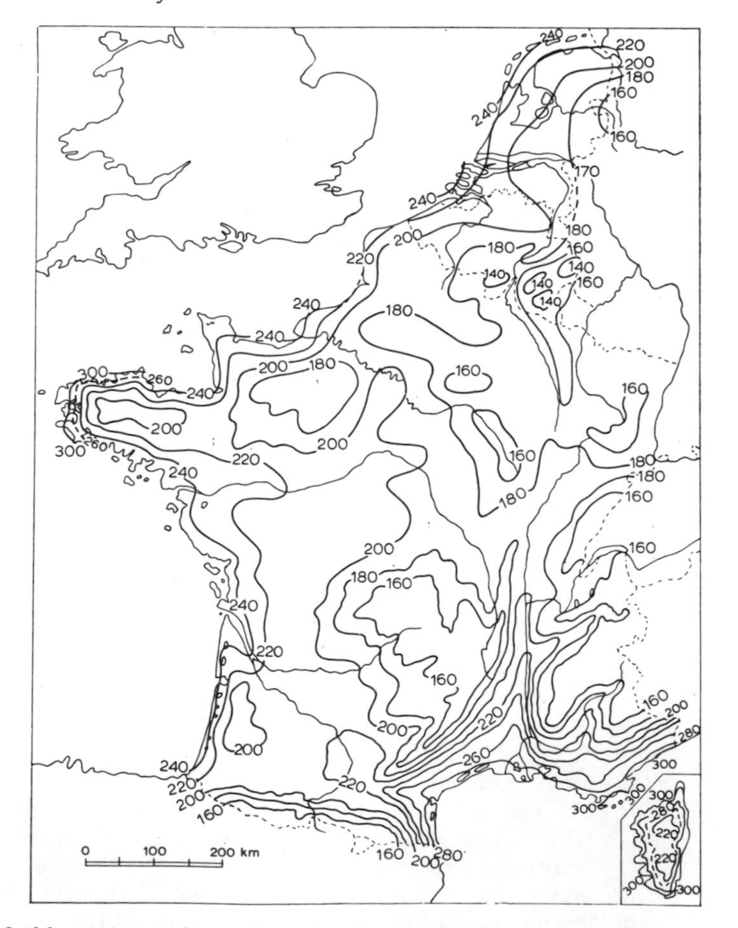

Fig. 3.11. Mean duration in days of the frost-free period for the reference years 1931-60. After Arléry (1970).

The mean durations of these periods for the reference years 1931-60 have been published for a number of European countries, see e.g. Figs. 3.11 and 3.12. Publications sometimes also include the mean dates of the beginning and ending of these periods.

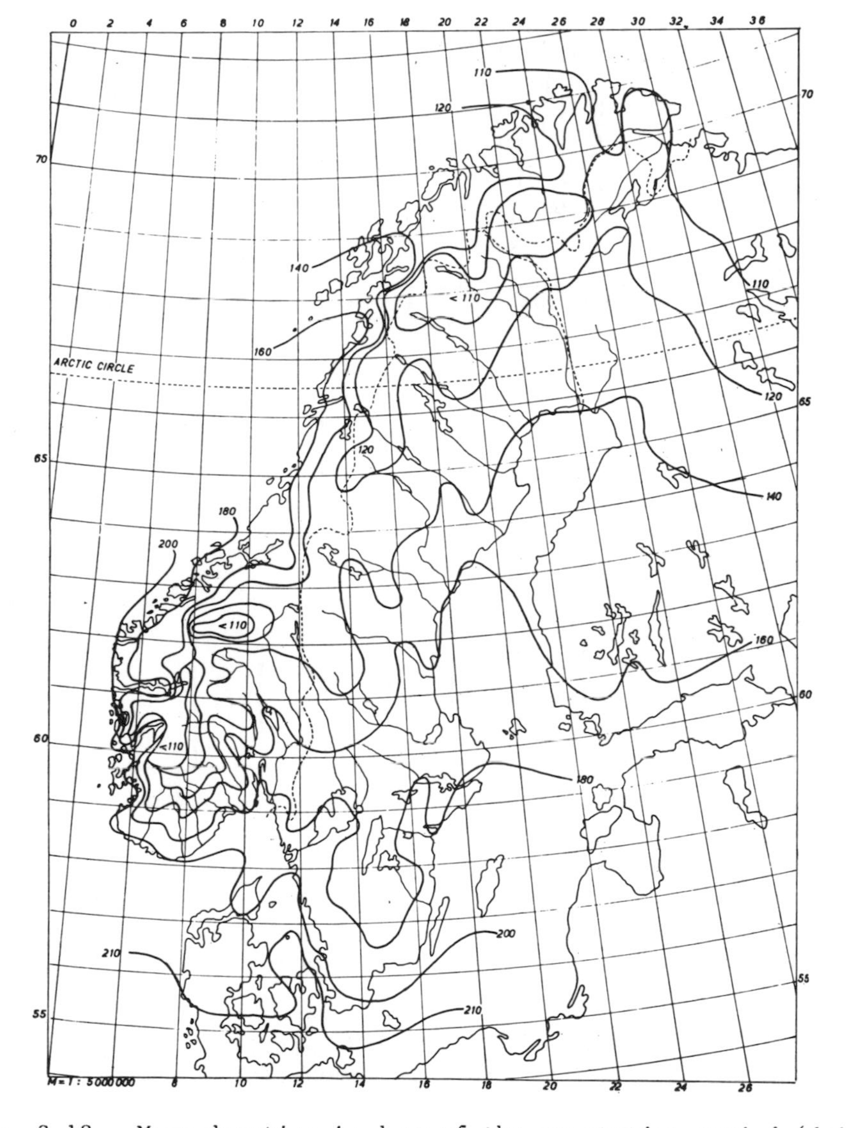

Fig. 3.12. Mean duration in days of the vegetation period (daily mean temperature above 6°C) for the reference years 1931-60. After Johannessen (1970).

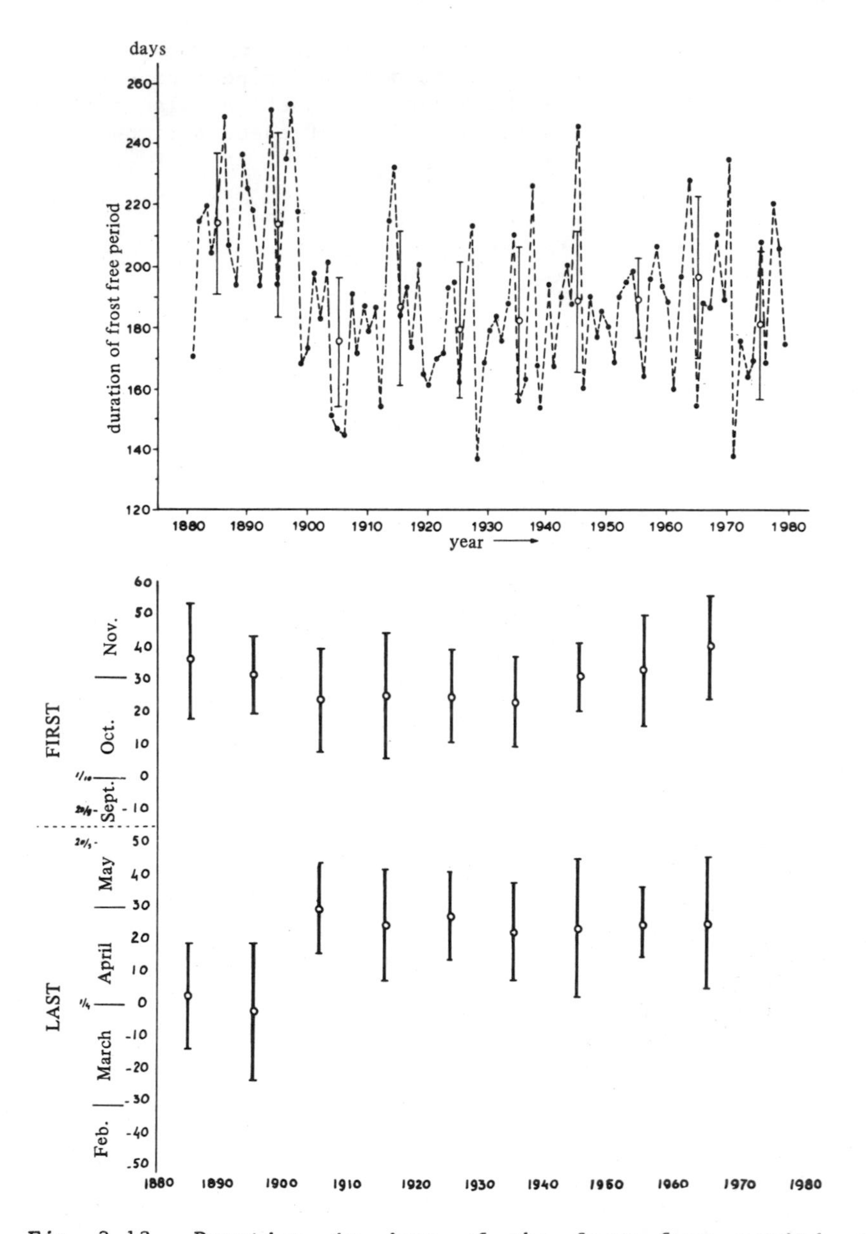

Fig. 3.13. Duration in days of the frost-free period at De Bilt (a), the data of the last frost in spring and of the first frost in autumn (b,c); circles = 10-year averages, bar = σ.

Less emphasis, however, has been given to the time change and variability of the duration and timing of such periods. As far as the growing season is concerned, Lamb (1977) presented some data for the lowlands of Central England (see his table 18.1), indicating that decade averages over the last century have varied by as much as 20 days, while for the minimum duration in each decade a variation of nearly 40 days has been observed. For the 1930-49 period the growing season in Central England seems to have been at an optimum with decadal averages up to 9 months and more.

According to Lamb, it seems likely that in the coldest decades of the Little Ice Age the average growing season was close to 8 months. This tremendous change to more favourable growing conditions has also been shown by Manley (1970). Time change and variability of the duration, beginning and ending of the frost-free period have been studied for De Bilt for the period 1881-1975. The results are presented in Fig. 3.13.

The figures show the rather unexpected result that the duration of the frost-free period was relatively short during the "warm epoch" 1900-1940. On a decadal basis the difference between the end of the last century and the more recent years amounts to 20-40 days. As is clear from the time changes in the first and last days of the frost-free period in The Netherlands, the largest contribution to the shortening of the period

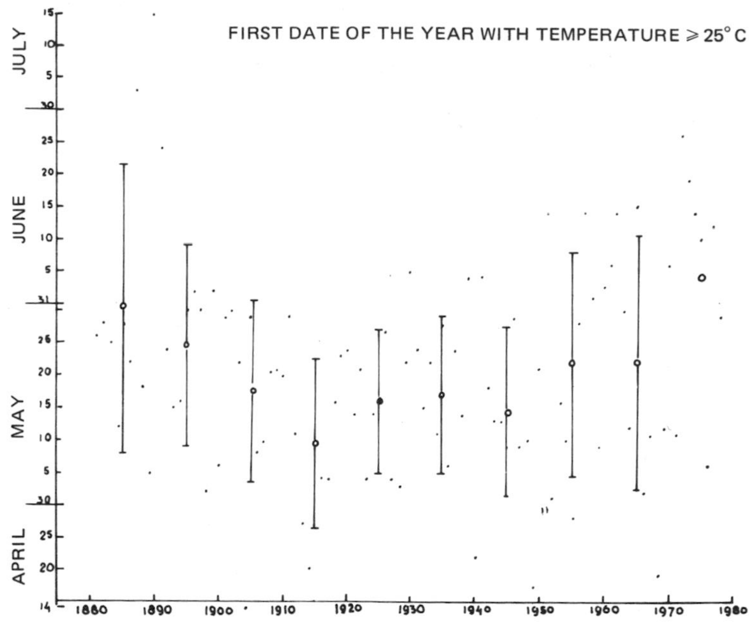

Fig. 3.14. First date of the year with temperature above 25°C at De Bilt. Ten-year averages and standard deviations are indicated.

stems from a shift of the last frost day in spring towards a later date. The increase in the frost-free period in recent years however seems to be due to a postponement of the first frost in autumn.

The standard deviation σ of the elements studied in Fig. 3.13 also varies with time, but not in a systematic way.

An interesting result in the present context is the time variation of the first date of the year that temperature at De Bilt exceeds 25°C (Fig. 3.14). Also in this graph, decadal averages over the last century appear to vary by some 20 days. The earliest dates occur in period 1900-1940, being also the period of lower variability (in terms of standard deviation) of the date of the first summer day as defined by a maximum temperature above 25°C in Holland.

3.1.3.4. *Yearly amplitude of temperature*

The yearly amplitude of temperature, defined as mean July temperature minus mean January temperature, is relatively small in the maritime areas, increases farther inland and reaches its highest values in the interior parts of the continents, at least in the temperate and high latitude areas. Time variations of this yearly amplitude have therefore often been interpreted as changes of the maritime or continental character of climate at the station concerned. Both for the long Central England (1679-1960) and De Bilt (1734-1944) temperature series graphs of the time variation of the yearly amplitude of temperature have been published, see von Rudloff (1967, p. 159), and Labrijn (1944, p. 108), respectively. Only the first-mentioned is reproduced here (Fig. 3.15).

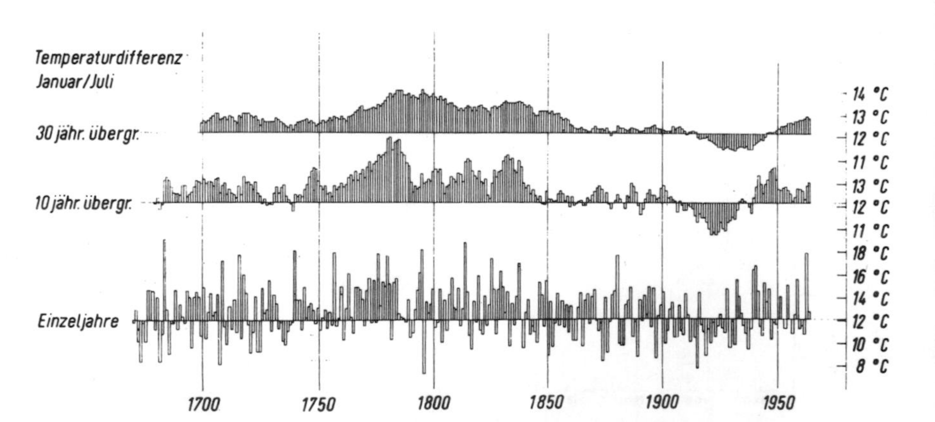

Fig. 3.15. Temperature difference between July and January for individual years and running 10 and 30-year periods in Central England. After von Rudloff (1967).

The two graphs agree to a very large extent in showing an increasing trend of temperature amplitude (more continental character of the climate) from 1730-1780 and a decreasing trend (more maritime conditions) during the whole nineteenth century up till 1930 and increasing again afterwards.

Since the two temperature series belong to nearly the same climate regime, one may wonder whether these tendencies also occurred elsewhere in Europe. Judging from Fig. 3.16 this is partly the case.

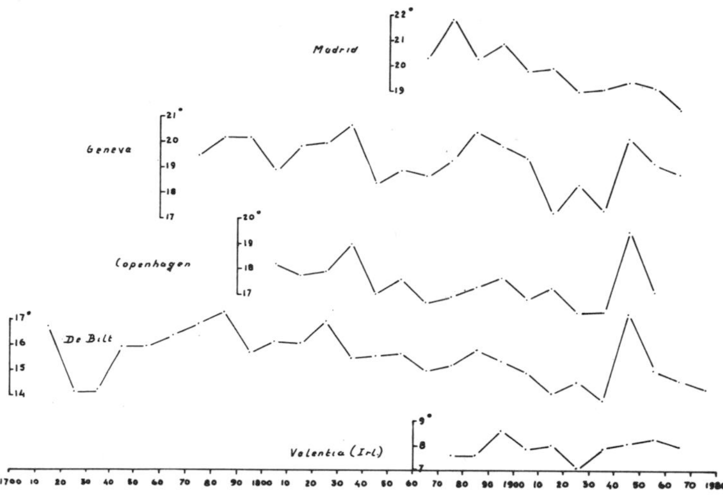

Fig. 3.16. Ten-year averages of temperature difference between July and January in different parts of Europe.

Even stations as far apart as Geneva and Copenhagen show a positive correlation of the time changes of the yearly temperature amplitude, at least on a decadal basis. In general, at these stations the pattern of time change is the same as that described above for Central England and De Bilt. In addition to Fig. 3.15, Fig. 3.16 shows that for the more recent decades the upwards jump in the 1930's or 40's did not continue afterwards. Instead a still decreasing trend is indicated. A general decrease of temperature amplitude is observed in the climate of Madrid. For Valentia (Ireland), the time changes of temperature amplitude have been relatively small, showing a minimum in the 1920's, which is in agreement with the general character of the time changes on the continent.

In general one gets the impression that the time changes of the yearly amplitude of temperature (and thus of changes in the continentality of climate) are coherent over a much larger space scale than for instance the time changes of temperature variability.

3.1.3.5. The city-effect on temperature

Cities are generally warmer than the surrounding countryside for a number of reasons which are well understood (see also Section 4.2.6.1). The effect increases with increasing size of the city and may even be expressed in a simple form as a function of total city population (Oke 1973). For various cities in Europe (London, Oslo, Berlin, Basle, Rome, Vienna, etc.) the effect on temperature has been investigated, though perhaps not in as much detail as the analysis of Dettwiller (1978) for Paris. In Fig. 3.17, annual mean temperatures at Paris-Montsouris are shown for the period 1800 to the present.

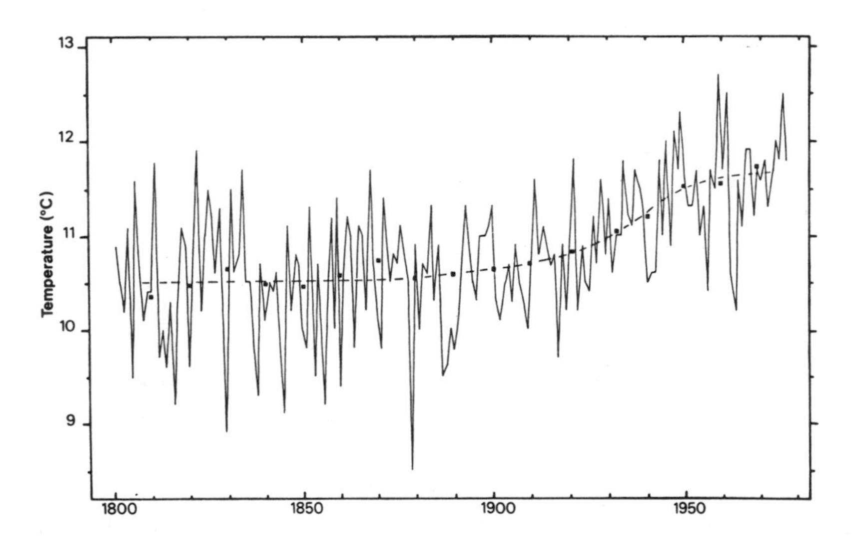

Fig. 3.17. Annual mean temperatures at Paris-Montsouris, with dots indicating median values over 20-year periods, overlapping 10 years with the foregoing period. After Dettwiller (1978).

An increasing trend during the first half of this century is evident, while after 1950 temperature is still increasing though less rapidly. The difference between the mean temperature over the nineteenth century (1801-1900, $\overline{T}$ = 10.50°C) and a recent period of 30 years (1948-1977), ($\overline{T}$ = 11.58°C) amounts to more than 1°C. Of course, as we know, temperature levels in general have been increasing during the present century. Therefore, a comparison with the behaviour of temperature at some nearby station not influenced by the city effect is necessary. Dettwiller made comparisons with temperature evolution at two much smaller cities in France (Besançon and Chartres). Figure 3.18 shows the difference between annual mean temperature at Paris-Montsouris and Besançon, from which it may be concluded that most of the temperature rise observed at Paris during the present century is probably due to the city effect.

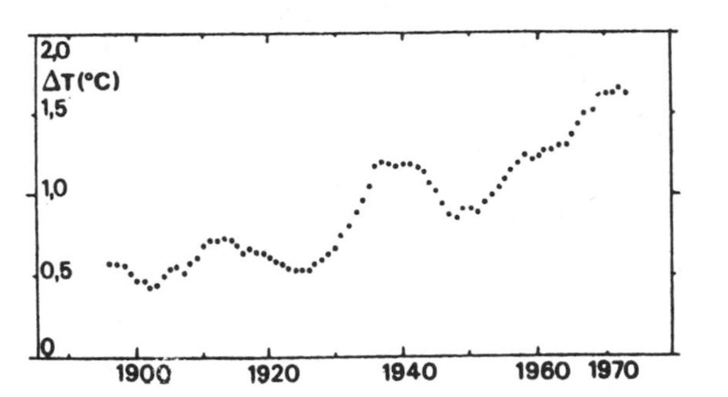

Fig. 3.18. Time evolution of 10-year running averages of the difference in annual mean temperature at Paris-Montsouris and Besançon (Besançon is located at some 300 km east of Paris). After Dettwiller (1978).

From this and many other examples, it has become clear that the city effect on temperature is very large and may seriously influence conclusions about climatic changes based upon an analysis of temperature records of growing cities. Dronia's (1967) warning about this falsifying influence is certainly still valid. A comprehensive study not only of the city-effect on temperature in Europe but of anthropogenic influences on European temperatures in general (e.g. due thermal pollution of inland waters in winter) is strongly to be recommended.

3.1.4. Relation between temperature, winds and ocean currents

The variations of the large-scale wind circulation patterns from the cold climate period of the Little Ice Age climax in the late seventeenth-early eighteenth century to the present day have been the subject of extensive studies, particularly from about 1750 onwards (Lamb and Johnson 1959, 1961) over the region extending from Iceland and eastern North America across the North Atlantic and Europe where the available observation network appears adequate. These studies show that the frequencies of vàrious types of pattern and the positions of main features such as the storm tracks vary not only from year to year but also from decade to decade and over still longer periods. The mean positions of the North Atlantic low pressure region, the "Iceland low", and of the subtropical anticyclones appear to have been about 2 degrees of latitude farther south in the eighteenth-early nineteenth century than in the first half of the present century; the prevailing westerly winds of middle latitudes across the Atlantic and Europe were less regular in their occurrence and more often interrupted by spells of northerly or easterly, and sometimes southerly winds. Since about the 1940s all these tendencies have apparently reverted to a pattern of behaviour more like that which occurred in about the last third of the nineteenth century. Some details of the history of the winds are illustrated in Chapter 2 (see Figs. 2.10, 2.13 and 2.14). There have been corresponding changes also in low latitudes and in the southern hemisphere, though the observation records are shorter there.

The variations of the wind circulation appear to be accompanied by variations in the ocean surface currents also, affecting the course and strength of the main flow of Gulf Stream-North Atlantic Drift water across the North Atlantic and the fluctuating position attained by the boundary between this warm saline water and the polar water flowing south from the area northeast-Greenland-Spitsbergen (see Section 2.3.3.1).

These variations of winds and ocean currents are clearly of importance in connection with many aspects of the climate of Europe. They are intimately connected with the variations of prevailing temperature and rainfall and the incidence of occasional long spells of one or another kind of extreme weather. They also affect the movements of fish stocks in the sea and hence the fisheries. More detailed investigations are needed, including the large-scale spatial correlations (teleconnections) between climatic anomalies in western and central Europe and other regions of the world, having in mind the economic role of such anomalies and their interdependence.

3.1.5. Conclusions

3.1.5.1. Mean temperature level

During the period of instrumental observations 30-year average seasonal mean temperatures in Europe have varied within a range of 0.5-2.5°C, depending upon location and season. Most of this variation is not random but part of a trend towards greater warmth which culminated during the first half of the present century. In contrast to other parts of the world, the slight decline of temperatures thereafter did not develop into a definite downward trend at any of the stations considered. For winter and autumn a still increasing trend is indicated. These temperature changes show a rather coherent pattern over the whole area of the European Community. Phase shifts of some tens of years occur between changes in different seasons.

3.1.5.2. Interannual variability

The magnitude of the interannual variability is of the order of 1°C. Its value, however, is not constant and on a 30-year time scale may vary by a factor of about two at all stations and for all seasons. Time changes of interannual variability are far less coherent in space and time than time changes of mean temperature level. The association between the two, on time scales of 10-30 years in rather weak, except perhaps for the winter season where warmer periods coïncide with lower, and colder periods with higher variability. Interannual variability of the seasonal mean temperatures in Europe does not show an increase at present; compared with the reference period 1931-60, variability over the last two decades has been declining at most stations.

3.1.5.3. Other aspects

The frequency of occurrence cf temperature extremes in Europe since 1940 has been somewhat higher than in the first 40 years of this century. The areal extent of extreme seasons comprises the whole of the European Community or a large part of it; maximum deviations from normal temperatures range from -8 to +5°C for winter and -3 to +4°C for summer.

Apart from the yearly cycle, temperature variations in Europe do not show any clear-cut periodicity except for the quasibiennial cycle (2-3 years).

Decadal mean duration of the vegetation season and the frost-free season in Mid-Western Europe has varied by several weeks, i.e. by some 10-15% of average duration. Shifts in the decadal mean date of the beginning or ending of these "seasons" may also amount to several weeks.

After a temporary increase in the 1940's the yearly amplitude of temperature (difference between the mean temperatures of July and January) has continued its decreasing trend which at most European stations started already in the foregoing century or even earlier.

For large European cities the city effect on temperature during this century amounts to about 1°C.

3.2. Rainfall and water budget

3.2.1. *Definition and measurements*

The water budget of an area is determined mainly by precipitation P, evapotranspiration E and runoff R, together with the storage of water ΔW in the soil or in the form of snow and ice at the surface. The balance equation on land reads simply

$$P - E - R = \Delta W.$$

Unfortunately, measurements of all these quantities are subject to many systematic sources of error, which have been partly corrected during the last 100 years; records before about 1850 are difficult to correct appropriately and can only serve as approximations. Particularly difficult are representative measurements of the actual evapotranspiration which consists of the evaporation from bare soil and the transpiration from plants. Simple instruments have a systematic bias and reliable measurements are possible only at special observatories.
Empirical formulae such as those developed by Penman and others usually give results which are - in spite of their deficiencies - more reliable than those from uncorrected instrumental series. The potential evaporation represents the evaporation from a permanently wet surface, e.g. a lake surface; its calculation is more reliable than that of the actual evapotranspiration which depends strongly on the availability of water to be evaporated. This quantity varies strongly with soil type, vegetation and other surface parameters.

River discharge can be directly measured, with an artificial channel, on small streams; in large rivers the relation between the measurable height of the water level and the total discharge R must be individually determined and frequently verified, due to changes of the profile. Since during the course of the year ΔW is by no means negligible, each parameter of the water budget must be determined separately, and only in undisturbed averages is $\Delta W = 0$ which enables a mutual verification of data.

The designing of instruments to measure precipitation seems to be quite easy; thus it is not surprising that rainfall measurements have been carried out as early as about 2500 years ago in India and in Israel, both countries where food and the economy depend critically on the vagaries of an unreliable seasonal rainfall.

However, the possible errors of such simple instruments are great, especially those caused by winds driving, away from the vertical, raindrops and more so particularly snowflakes falling into a horizontal gauge surface. In addition to this, rain and other hydrometeors including graupel and hail not infrequently fall in showers or thunderstorms with a characteristic diameter of a few kilometres, producing a swath of a size of the order of 20x2 km, in which the intensity may vary by a factor of 100 or more. The average density of rainfall stations in European countries is of the order 1-2 stations per 100 km^2 (locally in cities like West-Berlin up to 20 stations/100 km^2).

With a raingauge surface of 200 cm^2, only $2\text{-}4 \times 10^{-8}$ of the surface is represented in our data. This density is insufficient when considering the high spatial and time variability of precipitation, which is larger than for most other meteorological quantities. Microwave measurements with radar have promised to yield complete spatial coverage together with high time resolution; unfortunately the variability of the drop-size spectrum, together with other sources of error, here also prevents a greater accuracy.

Such deficiencies suggest using, for investigations of the time variability of precipitation, averages derived from a group of homogenized, well-distributed station records. Such area-averaged series are available from the Netherlands and have been recently prepared in the United Kingdom (Craddock 1976, Gray 1976) and in the Federal Republic of Germany (Hillebrand 1978, Ambs 1979). They are meaningful only if the area is small enough to ensure significantly positive correlations between monthly data from all the individual stations (cf. Dupriez and Sneyers (1978) for Belgium); this has been carefully checked for the German series and is doubtless true also for the south-east England series (Gray 1976). In order to obtain representative and comparable data, the sums for the summer tertial (May-August) and the winter tertial (November-February) have been used here (partly because of the relatively high intermonthly auto-correlations), thus excluding only one third of the total observations during the transitional seasons.

3.2.2. *Time changes and variability of rainfall*

In spite of their spatial separation, most series of homogeneous records (Table 3.7) and of area-averaged precipitation (Table 3.8) show an unexpected degree of coherence. The most remarkable common feature is, in general, an increase of winter precipitation indicated by the 30-year averages centred around 1900 to 1925, i.e. during the period lasting between about 1885 and 1940. Since the increase leads to a level which is now about 10-15 per cent higher than in the 19th century, this is by no means negligible; this matter has to be considered again when regarding runoff (see later). Only at Rome has winter rainfall (figs. 3.19 and 3.20) decreased significantly between 1905 and 1960; at British station high values like those around 1960 were also observed much earlier (Kew around 1760, eastern Midlands around 1810).

117.

TABLE 3.7 30-Year Averages and Interannual Variabilities (IAV) based on long precipitation records.

Station	Period of Observation	Summer Tertial (5-8) Average	Summer Tertial (5-8) IAV	Winter Tertial (11-2) Average	Winter Tertial (11-2) IAV
Edinburgh	1785-1960	231-272 mm	57-104 mm	188-232 mm	49- 85 mm
East Midlands	1726-1975	205-261 mm	51-100 mm	157-234 mm	47- 73 mm
London-Kew	1698-1970	188-242 mm	38- 95 mm	165-226 mm	47- 85 mm
Zwanenburg	1735-1978	223-289 mm	54-100 mm	205-272 mm	37- 97 mm
Utrecht-De Bilt	1849-1976	258-299 mm	74-107 mm	205-278 mm	42- 97 mm
Bruxelles-Uccle	1833-1976	259-291 mm	50-116 mm	225-282 mm	37- 93 mm
Münster (Westf.)	1819-1976	260-303 mm	61-103 mm	215-260 mm	42-102 mm
Stuttgart	1825-1975	283-330 mm	53-118 mm	140-188 mm	50- 84 mm
Lille	1801-1960	232-253 mm	57-107 mm	204-245 mm	30-103 mm
Paris(Observ. Cour(1))	1770-1972	191-241 mm	34-102 mm	152-208 mm	42- 75 mm
Dijon	1831-1972	242-268 mm	48- 95 mm	188-241 mm	56- 99 mm
Marseille	1821-1972	99-130 mm	44- 91 mm	193-257 mm	83-174 mm
Milan	1764-1978	286-383 mm	91-154 mm	263-345 mm	95-167 mm
Rome	1783-1978	108-162 mm	43- 86 mm	298-409 mm	105-183 mm

(1) 1770-1810 reduced from Observ. Terrace, IAV only since 1811.

The highest 30-year winter average appears in many German and Dutch series as late as around 1955, with earlier peaks centred around 1873/80 and 1916 (including S.E. England). Minima are centred around 1859 (a cluster of unusually dry years) and 1894.

TABLE 3.8 30-year Averages and Interannual Variability of Area-averaged Precipitation.

Area	Period	Summer Tertial (5-8)		Winter Tertial (11-2)		Author
		Average	IAV	Average	IAV	
Southeast England	1840-1969	216-241 mm	52- 81 mm	215-271 mm	56- 86 mm	Gray
Netherlands	1849-1978	244-283 mm	58- 85 mm	199-265 mm	35- 90 mm	KNMI
Germany West of the Oder	1850-1975	268-304 mm	41- 74 mm	179-222 mm	30- 76 mm	Baur
Lower Rhine (F.R.G.)	1806-1975	248-306 mm	47- 84 mm	207-258 mm	34- 85 mm	Hillebrand
Upper Rhine (North)	1837-1970	244-285 mm	56-103 mm	170-208 mm	44- 78 mm	Ambs
Upper Rhine (South)	1864-1970	325-372 mm	52-113 mm	189-252 mm	59- 98 mm	Ambs

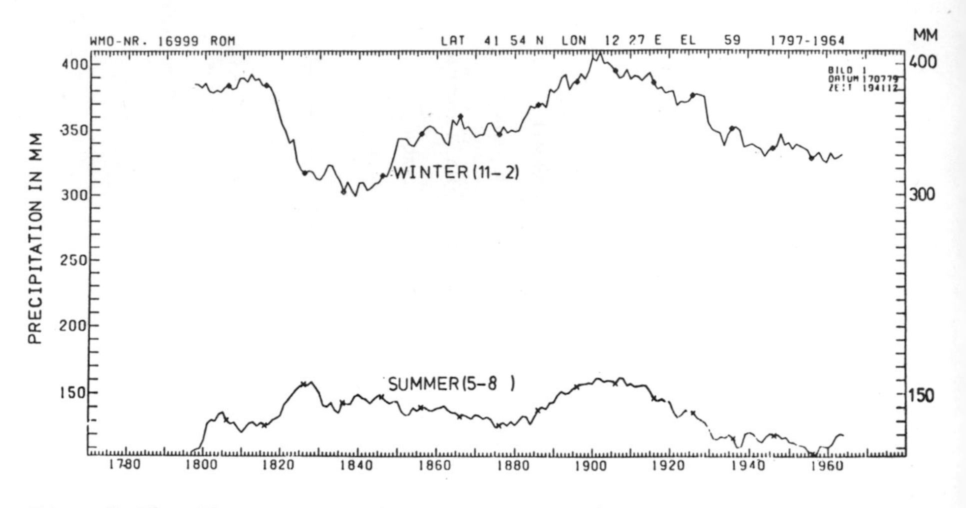

Fig. 3.19. 30-year running averages of precipitation: Rome, May-August and November-February. Scale much smaller than in the other figures; 1797 = average 1782-1811.

The interannual variability during the winter tertial reached a well-marked minimum around 1900, while the highest values occurred in most series except in England only quite recently (centred around 1955 when the values were higher than during the 19th century).

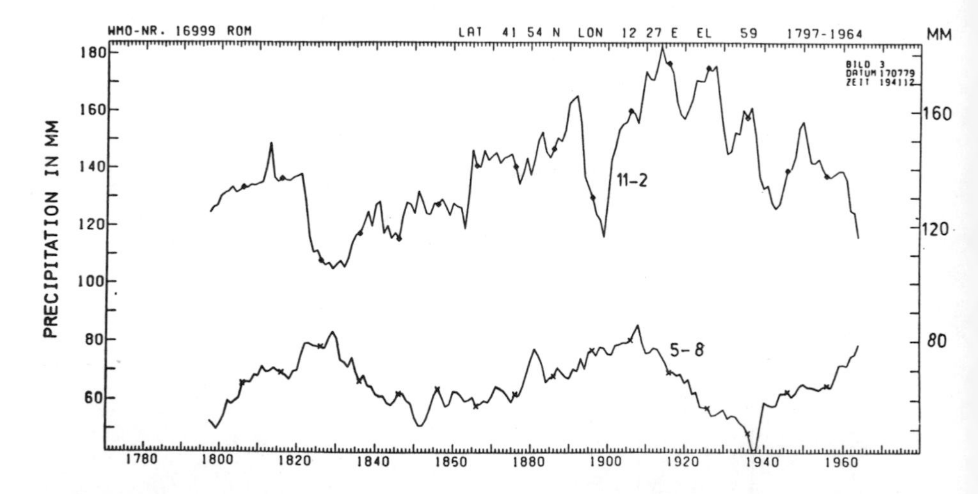

Fig. 3.20. Interannual variability of precipitation at Rome, May-August and November-February: 30-year running averages.

While the amplitude between the extreme 30-year averages in our series during summer reaches 10-15%, and during winter 20-25%, the amplitude between extreme interannual variabilities amounts to about 100% of the minimum (Fig. 3.20).

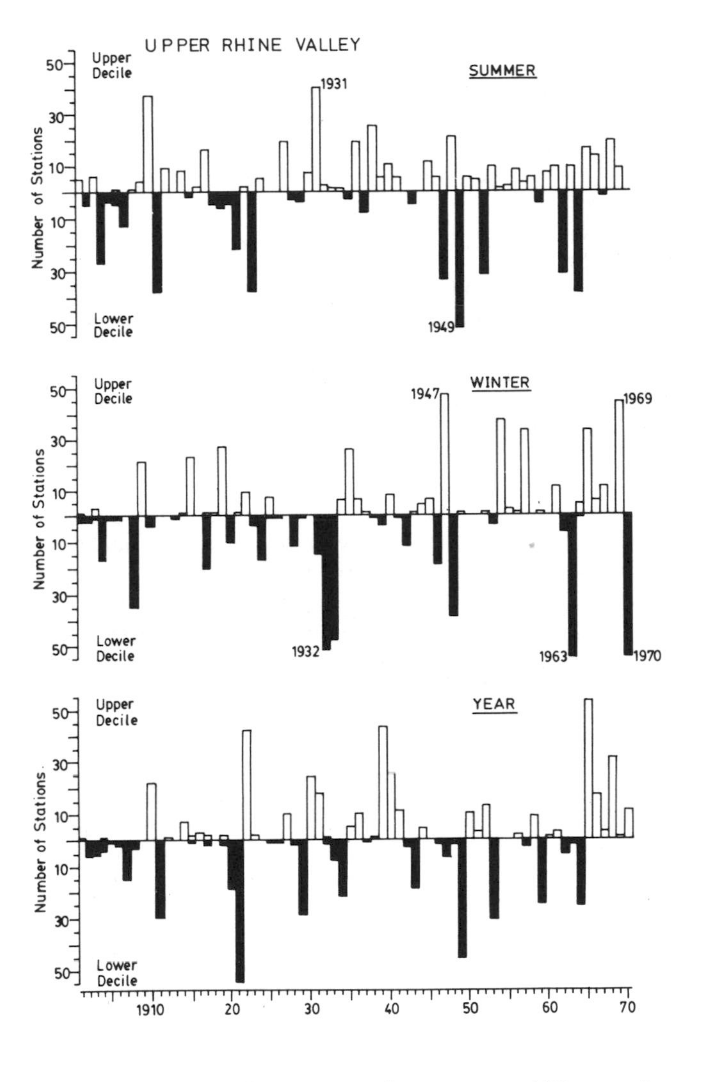

Fig. 3.21. Rainfall in the Upper Rhine area (55 station series, 1901-70). Numbers of stations in the upper decile in white, in the lower decile in black, meteorological seasons (June-August, December-February) and year (data: Ambs 1978)

Calculations of the time variations of the standard deviation σ show similar differences, but they were disregarded here as less representative, because very extreme single values are over-represented in σ due to the use of squared deviations. Any detailed study of the interannual variability shows that the frequency and intensity of extreme deviations is one of the most essential parameters. It can be expressed as the frequency of stations with values among the lowest (or highest) 10% of all measured data (lower or upper decile). The example of the Upper Rhine area (Fig. 3.21) indicates that droughts are less frequent especially during summer but more extended than wet seasons. There is a tendency for clustering of extremes, including some of opposite sign, e.g. the two years 1921/2 or 1947-49 in both summer and winter.

During the summer tertial the spatial coherence of area averages is only slightly weaker than during winter. This indicates the relatively minor role of "scattered showers", with the bulk of rainfall falling during large-scale weather events in much the same way as during winter. The highest averages are frequently also centred around 1955, but some longer records indicate similarly high values around 1840-1850 or 1867. Minima are frequently centred around 1907. The interannual variability shows also minima around 1895 and high values centred between 1950 and 1960. Together with similar winter results, this indicates a relatively high frequency of extreme seasons (especially of droughts) in the latest decades. Since droughts are usually caused by a sequence of anticyclonic periods, where dry conditions prevail over a much larger area (in the order of more than one million km^2) than heavy rainfall in cyclonic periods, they contribute most to the spatial coherence of precipitation variability. They tend also to last longer than sequences of rainy days; one reason is that after a long drought the first rain-producing situations are ineffective since most precipitation falling from high cloud levels evaporates in dry air below before reaching the ground.

Since individual station records show less spatial coherence due to unavoidable local inhomogenities, these results are only valid for the Rhine catchment from the Swiss Alps to the Netherlands, including south-east England. Table 3.7 gives a review of the variability of 30-year averages - such data cannot be taken without further verification as "normal". The surprisingly great time changes of the 30-year interannual variability indicate that the variations of this parameter must not be neglected when discussing climatic variability, as is also true for temperature (see Section 3.1).

The greatest variability of rainfall in the area of the European Community is observed in the Mediterranean winter rain belt; the long records of Marseille, Milan and Rome (Fig. 3.20) may be taken as examples.

If the tertials are representative for the half-year periods, Milan shows in some parts of its record a prevalence of winter rain, while in most periods the summer rainfall is much the greater, as in the Alps. At all three stations the interannual variability reaches 40-50% during winter, and during summer even 50-60% of the average (see also Section 6.4). The highest average rainfall figures are 30-40% greater than the lowest; for Rome this figure is even 50% for summer rainfall. In Milan and Rome the interannual variability reached its greatest summer values around 1910 and again in the most recent periods. The winter rainfall averages arount 1948 were the highest since the early 19th century, when they were about 10% higher still (Milan, Rome)

As regards the summer rainfall, unusually high values were observed at most stations around 1770 and again between 1830 and 1910 (Fig. 3.22). In both these periods, series of cool wet summers associated with blocking anticyclones were particularly frequent. Bad harvests and severe famines occurred around 1770 in central Europe. And in 1816-17 after the heaviest volcanic eruption in recorded history - that of Tambora in Indonesia in 1815 - there was a virtually world-wide harvest failure (see Fig. 6.7).

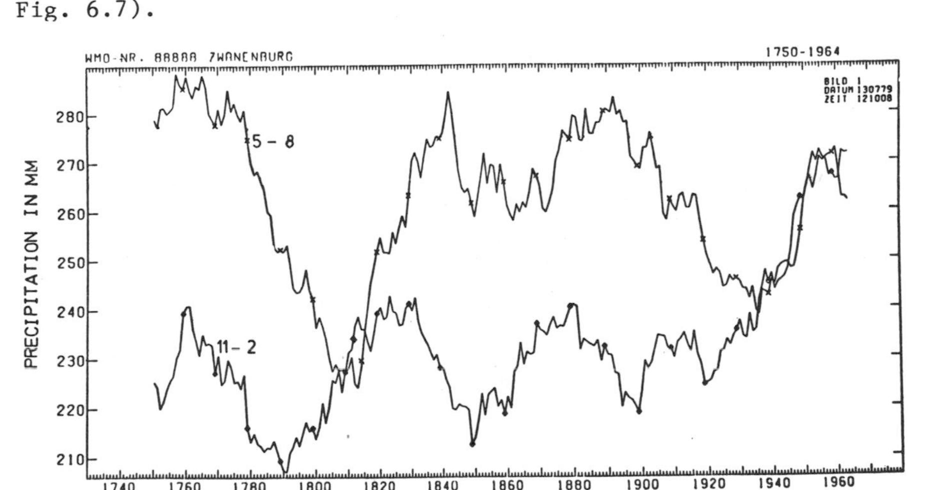

Fig. 3.22. 30-years running averages of precipitation, Zwanenburg (Netherlands). (cf. Fig. 3.21, data: KNMI)

The few rainfall records from the 18th century are less certain than those after responsibility for the station network had been taken over by newly founded state meteorological services. Nevertheless, a remarkable number of more or less coherent records for Great Britain exist (Craddock 1976) and also for France (Garnier 1974) and Italy.

3.2.3. *Time changes and variability of runoff*

Water budget terms other than rainfall data are rarely available in long records. Records of groundwater are, in European countries, nearly everywhere influenced by man and cannot be taken as representative. A long time series of soil moisture has been calculated for southern England from Penman's formula with remarkable results (Wigley and Atkinson 1977); comparable results exist from Edinburgh (Ledger and Thom 1977). Runoff records are available, but older data are only records of water level, unchecked with regard to changes of the profile of the river. River-gauge records from the Rhine have recently been checked and homogenized for five stations between Basle and Andernach, north of Koblenz (Figs. 3.23 and 3.24); two stations on the Lower Rhine are still under investigation.

The runoff records (Table 3.9) are certainly not without error, but the errors of monthly averages when properly checked can be estimated at hardly more than 5%. Their great advantage is that they represent large catchment areas (however, with different geological structures). The water budget equation for a catchment area has been given at the beginning of Section 3.2.

The actual evaporation depends, at least during the warm season, mainly on the available (net) radiation energy, which is more or less proportional to the incoming global radiation (i.e. the radiation from sun and sky on a horizontal surface); the role of temperature and wind is not negligible, but only of second order. For each month from July to September there is a correlation coefficient of from -0.62 to -0.71 between area-averaged rainfall (northern part of the upper Rhine) and global radiation (Karlsruhe, 1895-1977). For all other months except November/December the correlation is also negative and significant at the 0.1% level. Such a negative correlation means that the differences between rainfall extremes are amplified in runoff records: during droughts higher evaporation reduces infiltration and discharge, while during wet and cloudy periods evaporation is diminished. The role of temperature is here included: the correlation between global radiation and temperature (both at Karlsruhe, 1985-1975) between June and September amounts to between +0.62 and +0.84, but becomes negative during winter. The global radiation has been computed (Wacker 1979) with an empirically verified formula, which depends mainly on homogenized records of sunshine duration.

Table 3.9 gives the ranges of 30-year averages and the interannual variabilities of runoff for similar periods but for hydrological half-years. The relation between summer and winter remains approximately constant in spite of the increasing contribution of winter rains from south to north. For summer the highest 30-year average is 10-15% greater than the lowest, while for winter the greatest is 15-20% higher than the lowest.

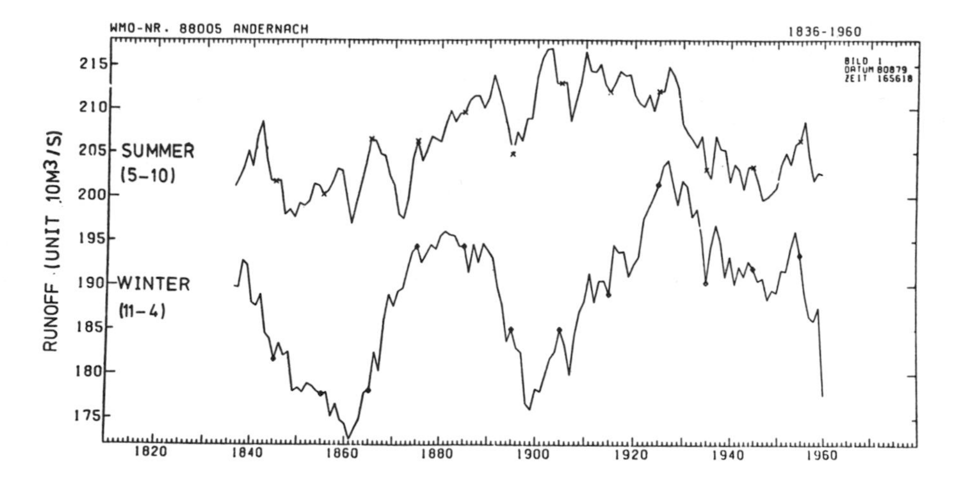

Fig. 3.23. 30-year running averages of runoff (Rhine river at Andernach, catchment area 140 x 10^3km^2) for hydrological half years (crosses May to October, diamonds November to April), units 10 m^3/s.

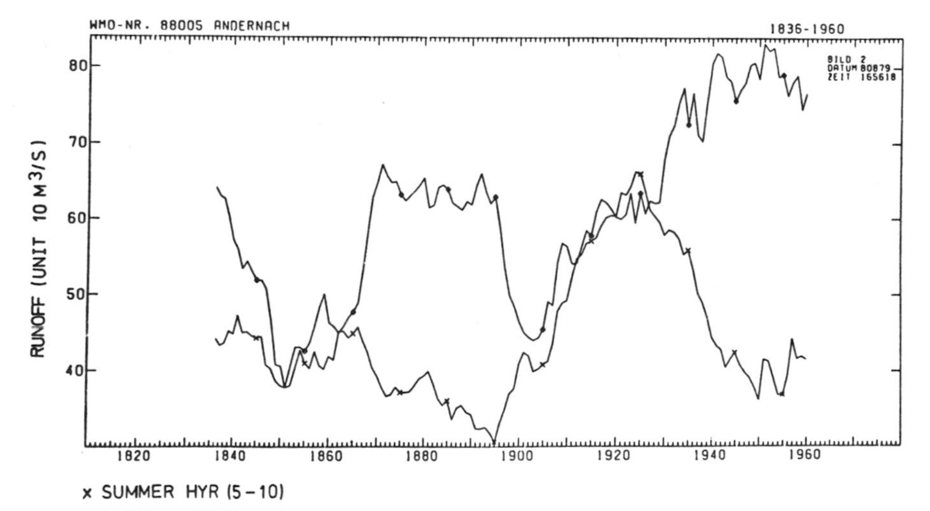

Fig. 3.24. 30-year running interannual variability of runoff (Andernach, data as in Fig. 3.23)

Regarding the low range of the variations at Basle, the storage provided by Lake Constance and other Swiss lakes should not be overlooked; however, the nearly constant ratio at the other stations indicates no long-term storage efficiency. The time variations of the interannual variability (Fig. 3.24) are much great-

TABEL 3.9 30-years averages and interannual variabilities of runoff (River Rhine)

	Period	Catchment	Summer Half-year (5-10)		Winter Half-year (11-4)	
			Average	IAV	Average	IAV
Basle	1809-1975	35,900 km^2	1160-1298	157-321	767- 883	150-304
Karlsruhe-Maxau	1821-1974	50,300 km^2	1360-1504	167-336	986-1144	168-390
Worms	1820-1974		1512-1657	206-449	1112-1321	222-489
Kaub	1821-1974	103,700 km^2	1668-1830	248-521	1372-1575	297-571
Andernach	1821-1975	139,800 km^2	1969-2170	306-663	1726-2041	378-833

All values in m^3/s; data from Bundesanstalt für Gewässerkunde, Koblenz (courtesy H. Liebscher).

er; the greatest during summer are nearly twice, during winter more than twice as high as the lowest. Somewhat surprising is the increase of amplitude with increasing size of catchment; this indicates that the individual events (floods and droughts) are by no means locally restricted but all of regional character and tend to become amplified with increasing area of the catchment because fo the negative correlation between evaporation and precipitation.

These time series reveal a remarkable coherence: maxima and minima of the 30-year averages and variabilities nearly coincide along the course of the river, in spite of the nearly fourfold increase of the catchment area. The highest summer averages occurred around 1860-70, 1910 and 1956 (= period 1941-70), the lowest around 1870-80 and 1947. The peaks of the interannual variabilities tend to occur between the times of highest average rainfall, a relationship which appears to be accidental; the greatest variability was centred around 1924, the lowest around 1954. The greatest winter averages occurred around 1840, 1882 and 1927, the lowest were centred around 1860, 1907 and 1960. The greatest variability winter rainfall was centred around 1952 nearly coinciding with the lowest for summer, while lowest variability of winter rainfall was about 1855 and 1910.

The general rise of winter precipitation (see Fig. 3.25) between 1900 and 1925 is weakly reflected in the runoff data; however, after this maximum it falls again after 1927 to levels as low as earlier, in contrast to most of the area-averaged precipitation series.

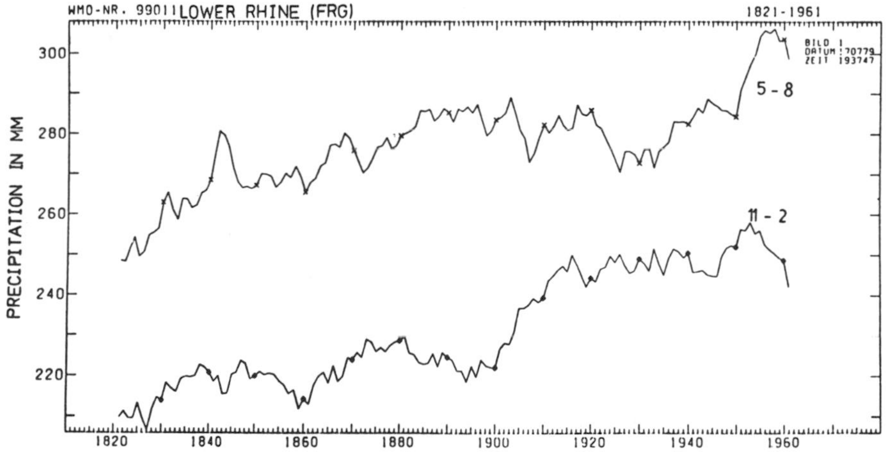

WMO-NR. 9901 LOWER RHINE (FRG) 1821-1961

Fig. 3.25 Lower Rhine (Federal Republic of Germany). 30-year running averages of area-averaged precipitation, summer tertial (5-8) crosses, winter tertial (11-2) diamonds, unit mm/4 months (data: Hillebrand 1978).

Most probably this partial lack of coherence can be interpreted as caused by increasing accuracy of rainfall measurements. The apparent precipitation trend may thus be partly biased; the averages before 1890 appear to be about 10% too low. Unfortunately no independent parameter is available to verify this discrepancy. Table 3.10 shows some examples of various water budget data from long records: the most important result is the low variation of evaporation and of global radiation, in the order of 5-7% in contrast to precipitation with 15-20% or more. Parameters as runoff or soil moisture (proportional to ΔW of the above-mentioned water balance equation) depend on the difference $P - E$; with the negative correlation between these two terms, their variability is even greater than that of rainfall. The result of this is also seen in the extreme observed maxima and minima. The drought of 1976 (see Fig. 6.4), which was centred on both sides of the English Channel, but extended over large areas of central and western Europe, was probably the most intense drought in the London area since the beginning of instrumental observations (Wigley and Atkinson 1977).

In an area as large as that of the European Community, which embraces at least two of the large-scale climatic belts - the temperate zone with all-year rains and the Mediterranean zone with prevailing winter rains - no uniform behaviour of the climatic fluctuations of the past can be expected. But the fluctuations of average seasonal rainfall are of the order 20%, in the Mediterranean even 30-50%, and there exist irregular periods of high and low variability, indicating greater or lower frequency of extreme anomalies. These extreme anomalies of individual years and seasons are much greater than experience during periods of low variability seems to demonstrate; in some areas we live in a time of high variability with extremely wet years in the late 1960's and dry spells during the 1970's. Such anomalies are, at present, unpredictable: since the whole time-scale beyond one month has been badly neglected in meteorological research, more intense investigations of individual anomalies and their teleconnections (spatial correlations) are needed, as well as their relations with anomalies of the temperature and pressure fields. A capacity to predict such anomalies, which usually last between a few months and a few years, would be of very great economic importance. In this the interactions between the atmosphere and the slower, but highly efficient members, of the climatic system (the upper ocean mixing layer, drifting sea ice, snow-cover and soil moisture (see Chapter 1) must be taken into account. Due to the large extent of the global climatic system and to the great number of its "degrees of freedom", the problem is highly complex and progress towards its solution is still in its infancy.

TABLE 3.10 Long record parameters of the water budget and its variations

	Average	CV[1)]	Maximum	Minimum	Period
Netherlands: annual precipitation	677 mm	14.5%	877 mm	423 mm	1849-1908
Netherlands: annual precipitation	749 mm	14.8%	1038 mm	434 mm	1909-1978
London-Kew: annual precipitation	613 mm	17.1%	970 mm	308 mm	1876-1970
London-Kew: annual potential evaporation	547 mm	5.6%	658 mm	439 mm	1876-1976
London-Kew: Soil moisture deficit (May-August)	77.9 mm	35.6%	116 mm	0 mm	1871-1925
London-Kew: Soil moisture deficit	86.3 mm	29.4%	126 mm	24 mm	1926-1976
Karlsruhe: annual actual evaporation[2)]	454 mm	5.5%	487 mm	389 mm	1931-1970
Karlsruhe: global radiation[3)], annual	129 W/m^2	5.1	147 W/m^2	111 W/m^2	1895-1977
Karlsruhe: global radiation, May-August	224 W/m^2	5.7	256 W/m^2	192 W/m^2	1895-1977

[1)] CV= coefficient of variation $=\sigma/M$ (standard deviation in per cent of average M)

[2)] computed from empirical formula (1944-45 no data)

[3)] global radiation from sun and sky, main source of energy (in W/m^2) for warm season evaporation.

An increasing demand for freshwater for irrigation, for new industries and for domestic use, together with the sealing of the soil in expanding cities are having a strong influence on the water budget; the percentage of precipitated water used by man will further increase, while runoff is expected to decrease. In dry years, such as 1921 or 1976, the potential evaporation in many areas is greater than the precipitation, and the water supply becomes marginal. This is not only true for Mediterranean areas with a regularly dry summer period (Section 6.4); an increase of the demand for water by 2-3 percent per year will lead, even in apparently humid countries, to the brink of repeated water shortages.

Rainfall is mainly concentrated in relatively small areas, along fronts or similar disturbances, whereas anticyclonic (dry) conditions or the transport of warm or cold air ahead of or behind an upper tropospheric wave cover much greater areas. Accordingly the spatial variability of rainfall for individual months, seasons or years is greater than that of temperature. Nevertheless, the example of runoff along the Rhine indicates a higher degree of coherence than expected.

3.2.4 *City effects on precipitation*

As outlined in Section 3.1.3.5, cities tend to be, in all seasons, slightly warmer than their undisturbed surroundings. Such a heat island increases locally the duration and intensity of convective activity; this effect is enhanced by the water vapour output of all combustion processes. If a city effect on precipitation exists, convincing evidence could be found at first in the frequency of heavy showers during the warmer season. Single cases above large cities have been frequently described, but they are not conclusive since similar events can happen everywhere without perceptible local triggering. Systematic studies as in the METROMEX experiment around St. Louis (Ackermann and Changnon et al. 1978) with up-to-date instrumentation are as yet unavailable in Europe.

A first climatological study on the relative frequency of heavy showers has been made by Reidat (1971) around Hamburg. He selected 100 cases during the warm season, in which more than 20 mm/d fell at least on one station of a rather dense grid of 35 stations around Hamburg. Then at each station the relative frequency of days with above 20 mm/d was evaluated. While in the rural surrounding this frequency varied only between 25 and 30%, a narrow band with a maximum frequency above 55% was observed within the city, elongated from S.W. to N.E. (which is the main direction of wind) and starting about 5 kms N.E. of the industrial centre in the harbour region.

A similar investigation has been made by Seibel (1980) for the "Ruhrgebiet", where several industrial centers triggered similar bands of increased frequency of heavy showers elongated in the direction of prevailing winds. While the average relative frequency (at about 160 stations) was about 10 percent of the selected days, it increased to 18-20% within the bands. This increase occurred simultaneously with a decrease to 4-6% only at the forest area at the northwest; the total result appears to be not an increase but a redistribution of strong convective rainfall.

A similar redistribution has also been found above the Netherlands (Witter 1982). Here, the frequency of heavy summer showers increased significantly between 1949 and 1978, above industrial areas and downwind, while on the northern and northwestern (left-hand) flank over the coastal regions the shower frequency decreased.

In the area of Naples, the difference between the average rainfall at 4 undisturbed stations in the vicinity of the city and in the city itself decreased from 86 ± 25 mm (1885-1915) and 83 ± 14 mm (1916-45) to -66 ± 20 mm (1946-75) (Palumbo and Mazzarella 1980).

3.2.5. *Conclusions*

(1) In western and central Europe the 30-year seasonal averages based on long rainfall series vary mostly by about 20-30%, in the Mediterranean up to 50-60 per cent.

(2) The interannual variability as evaluated from overlapping 30 years series varies much more with time: the greatest values are 60-100% or more higher than the lowest. The spatial coherence of this parameter is weak.

(3) Due to a significant negative correlation between evaporation and rainfall, runoff data from catchment areas up to 10^5 km^2 are subject to even greater variability. A similar result can also be derived from the few existing soil moisture series.

Chapter 4

CARBON CYCLE, GREENHOUSE EFFECT AND OTHER ANTHROPOGENIC IMPACTS ON CLIMATE

4.1 The carbon cycle and the accuracy of its modelling

4.1.1. Introduction

Throughout most of the millions of years of Man's existence on Earth, his fuel consisted mainly of timber and plant remains which had grown only a few years before they were burned. The effect of this burning on the amount of carbon dioxide in the atmosphere was negligible, because it accelerated only slightly the natural processes of decay which continually recycle carbon from the biosphere to the atmosphere (by biosphere we refer to the carbon reservoir which constitutes living and dead organic matter, providing this remains subject to oxidation in natural conditions). From about 1850, man began to burn intensively the fossil fuels that were locked in the sedimentary rocks, and thus significantly changed the atmospheric carbon dioxide concentration. The amount of CO_2 accumulated in the atmosphere is about half the total released from fossil fuels (Fig. 4.1).
A considerable transfer of CO_2 into other natural reservoirs must therefore have occurred and it seems likely that the oceans have played the most important role in this regard. Increased assimilation by land plant biota cannot be excluded however, since it is well established that plants grow faster in atmospheres richer in CO_2. On the other hand, land biota could also serve as a source of CO for the atmosphere as a consequence of deforestation and cultivation of land for agricultural purposes.

Since most pools and fluxes of the carbon cycle are large compared to the carbon flux produced by human activities, it is not easy to evaluate the importance of the perturbation of the natural carbon cycle by Man. This section reviews the present state of the problem of modelling the carbon cycle (see also Section 4.2.3).

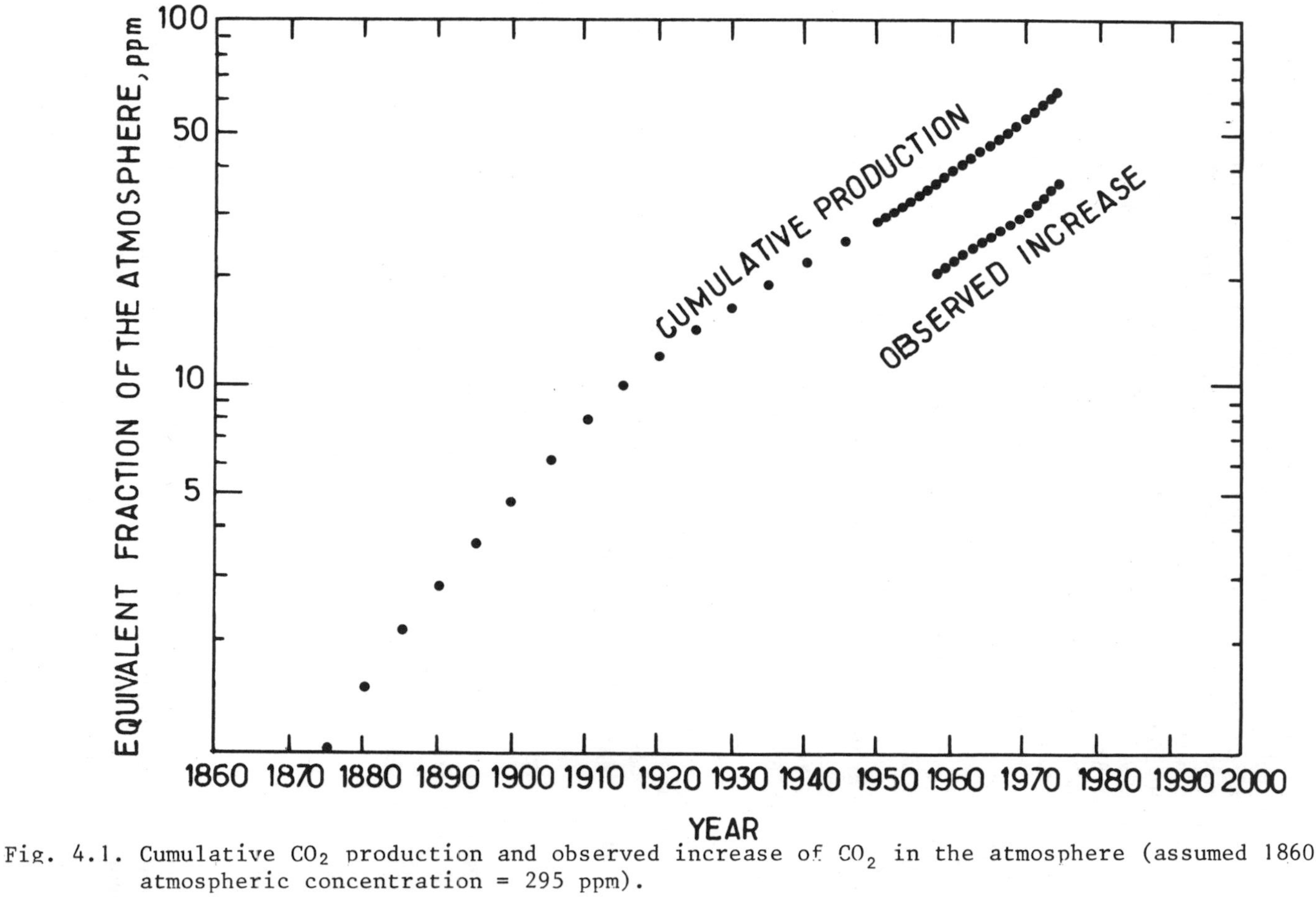

Fig. 4.1. Cumulative CO_2 production and observed increase of CO_2 in the atmosphere (assumed 1860 atmospheric concentration = 295 ppm).

4.1.2. *The various carbon pools*

In the geological past, very large quantities of carbon dioxide entered the atmosphere from volcanoes and other natural ventings. The total amount was at least fifty thousand times the quantity of CO_2 now present in the atmosphere. Most of it (82%) became combined with calcium or magnesium and was precipitated on the sea floor as limestone or dolomite constituting a reservoir, pratically isolated, of about 30 x 10^{15} tonnes of carbon. The remaining 18% (or 6.6 x 10^{15} tonnes of carbon) was reduced by plants to organic compounds and became buried as organic matter in the sediments. A small part of this organic matter (about 0.15% or 10^{13} tonnes of carbon) was concentrated in the deposits we call coal, petroleum, oil shale or natural gas. Only a very small fraction of the emitted CO_2 (less than 0.2%) has remained in the rapidly exchangeable reservoirs: atmosphere, biosphere or ocean (Baes et al. 1976).

Compared to the present annual release of about 5 x 10^{9} tonnes of carbon as CO_2 from the burning of fossil fuels (Rotty 1977), the rate of natural outgassing of CO_2 from volcanoes (0.01 - 0.05 x 10^{9} tonnes of carbon per year is insignificant (Baes et al 1976). Although long-term natural variations of atmospheric CO_2, e.g. those linked to climatic variations, are poorly known, the carbon dioxide problem concerns mainly the flux changes between atmosphere, biosphere and ocean as a consequence of the injection into the atmosphere of CO_2 produced by the burning of fossil fuels and other human activities like deforestation and agricultural practices.

4.1.2.1. *The atmosphere*

Measurements of atmospheric CO_2 are at present made as a routine at different locations from 78°N to 90°S. Observations at all these stations show the same trend. The Mauna Loa observatory (Hawaii, ca. 3400 m), at which accurate and regular measurements have been made since 1958, provides the longest continuous monitoring of the evaluation of the CO_2 content of the atmosphere. These measurements show an accelerating increase of the CO_2 concentration upon which are superimposed annual fluctuations the amplitude of which depends on the location of the station (Fig. 4.2).

A provisional estimate of the seasonally adjusted CO_2 concentration for Mauna Loa observatory as representative of the Northern hemisphere on 1 January, 1978, is 334.2 ppm (ppm = part per million volume: 10^{-6}) (Keeling et al. 1979), corresponding to an atmospheric total of nearly 709 x 10^{9} tonnes of carbon. This value represents a significant increase since 1958: at that time the total atmospheric CO_2 content was only 670 x 10^{9} tonnes of carbon and its concentration 315.5 ppm.

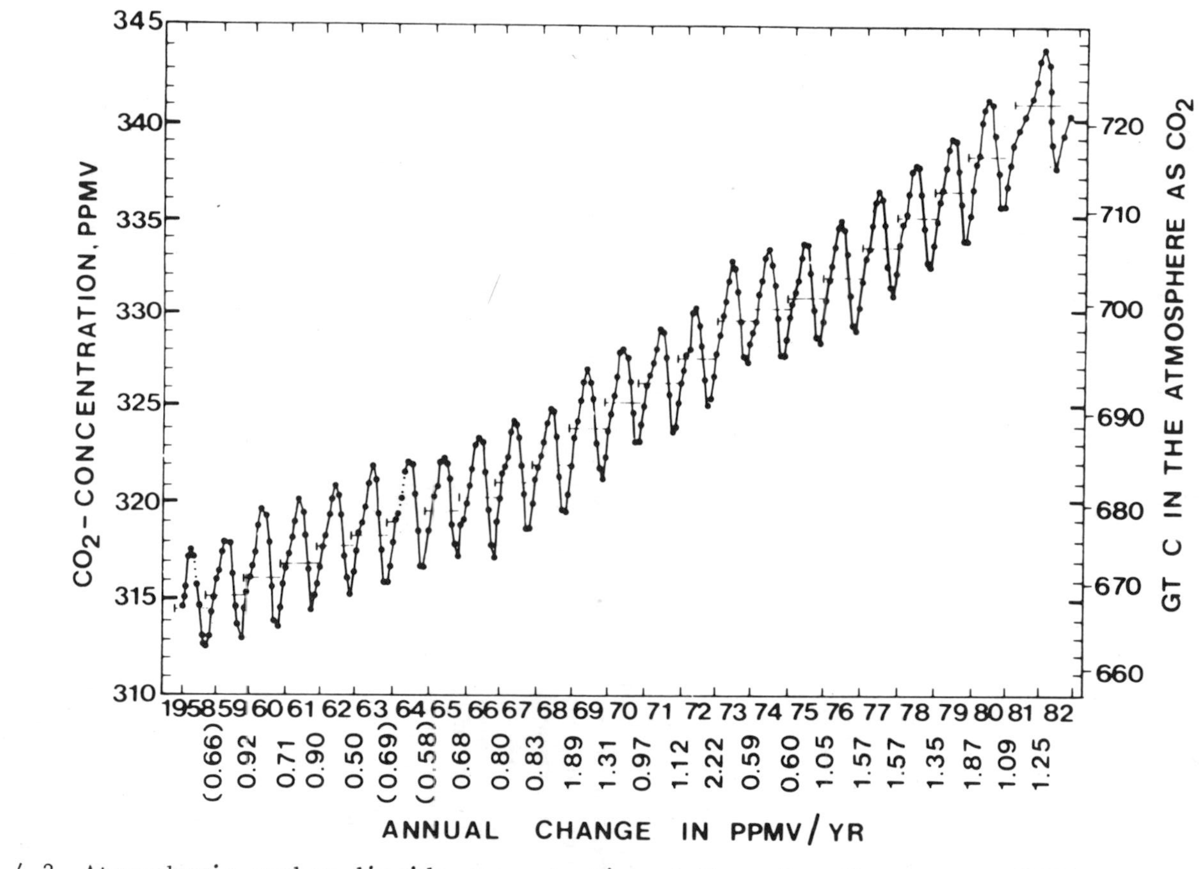

Fig. 4.2. Atmospheric carbon dioxide concentration at Mauna Loa Observatory (1958-1971 data from Keeling et al. 1976; 1972-1974 data from Keeling, personal communication).

Since this increase is directly related to the burning of fossil fuels, the atmospheric CO_2 partial pressure was much lower in pre-industrial times, but its value is not accurately known. Estimates fo atmospheric carbon dioxide in the 19th century vary from 268 to 290 ppm (Keeling and Stuiver 1978). This uncertainty is one of the major difficulties for the evaluation of the various human activities in perturbating the natural CO_2 cycle.

The seasonal change in atmospheric CO_2 is assumed to reflect primarily the biotic metabolism in the Northern Hemisphere. During summer, primary productivity is at its highest level. Plants take up CO_2 from the atmosphere and fix it as organic carbon. The total amount of carbon fixed in that manner per unit area of ground, integrated over time, is the *gross primary production*. However, this gross primary production does not represent the real amount of carbon stored annually by plants. In fact, all the life processes of green plants require energy which is produced by the oxidation of some organic carbon. The *respiration* reaction releases CO_2 into the atmosphere so that the net amount of carbon fixed per unit area of ground, integrated over time -*the net primary production*-is the difference between the gross primary production and the respiration of green plants. Seasonal changes of the net primary production are maximum in forests of the middle latitudes. These forests are extensive in area (Table 4.1) and have the potential for storing carbon in quantities that are sufficiently large to affect the CO_2 content of the atmosphere. During summer the net primary production increases sharply and from May to September the atmospheric CO_2 content decreases in the Northern Hemisphere. During autumn, winter and spring, the net primary production becomes negative (the plant respiration dominates) and the atmospheric CO_2 content increases (see Fig. 4.2.).

The variation in the amplitude of the difference between the winter and late summer concentrations is consistent with this model. This difference ranges from about 15 ppm at Pt. Barrow, Alaska (73°N) and other middle latitude sites (Long Island, N.Y. -Ocean Weather Station P, 50°N - 145°W) to 6 ppm at Mauna Loa (19°N) (Machta et al 1977; Wong 1978). The difference thus drops towards the tropics where the seasonal pulse of the metabolism is less pronounced. This model is also supported by the observation that oscillations damp out with a monthly lag at an altitude of 7 - 9 km; in the stratosphere the seasonal variation almost disappears (Bishoff and Bolin 1966). However, purely physical processes, such as annual variations in sea surface temperature which modulate the oceanic CO_2 partial pressure, may contribute to the seasonal oscillation and to the total flux from or into the atmosphere at all latitudes. Extra atmospheric monitoring stations are needed to clarify the influence of these physical processes.

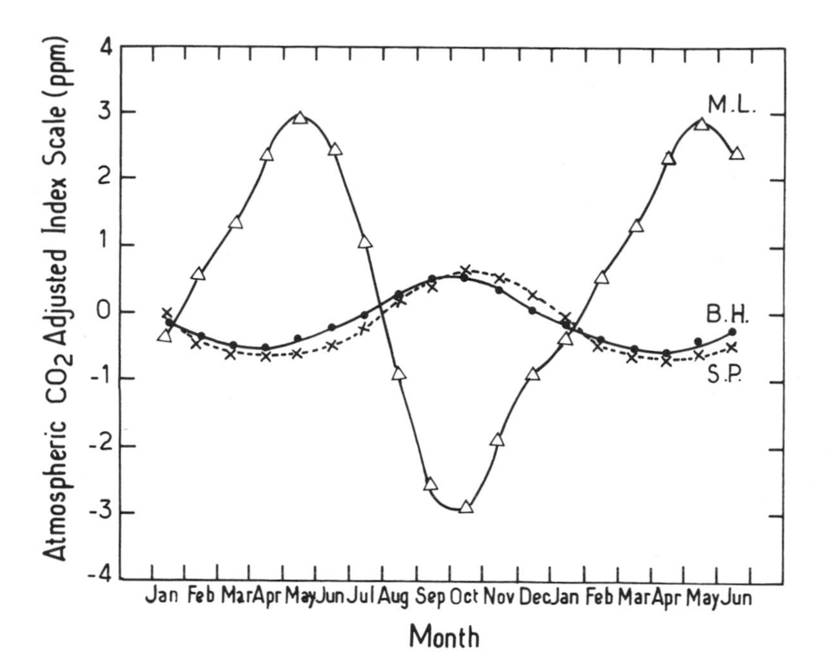

Fig. 4.3. Average seasonal variation of the atmospheric CO_2 concentration at Baring Head (New Zealand), Mauna Loa and South Pole, from 1973 to 1976. The solid line (B.H.) indicates the Baring Head seasonal variation, the broken line (S.P.) the South Pole seasonal variation, and the (M.L.) the Mauna Loa seasonal variation. The seasonal variation was computed for each station by removing a linear trend from the monthly averages of CO_2 concentration over the period January 1973 to December 1976. The resulting monthly averages were reduced to 12-monthly averages for the plot, by averaging each set of January, February, ... December. January-June values have been plotted twice to show the seasonal variation more clearly. The Mauna Loa and South Pole variations are computed from unpublished data.

In the Southern Hemisphere, the seasonal variation is smaller. The New Zealand and South Pole variations, with an amplitude of 1.1 ppm are almost identical. The maximum concentration occurs in October (Fig. 4.3), which is two months earlier than would be expected for a change caused by summer growth in the Southern Hemisphere (Loewe et al 1979). The phase of the oscillation cannot be explained by seasonal changes in sea surface temperature since the solubility of CO_2 increases when the temperature decreases and since the surface water is still cold in October.

It is thus possible that part of the seasonal oscillation observed in the Southern Hemisphere originates from the oscillation in the Northern Hemisphere (Bolin and Keeling 1963).

4.1.2.2. The terrestrial biosphere

Many attempts have been made to estimate the plant biomass for each ecosystem type. Details of such estimates are given in Tables 4.1 and 4.2 (from Olson et al. 1977). Table 4.3 compares, in a simplified way, the partitioning estimates given by various authors. All the estimates recognize that wood complexes account for more than 80% of the live carbon pools. However, estimates of wood areas and live carbon pools put forward by different authors differ by a factor of 2, whereas estimates of dead organic carbon differ by a factor of 5. Satellite data are thus necessary to measure more accurately the present extent and character of land vegetation and to determine their changes with time.

Non-fossil organic carbon pools are a mixture of materials characterized, respectively, by rapid and slow recycling of CO_2. Material featuring rapidly exchanging carbon consists of plant leaves, small branches, roots, litter and agricultural products. The largest estimate of its extent - 160×10^9 tonnes of carbon - represents 10% or less of the total carbon content of the terrestrial biomass, which is thus dominated by materials with slowly exchanging carbon. A mean residence time for each carbon pool could be calculated by dividing the size of the pool by the flux into or out of the pool. For live materials, the flux into the pool is the net primary production. The extimates of net primary production vary between 41 and 78×10^9 tonnes of carbon per year. Table 4.1 reproduces those given by SCEP (1970). These values lead to a residence time of a few years for the rapidly recycling to CO_2 material. Carbon in soils, peat and other sediments consitutes an additional pool of $700 - 3000 \times 10^9$ tonnes of carbon with slower turnover: it means residence time exceeds 1000 years.

The biosphere thus exchanges carbon with the atmosphere and, in equilibrium conditions, the fluxes inwards and outwards must balance. This balance is obtained when the net primary production is equal to the respiration by heterotrophic organisms and the fires from natural processes and from accidental or deliberate burning of wood by man.

4.1.2.3 The ocean

In the oceans, carbon can be found in four main forms: living organisms, dead organic carbon, dissolved inorganic carbon and solid carbonate.

4.1.2.3.1 Marine biosphere

Estimates of this carbon pool vary between 0.4×10^9 and 2×10^9 tonnes of carbon. The lowest estimate takes account only of the concentration of phytoplankton in the open sea (Strickland 1965). the highest one makes allowance also for the most productive areas of the oceans such as upwelling, coral reefs, etc.

The growth of phytoplankton is limited by the availability of nitrogen, phosphorus and silicon in the oceanic surface water. This pool is likely to vary in response to any change in the chemistry of the ocean (Broecker, 1981).

4.1.2.3.2 Dead organic matter

Estimates of *marine productivity* (Ryther 1969) are close to 20×10^9 tonnes per year, at least ten times higher than the marine biomass. This reflects the fact that nutrients are recycled many times before they return at depth. As a consequence to this heavy recycling, dead organic matter is quite variable in the various surface water masses. This pool is estimated at about 30×10^9 tonnes of carbon, while the whole ocean contains about 1650×10^9 tonnes of carbon. Both estimates are quite uncertain. Measurements of the content of organic matter in the ocean are neither accurate nor sufficiently extended in time to detect any evolutionary trend of this reservoir in response to human activities.

4.1.2.3.3 Dissolved inorganic carbon

The most abundant exchangeable forms of carbon in the oceans are inorganic: bicarbonate (HCO_3^-, 89%); carbonate (CO_3^{--}, 10%) and dissolved CO_2 ($CO_2 + H_2CO_3$, 1%).

The concentration of total dissolved inorganic carbon (CO_2) is about 2 mMol/ℓ in surface water and 2.4 mMol/ℓ in deep water, reflecting the fact that deep waters are progressively enriched in CO_2 through oxidation of dead organic matter and dissolution of carbonate initially precipitated in surface water by living organisms. The total amount of dissolved inorganic carbon is 580×10^9 tonnes in the wellmixed surface waters and $39\,000 \times 10^9$ tonnes in the entire ocean reservoir.

Surface water exchanges its CO_2 with the atmosphere (Kanwisher 1963; Hoover and Berkshin 1969; Quinn and Otto 1971). Since surface waters are separated from deeper waters by a steep gradient of temperature and density (the thermocline), the size of this reservoir is well defined; its average depth is 75 m and exchanges with the deep water masses are slow. The capacity of this reservoir to absorb any excess of atmospheric CO_2 is limited by the buffering factor of the carbonate-bicarbonate system

TABLE 4.1. Estimated partitioning of Terrestrial Ecosystem Area Trends, Organi Carbon Pools, and Production Rates

	Estimated Area (10^6 km²)			Estimated Carbon Pool								Net Primary Production (Gtons yr^{-1}) 1970 estimate		
				liv				Dead[e] Gtons	Total Gtons	Rapid[f] Gtons	Slow[g] Gtons			
	Preagri-cultural[a]	1860[b]	1970[c]	Preagri-cultural[a]	1860[b]	1970[c]	ktons/km²[d]					Rapid[f]	slow[g]	Total
1. WOOD COMPLEXES														
A. Boreal + Temperate														
Boreal (taiga)	10.19	9.5	9	101	88	81	9							
Semiboreal	6.91	5.6	5	64	48	40	8							
Cordileran	3.77	3.5	3	68	61	43	15							
Other cool temperate	3.76	3	2	68	33	20	10							
Warm temperate	5.76	4.2	3.8	108	57	38	10							
Semiarid	3.83	2.5	2	25	15	10	5							
Arid moistland	1.07	0.4	0.2	13	2	1	5							
TOTAL	35.2	28.7	25.0	447	304	235	(9.4)							
SUBTOTAL N of 30°N (exclu-				24	284	226		384	610	70	540	10	8	18
ding S of 30°N: add below)				1	20	(9)								
B. Tropical + Subtropical														
Wet site, rainforest	4.56	4.3	3.3	84	68	60	18							
Other tropical moist	8.83	7	5.3	216	126	90	17							
Montane, seasonal	1.18	1	0.5	38	16	6	12							
Montane, humid	2.42	2.2	2	60	26	20	10							
Arid moistland	0.32	0.2	0.1	3	2	1	10							
Woody savanna, scrub	14.05	13	11	139	91	77	7							
TOTAL	31.36	27.7	22.2	340	329	254	(11.4)							
SUBTOTAL S of 30° N			∿ 23.2		349	263		307	570	50	520	11	9	20
WOODS TOTAL			∿ 47.2	987	633	489		691	1180	120	1060	21	17	38
2. NONWOODS COMPLEXES														
Agro-urban														
Crops[i]	0	5	12		4	12	1							
Fringe area[j]	1	3	7	2	6	14	2							
Buildings, etc.[k]	0	1	3		0	0	0							
		9	22		10	26	(1.2)							

Other Land														
Tundra-like bogs	13.53	13	12	21	17	12	1.0							
Grasslands	22.96	21	20	31	21	14	0.7							
Desert, semidesert	29.35	30	29	9	15	17	0.6							
	65.84	64	61	61	53	43	(0.7)							
NONWOODS TOTAL	67	73	83	63	63	69		311	380	40	540	10	8	18
EARTH minus water, ice			130.2			557		1203	1740	160	1600	31	25	56
3. LAKES, RIVERS (including reservoirs)	3		3.1			0.03		60?	60?		60			
4. GLACIERS	15+		15			0.0								
EARTH minus oceans			148.3											
5. OCEANS	350+		361.8			1		1650	1651	2	1649			
						558		2913	3471	162	3309			
EARTH TOTAL			510.1											

[a] After Rodin, Bazilevch and Rosov (1970, and Olson (1974, 1975).

[b] Very Preliminary judgement of forest clearing before and after 1860 (subject to revision).

[c] After Olson (1970a, and new estimates).

[d] 1970 estimate only; previous value higher; parenthetical averages are weighted.

[e] Dead of relatively active dead pool, probably excluding significant amounts of past (histosols) and other resistant humus (estimated as having residence time near or greater than 1000 years).

[f] Includes materials with a probability distribution of fairly short residence times (such as living and dead stages of leaves, flowers, fruits, small roots, most animals, and their unstable residues.

[g] Most woody parts of live plants and dead residues having residence times averaging many years.

[h] From SCEP, 1970; includes photosynthesis minus respiration of all live plant parts; subject to revision.

[i] After Ryabchikoff (1975)

[j] Fringe areas include decorative and wild vegetation, abandoned fields, roadsides, and other more or less vegetated areas around towns or other settlements.

[k] Relatively unvegetated areas in towns or industrial areas, mines, queries, highways, and other disturbed areas besides agricultural fields.

TABLE 4.2. Alternative Estimates of Primary Production and Biomass for the Biosphere (from Whittaker and Likens 1973)

1	2 Area 10^6 km^2 = 10^{12} m^2	3 Mean net primary productivity, g C/m^2/year	4 Total net primary production, 10^9 metric tons C/year	5 Combustion value kcal/g C	6 Net energy fixed 10^{15} kcal/year	7 Mean plant biomass, kg C/m^2	8 Total plant mass, 10^9 metric tons C
Tropical rain forest	17.0	900	15.3	9.1	139	20	340
Tropical seasonal forest	7.5	675	5.1	9.2	47	16	120
Temperature evergreen forest	5.0	585	2.9	10.6	31	16	80
Temperature diciduous forest	7.0	540	3.8	10.2	39	13.5	95
Boreal forest	12.0	360	4.3	10.6	46	9.0	108
Woodland and shrubland	8.0	270	2.2	10.4	23	2.7	22
Savanna	15.0	315	4.7	8.8	42	1.8	27
Temperate grassland	9.0	225	2.0	8.8	18	0.7	6.3
Tundra and Alpine meadow	8.0	65	0.5	10.0	5	0.3	2.4
Desert scrub	18.0	32	0.6	10.0	6	0.3	5.4
Rock, ice, and sand	24.0	1.5	0.04	10.0	0.3	0.01	0.2
cultivated land	14.0	290	4.1	9.0	37	0.5	7.0
Swamp and marsh	2.0	1125	2.2	9.2	20	6.8	13.6
Lake and stream	2.5	225	0.6	10.0	6	0.01	0.02
total continental	149		48.3	9.5	469	5.55	827
Open ocean	332.0	57	18.9	10.8	204	0.0014	0.46
Upwelling zones	0.4	225	0.1	10.8	1	0.01	0.004
Continental shelf	26.6	162	4.3	10.0	43	0.005	0.13
Algal bed and reef	0.6	900	0.5	10.0	5	0.9	0.54
Estuaries	1.4	810	1.1	9.7	11	0.45	0.63
Total marine	361		24.9	10.6	264	0.0049	1.76
Full total	510		73.2	9.9	723	1.63	829

* All values in colums 3 to 8 expressed as carbon on the assumption that carbon content approximates dry matter x 0.45.

TABLE 4.3. Various estimates of organic carbon reservoirs

		Whittaker et Likens (1973) ESTIMATES FOR 1950	Olsen et al (1977)			Reiners et al (1973)	Bohn (1976)
			Preagricultural	1860	1970		
WOODS	AREA $10^6 km^2$	71.5	31	27	22		
	POOL 10^9 T of C	792	540	329	254		
NON WOOD	AREA $10^6 km^2$	77.5	67	73	83		
	POOL 10^9 T of C	45	63	63	69		
DEAD ORGANIC CARBON	POOL 10^9 T of C				1203	700	3000

in the ocean, the consequence of which is that, if the atmospheric carbon dioxide increases by x%, the partial pressure of the dissolved CO_2 will also increase by x% but the total dissolved CO_2 will only increase by x/B%. Theoretical values of B are close to 10 (Broecker et al. 1971) but experiments by Rebello and Wagener (1976) yielded a value of only 7. Finally, since the surface water carbon pool is of a size similar to that of the atmosphere, its efficiency to absorb CO_2 remains small as a consequence of the buffer effect, despite its fast reaction to the increasing atmospheric concentration.

Clearly oceans can be a sink for the fossil carbon flux only because mixing occurs between surface and deeper waters. Transfer of surface water directly to the deep ocean (deeper than 1000 m) takes place in restricted areas, mainly the Norwegian and Labrador Seas in the North Atlantic and the Weddell Sea in the Southern Ocean. This process is particularly slow: the residence time of the deep ocean is in the range 500 - 1500 years (Broecker 1963) indicating an annual transfer of only 2 - 6% of the surface water into the deep sea. Another mode of mixing of surface water with deeper waters occurs with much efficiency: the formation of intermediate waters and the eddy diffusion in the upper 1000 m (Broecker et al 1971). At present, this process is not known with sufficient accuracy, but the penetration in the oceans of radioactive tracers produced by nuclear explosions (such as carbon 14 and tritium) suggests that it could be responsible for most of the CO_2 uptake by the oceans (Stuiver 1978; Siegenthaler and Oeschger, 1978).

4.1.2.3.4 Solid carbonate

Carbonate shells are secreted by zooplankton and phytoplankton in surface waters. After the death of these organisms their shells fall to the sea floor. As these shells settle from the surface, they encounter undersaturated waters below a certain depth (depending on the ocean). Thus, the net flux of carbonate arriving at the sea floor is lower than the carbonate flux leaving the surface water. At present, the net flux of carbonate arriving at the sediment is not accurately known. Generally, the carbonate content of the sediment is fairly high (often over 50%) above a water depth of 4 km.

Most carbonates in the sediment are definitely stored and only the superficial layer remains in contact with deep water. However, the activity of benthic organisms continuously mixes the upper 5 - 15 cm of the sediment (Peng et al. 1977). Broecker and Takahashi (1977) estimate the total amount of $CaCO_3$ maintained in contact with bottom water to be 0.5×10^{13} tonnes of carbon, a quantity equivalent to the amount of carbon locked up in recoverable fuels. This carbonate will ultimately neutralize the CO_2 generated by fossil fuel combustion via the reaction

$$CaCO_3 + CO_2 + H_2O \rightleftharpoons 2\ HCO_3^- + Ca^{++}.$$

However, we do not know where and at what rate "enhanced" dissolution of $CaCO_3$ will occur.

4.1.3 *Impact of human activities*

The perturbations of the natural carbon cycle are difficult to estimate because the biosphere is at present poorly known and the fate of fossil fuel CO_2 in the oceans itself is not determined in detail. We shall summarize here what we know about changes in the natural carbon cycle, what we do not know and what indications we have through proxy-indicators and models.

4.1.3.1 *What we know*

- The quantity of fossil fuels annually burned by Man's activities.
- The annual increase in the atmospheric CO_2 concentration at Mauna Loa and a few other monitoring stations durings the past twenty years or so (Fig. 4.1).

The discussion of these two points in the foregoing shows that our knowledge is not sufficient to determine accurately the relationship between anthropogenic release and atmospheric reservoir content.

4.1.3.2 *What we do not know*

Is there another important source of CO_2 resulting from human activities?

A number of authors have pointed out that the amount of wood consumed in deforestation to increase agricultural land and as firewood in underindustrialized countries could constitute another important source of CO_2, perhaps of the same order of magnitude as the fossil fuel emission (Adams et al. 1977, Bolin 1977, 1979, Woodwell ct al. 1978). Table 4.4 presents various estimates of man's contribution to the fluxes to and from the atmosphere. These estimates depend often on limited data gathered from relatively small geographical areas and then extrapolated to a global scale. Moreover, contradictory data are sometimes to be found in the litterature: the belief that tropical soils lose their organic matter after forest cutting is still controversial (Nye and Greenland 1965; Lemon 1977). Estimates for the biospheric source thus vary from a few per cent to two hundred per cent of the fossil fuel emission.

TABLE 4.4

Estimates of anthropogenic injections and of their consequences (10^9 T.C/yr.)

	Bolin 1979	Broecker et al. 1979	Stuiver 1978	Woodwell et al. 1978
Fossil fuel combustion	5	5.2	5	5
Deforestation soil	1	0	0	2-18
decomposition	1	0	0	
Annual anthropogenic injections	7	5.2	5	7-23
Increase in the atmosphere	2.5	2.7	2.7	2.5
Dissolution in the ocean	1-3	1.9	2.3	?
Increased plant assimilation	1-3	0.6	0	?

Is the biosphere a sink for man-made CO_2?

The biosphere might also be a sink for fossil fuel CO_2 since laboratory experiments show that photosynthesis increases with increasing CO_2, provided enough water, light, nutrients and a proper temperature are provided for the plant. However a detailed discussion of this factor by Lemon (1977) which takes into account the complex processes which effect the final yield of carbon concludes that the final biomass is not necessarily increased by added CO_2. The main effect could be a decrease of the duration of the plant life cycle.

What is the quantity of man-made CO_2 already taken up in the sea?

GEOSECS data for the Atlantic Ocean reveal directly the increase of the CO_2 concentration of Antarctic Intermediate Water and freshly formed North Atlantic Deep Water (Brewer 1978; Chen and Millero 1979).

However, at present, this increase is small and its amount can be calculated only with an accuracy of 50%. Similar data for the other oceans, particularly for the Indian Ocean, are completely lacking. It is thus impossible to obtain an accurate estimate of the quantity of man-made CO_2 already entered in the whole ocean, which would be of primary interest to balance the world carbon budget.

4.1.3.3 What we know through proxy indicators and models

Hall et al. (1975) presented a model which interpreted annual variations of the month-by-month CO_2 concentration at Mauna Loa as reflecting only changes in net photosynthesis and respiration of the biosphere. Using the estimate of Broecker et al. (1971) that 58 ± 7% of man-made CO_2 remains in the atmosphere, Hall et al. conclude that no change of biotic metabolism in the Northern Hemisphere emerges from 1958 to 1972 and that the biosphere has remained stable.

Siegenthaler and Oeschger (1978) used two different types of ocean modelling and determined the exchange coefficients between the atmosphere and the sea from the preindustrial carbon-14 distribution. In these models, the biosphere can be a sink or a source of CO_2 and the final atmospheric CO_2 concentration is calculated year by year from 1958 to 1974. The best fit between calculated and observed atmospheric CO_2 concentrations corresponds to a biosphere in equilibrium or slightly increasing.

Another approach used the isotopic ratios of the carbon. Biospheric and fossil fuel carbon is strongly depleted in carbon-13 and about 18 ‰ isotopically lighter than the atmosphere. The atmospheric $^{13}C/^{12}C$ variations of the cellulose fraction in well-dated tree rings reflect those of the atmospheric carbon and thus the changes of the CO_2 atmospheric concentration due to the injection of fossil fuel CO_2 *and* biospheric CO_2.

By contrast, fossil fuel carbon and biospheric carbon differ in their carbon-14 content: fossil fuel carbon in old and contains no carbon-14 whereas biospheric carbon is young and has a $^{14}C/^{12}C$ ratio close to that of the atmosphere. Changes in the $^{14}C/^{12}C$ ratio of the atmosphere (also recorded in tree rings) reflect thus *only* the fossil fuel injection.

Using this approach, Stuiver (1978) concluded that the net release of CO_2 from the biosphere to the atmosphere between 1859 and 1950 was about 1.2×10^9 tonnes of carbon per year and has decreased over the last few decades. A model of ocean-atmosphere exchange taking into account these calculations and the present distribution of the carbon-14 produced by nuclear explosions in the Atlantic Ocean suggests that 47% of the man-made CO_2 has been taken up by the sea. However, Freyer (1978) using a similar approach but different model assumptions calculated a higher biospheric injection.

4.1.4 Conclusion

The role of the biosphere as a source or sink for increasing carbon dioxide is not fully established. The mechanisms of assimilation of the anthropogenic carbon dioxide by the ocean are not described quantitatively enough and the nearly even division of this pollutant into the two main sinks (atmosphere and ocean) may change in the future.

As a result of the greenhouse effect (see Section 4.2.3.1) anthropogenic carbon dioxide could be the cause of a major climatic change the impact of which on our environment would be difficult to predict. At present, the world carbon budget is still unbalanced. As a consequence of the great amount of uncertainty, any projection of the future increase of the atmospheric carbon dioxide beyond some decades ahead can only represent a crude estimate. However, some of them (see Baes et al. 1976), assuming an ever increasing use of coal and oil as a source of energy, predict that the carbon dioxide content of the atmosphere might rise by as much as a factor of 4 to 8. Such a heavy pollution could not fail to influence the thermal equilibrium of the atmosphere and, thus, our climate.

Accordingly a major research effort is required, as a matter of high priority, to obtain a better understanding of the carbon cycle and to enable an accurate prediction of the future atmospheric concentration of carbon dioxide for various scenarios of fossil fuel burning.

4.2 Man's Impact on Climate

Serious assessment of human impact on the global environment started in 1970 with a workshop concerned with the study of Critical Environmental Problems (SCEP 1970). In the followup report of the Study of Man's Impact on Climate (SMIC 1971), it was concluded that in many critical areas the data were incomplete and even contradictory, and that without additional research and monitoring programmes the scientific community will not be able to provide the firm answers which society may need of large scale, and possibly irreversible, inadvertent modification of the climate is to be avoided.

Among all these impacts, the carbon dioxide question has become the subject of most extensive concern in recent years. In screening the existing knowledge, a CO_2 - Climate Review panel of the US National Research Council (1982) concluded that previous results, which interferred a relationship between man-made changes in atmospheric composition and substantial climate effects, remain unchanged: an increase of carbon dioxide in the atmosphere by a factor of 2 would cause the average global surface temperature to increase by $3° \pm 1.5°C$ and no overlooked or underestimated physical effects were found that could reduce this currently estimated global warming to negligible proportions or reverse them together. However, some important questions remain open such as: (i) how much carbon dioxide and other infrared absorbing trace gases will be really added to, and will remain in, the atmosphere in future years and at what rate? (ii) will it be possible to detect the first signal of a human impact on climate or will the CO_2 -induced warming be masked by "natural" climate changes caused by other factors such as the secular variations of atmospheric aerosols and solar irradiance? (iii) what would be the major consequences of possible climatic changes for human societies?

The purpose of this chapter is only to summarize briefly the most significant results obtained up to January 1983. Indeed not only is there an increasing number of papers being published on this subject, but also many reviews, books and proceedings; only among those which appeared between July and December 1982, I will cite Bach (1982a), Revelle (1982), Schlesinger (1982), Bach (1982b), MacDonald (1982), Clark (1982), Bach et al. (1983) This is probably sufficient to show that this is a field of real concern and in rapid motion and that this review can not be exhaustive.

4.2.1 Introduction and energetics of man's impact on climate

As shown in previous chapters, the facts are that climate everywhere fluctuates quite noticeably from year to year and that there are gradual changes in climate that make one decade or one century different from the one before. These yearly fluctuations and longer-term changes have been the result of natural processes or external influences at work on the complex system that determines the Earth's climate (Berger, 1980). However, it now appears that man is becoming another significant factor in the climatic balance not only at a local scale (e.g., city climates, cf. 3.135) but increasingly towards a global scale. The magnitude of the climate change that mankind will bring about, or its exact timetable, is not yet certain and there are many important implications of the change that are still merely based upon speculations. Nevertheless, it seems possible now to sketch a rough scenario of the anthropogenic changes that will probably occur in the next decades.

As a first step, it is important to know what is the relative contribution of man to the global pollution of the atmosphere. From Table 4.5, it can be seen that we are progressively becoming competitive with Nature, at least as far as sulphur compounds, CO_2 and aerosol production are concerned.

However, it is not necessarily the total amount of emitted pollutants which is important but their individual impacts on climate as summarized in Tables 4.6 and 4.7. From these tables and a comparison between the energetics of large-scale natural and man-made climatic changes (Flohn 1975), it can be seen that mankind may be able to exert an influence and alter the energy balance on a global scale. When these various man-made effects are tested by experiments with climate models, the influence on the global mean air temperature of increasing carbon dioxide and other atmospheric trace gases emerges as the largest single factor and the overall tendency is towards warming.

TABLE 4.5. Ranges of emissions for world-wide natural and anthropogenic pollutants in the atmosphere.

	Residence Time in the atmosphere	Current background concentration (Bach 1982b)	Natural Production (Gt/yr)	Anthropogenic input (Gt/yr)	Mean Contribution of man (%)
S Compounds (Georgii 1979)	1h-several dy		0.08 - 0.11	0.06 - 0.08	40
Sulphur dioxide	4 dy	0.05 - 1 ppb			
CO_2 (Bach 1976)	4 yr (trop) 2 yr (strat)	335 ppm	72 - 907	12 - 16	1 - 20
C (as CO_2) (Bolin et al.1979a)	4 yr (Zimen 1979)			FFC 5 NFFC 1-6	
Particulates (diam < 5 μm) (Dittberner 1978a, Bach 1982b)	- trop. 1-20 dy - strat.1- 5 yr	10-50 $\mu g/m^3$	1.2 - 1.5	0.7	35
N Compounds (Bach 1976)	5 days to tenths of years (N_2O; Hahn 1979)	310 ppb	1.4 - 7.4	0.02 - 0.1	1
CO Bach (1976)	0.1 - 3 yr	0.05 - 0.2 ppm	0.07 - 5	0.19 - 0.6	10
CO Zimen (1979)	1 month		4.5	0.6	12
Water vapour (evaporation)	10 dy (trop)	10-3x10^4 ppm(trop) 3 ppm(strat)	71500 (Lvovitch 1977)	1800 (1965) (Flohn 1973)	3

ppm = parts per million; ppb = parts per billion; FFC = fossil fuel combustion; NFFC = non-fossil fuel combustion
Gt = 10^{15} g; trop = troposhere; strat = stratosphere.

TABLE 4.6. Direct impact of man's activities on sensitive climatic factors.

ACTIVITIES	CO_2	α	dust	H_2O	heat	^{85}Kr
LAND USE						
Urbanisation		x		x		
Deforestation	x	x	x	x		
Overgrazing		x	x	x		
Savanna bushfires and slash-and-burn practices in agriculture	x	x	x		x	
Agriculture-fertilizer	x		x			
Large scale solar energy conversion systems		x				
Reservoirs-irrigation		x		x		
DOMESTIC ACTIVITIES						
Industries			x		x	
Burning of fossil fuel-energy consumption	x				x	
Freons	x					
Nuclear energy				x	x	x

CO_2 = carbon dioxide and trace gases; α = albedo; dust = dust and aerosols; H_2O evaporation and release of water vapour; heat = direct heat release; ^{85}Kr = Krypton-85.

4.2.2. *Energy demand and climatic implications of energy use*

4.2.2.1 *Energy demand*

Data show that since industrialization on a world scale started in the middle of the nineteenth century, the demand for energy has risen sharply, especially since the end of World War II: the global energy use rate was 0.5 TW* in about 1900, 2 TW around 1945 and has now reached 10 TW (including 2 TW from use of fire-wood in the Third World). In fact, during the 1860-1975 period, the growth rate has been almost constant, about 5.3 percent per year, except for the period of world wars and economic crises. Based upon consideration of demographic, economic, social and political factors, a current estimate of world energy demand in the year A.D. 2925 was 32 TW (Rotty 1979) but other more recent estimates of future world energy use have been derived from different assumptions (Fig. 4.4).

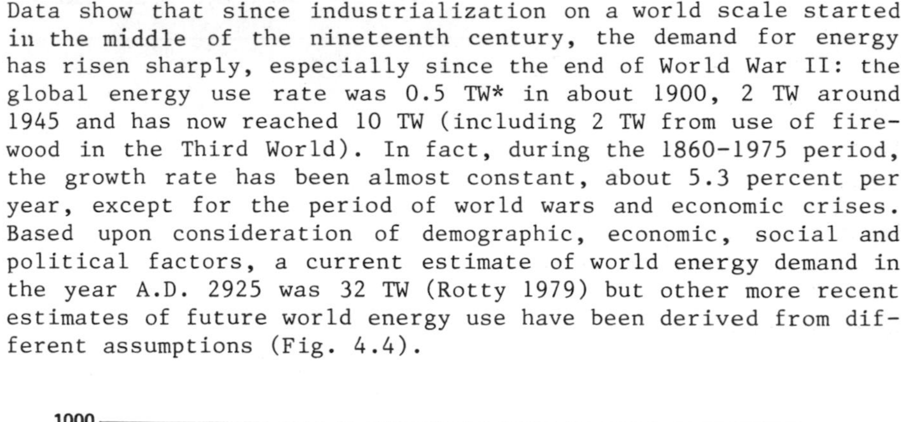

Figure 4.4. World energy use and projections (Rotty 1979).

*TW = 10^{12}

TABLE 4.7. Man's Impact on Climate

IMPACT	CLIMATIC EFFECT	SCALE AND IMPORTANCE OF THE EFFECT
(1) Release of carbon dioxide;	Increase the atmosphereic absorption and emission of terrestrial IR radiation (greenhouse effect) resulting in warming of lower atmosphere and cooling of stratosphere	- Global; potentially a major influence on the climate
Release of other trace gases		- Global; potentially significant
(2) Changes of surface albedo and evapotranspiration	Modification of the heat balance	Regional; importance speculative
(3) Release of particles or aerosols: from industry and slash-and-burn practices in agriculture	These sunlight-absorbing particles probably decrease albedo over land, causing a warming; they also change stability of lower atmosphere	Regional, since aerosols have an average lifetime of only a few days; stability increase may suppress convective rainfall
that act as condensation and freezing nuclei	Influence growth of cloud droplets and ice crystals; may affect precipitation in either direction	Local or (at most) regional influence on precipitation
(4) Release of heat (thermal pollution)	Warms the lower atmosphere directly; increase of convective precipitation	Locally important now (heat island), will probably become significant regionally
(5) Release of radioactive Krypton-85 from nuclear reactors and fuel reprocessing plants	Increase conductivity of lower atmosphere, with implications for electric field and precipitation from convective clouds	Global; importance of influence is highly speculative

Adapted from KELLOGG (1978)

Especially since the 1973/74 world energy crisis, the reduced energy use has resulted in regular downscalings of energy projections. It further appears that energy projections made by more independent bodies, and in more recent years, tend to be lower in magnitude than those carried out by organizations with vested interests. In Table 4.8(a), some recent examples of world energy projections of the global primary energy consumption are given (comments and details are available in Bach 1982a). For the year 2030, IIASA (International Institute for Applied Systems Analysis, Laxenburg, Austria) projects a use of 36 TWyr/yr for its high scenario and 22 for its low one, the CEC (Commission of the European Communities) Zero Growth Scenario (assuming an unchanged average global per capita energy consumption of 2 kW and a doubling of world population from 4 to 8 billion) lead to 16 TWyr/yr and the Efficiency Scenario (Lovins et al. 1982) to 5.2 TWyr/yr.

TABLE 4.8(a). Global primary energy consumption (TW) for different energy sources and a variety of energy scenarios 1975-2030 (BACH 1982a).

		IIASA Scenarios high		IIASA Scenarios low		Zero-Growth Scenario		Efficiency Scenario	
Sources	1975	2000	2030	2000	2030	2000	2030	2000	2030
Oil	3.83	5.89	6.83	4.75	5.02	4.26	3.58	1.77	0.24
Gas	1.51	3.11	5.97	2.53	3.47	2.27	2.48	1.51	0.34
Coal	2.26	4.94	11.98	3.92	6.45	3.51	4.60	1.77	0.38
Nuclear	0.12	1.74	8.08	1.29	5.17	1.15	3.71	0	0
Others*	0.5	1.15	2.76	1.09	2.28	0.98	1.60	2.02	4.27
Total	8.22	16.83	35.63	13.58	22.39	12.17	15.97	7.07	5.23
% 1975	100	205	435	166	273	146	195	86	63

* Other energy sources: hydro-solar-biogas-geothermal-commercial wood use. the estimated global potential of renewable energy sources for 2030 ranges from a practical value of 6 to 10 TW, to a theoretical one of 20 TW and a maximum available one of 26 to 283 TW.

TABLE 4.8(b) Projected energy consumptions by major regions of the world for the scenario leading to an energy consumption of about 40 TW in 2025.

Region	1980 (in TW)	1985 (in TW)	2000 (in TW)	2025 (in TW)
United Stated	2.44(10.1)	2.82	3.90	6.56(10.6)
North America (less US)	0.22	0.25	0.41	0.63
Western Europe	1.62(3.7)	1.81	2.70	4.12(7.8)
U.S.S.R.	1.42(5.0)	1.87	3.23	8.15(10.1)
Communist Eastern Europe	0.57	0.67	1.36	2.57
Japan	0.51	0.63	1.05	2.28
Other non communist Asia	0.57	0.73	1.84	5.52
Communist Asia	0.60	0.73	1.27	2.47
Africa	0.16(0.2)	0.22	0.63	2.60(1.2)
Latin America	0.41(0.8)	0.54	1.30	3.81(4.6)
Oceania	0.10	0.13	0.22	0.51
TOTAL WORLD	8.62(1.9)	10.40	17.91	39.42

After Perry en Landsberg (1977); ( ) per capita energy use in kW (Rotty 1979).

TABLE 4.8(c). Possible World Primary Energy Demand of the Efficiency Scenario (Bach 1982a).

Regions	Primary energy (TW)	Primary energy demand (TW)	
	1975	2000	2030
1. North America (USA and Canada)	2.65	1.78	1.06
2. USSR and East Europe	1.84	1.89	1.55
3. West Europe, Japan, Australia, New Zealand South Africa, Israel	2.26	1.51	0.94
Developed countries	6.75	5.18	3.55
Developing countries	1.46	1.89	1.68
World	8.22	7.07	5.23

This last scenario considers a population growth assumed to be the same as in IIASA and CEC studies, the per capita growth rates of the low IIASA scenario, some other specific parameters such as a structural energy service intensity and a technical energy efficiency, and further assumptions such as no changes in lifestyle, the use of only presently available technologies and the purchase of greater energy productivity at a level cost-effective at current German fuel prices. The results of this scenario where energy is used efficiently in a cost-saving manner were quite astonishing: its application reduces the global primary energy demand by 14% in the year 2000 and by 37% in 2030 as compared to the base year 1975. Moreover, this applies to a world with an almost five-fold rise in economic output and a world in which the present global average per capita energy consumption of 2 kW is raised in 2030 to the 1973 German level of 5.2 kW/cap.

To assume such a reduction in energy consumption and consequent fossil fuel release (cf. Section 4.2.3) is apparently not unreasonable when considering recent developments. The rise in fuel prices appeared indeed to have affected a more rational use of energy leading already to some substantial energy savings even without specific efficiency measures (Rotty 1981):

(i) average annual CO_2 emission rate from oil use decreased from 7.1% in 1959-73 to 1.7% after the energy crisis;

(ii) that of gas from 8% to 2.8%;

(iii) that of coal remained constant with 1.9% per year between 1950 and 1980;

(iv) that of all fossil fuels went down from the historical growth rate of 4.3% per year to 2.3% per year after 1973.

A further feeling about how this energy use is shared between the eight main world segments can be obtained from Table 4.8(b) where, using the 32-TW scenario as *an example*, Ncrth America and Eastern Europe-USSR are supposed to have the highest per capita energy use to at least the year 2025.

However, other scenarios are now available and Table 4.8(c) illustrates the possible world primary energy demand per world regions with efficient use, strong economic growth and constant urban fraction (Bach 1982a).

4.2.2.2 Energy release and climatic implications

The overall temperature of the Earth is determined by a mean balance over the course of the year between the solar energy that is absorbed and the infrared energy radiated to space and it is clear that the direct impact of injecting an appreciable quantity of heat into the atmosphere will contribute to change the Earth's climate. Table 4.9 enables a comparison between some man-made and natural energy releases.

The influence of this heat released will mainly depend upon the region. Cities at high latitudes and those with a compact layout, either already equal or even considerably exceed the local natural radiation receipt* (energy consumption density in

TABLE 4.9. Comparison of man-made and natural energy releases.

A. EXTERNAL PARAMETERS	Related to the globe TW(10^{12}W)	Related to unit area W m^{-2}
Solar energy intercepted by the Earth	173000	1360
One-fourth of the solar constant	--	340
Solar input to Earth surface + atmosphere	121000	240
Radiation balance at the Earth surface	52000	102
Geothermal heat	32	0.061
Radiation from the full moon	7	0.061
Krakato explosion of August 1883	1.7	---
B. INTERNAL PARAMETERS AT THE GLOBAL SCALE		
Production of available potential energy	1200	2.4
1% change in cloudiness	350	0.67
Change of evaporation equatorial oceans 10^7 km^2	300	0.59
Photosynthetic processes	90	0.18
Kinetic energy of general circulation	1.7	0.003

C. ENERGY RELEASE BY NATURAL LOCALIZED PROCESSES	area (km^2)	Total Power (TW)	Power Density ($W\ m^{-2}$)
Melting of available winter snow during spring	---	17000	---
Wet cooling tower of 1000 MW nuclear power plant (closed cycle)	0.02	1.68×10^{-3}	84000
Oil refinery ($5\text{-}6 \times 10^6$T of oil/yr)	0.1	0.4×10^{-3}	4000
20000 MW power park	26	0.04	1540
1 km freeway section with daily traffic density of 30000 cars and 7000 trucks	0.05	2.3×10^{-6}	46
One family residence (4 persons)	0.1×10^{-3}	5×10^{-9}	56
Street lightning for an average night in New York City	---	10^{-6}	---
Monsoon circulation	---	1700	---
Average cyclone (1 cm of rain/day)	10^6	100	100
Average hurricane	---	15	---
Great Lakes snow squall (4 cm of snow/hr)	10^4	10	1000
Thunderstorm (1 cm of rain/ 30 min)	100	0.1	1000
Tornado (kinetic energy)	10^{-3}	10^{-4}	100000
Average lightning stroke	---	10^{-8}	---
D. WASTE HEAT FROM URBAN AREA			
Ruhr district	6500	0.111	17
Chicago	1800	0.0527	29.3
St. Louis	250	0.0161	64.4
Berlin (West)	230	0.005	21.3
E. WASTE HEAT FROM SELECTED AREA			
One megaton nuclear device (heat dissipated over 1 hour)	---	1	---

After Sellers (1965), Flohn (1975), Hosler and Landsberg (1977), Bach (1982b).

New York city, Manhattan, is 630 W m^{-2}), but for areas on a regional or global scale, the heat release is only 0.002-0.5%* of the net radiation (Western Europe consumes around 1 W m^{-2}, Africa, 0.005, South America, 0.02, and world continents 0.06).

If we assume the world population to grow to 20 billion (20 x 10^9) inhabitants, whose energy use will be around 20 kW per capita (around 10 times the present world average and twice that of the US), the direct impact of the heat injected into the atmosphere will reach 0.8% of the total solar energy absorbed at the Earth's surface. Most will be released in the planetary boundary layer, and if we assume that this heat will end up being more or less evenly distributed in a given hemisphere, the additional heat can be considered as if it were an increase in the total amount of solar radiation reaching the surface. As the current set of models seem to converge quite well on the conclusion that a 1% increase in the heat available to the system would result in about 1.5°C increase in the mean surface temperature, the average surface temperature increase might be about 1°C by the end of the next century due to the direct release of heat related to this scenario.

As these assumptions are unlikely to be fulfilled, let us consider a slightly more realistic scenario as presented by Häfele et al. (1976). Assuming that there would be a population growth from today's 4 billion to 12 billion and that provision must be made for an average per capita energy consumption of 5 kW towards the year 2040, the world reserves of oil and gas and 20% of the present world's coal resources would be consumed. Such a direct heat release will raise the temperature of the globe by 0.15°C, which is negligible compared to the 2°C coming from a doubling of the atmospheric CO_2 content (Jason 1979) due to the burning of such fossil fuels.

4.2.3 The carbon dioxide problem

For over a century, concern has been expressed that atmospheric trace gases play a major role in controlling the Earth's heat balance. J. Tyndall (1863) clearly described the greenhouse effect. S. Arrhenius (1896), a Swedish chemist, and Th. C. Chamberlain (1899), an American geologist, indepently advanced the hypothesis that changes in the abundance of CO_2 in the atmoshere would affect the Earth's surface temperature. Arrhenius estimated that a doubling of the concentration would cause a global warming of about 6°C. In 1931, G.S. Callendar suggested that the global warming observed over the previous 60 years might have been caused by an increase in atmospheric carbon dioxide

* For comparison, the natural net radiation at the surface, averaged over the whole globe, is 102 W m^{-2} and in Europe mostly 50 - 80 W m^{-2}.

from the burning of fossil fuels, an argument which was taken back in the early 1950's by G.S. Plass. Neither suggestion was taken very seriously because of the difficulty to measure and relate the atmospheric CO_2 content and the global temperature variations, and also because many scientists assumed that all of the industrially produced carbon dioxide would be taken up by the oceans.

Since the 1960's, measurements and our understanding of the physical processes governing the carbon cycle and climate advanced markedly. Increasing the sense of urgency is the firmly established ovservational evidence of steadily increasing CO_2 concentrations in the global atmosphere. Internationally, the World Climate Conference (WMO 1979) highlighted the CO_2 problem, and its study became a major objective of the World Climate Program (WMO 1981). Comprehensive international studies in the context of energy-climate interactions were also conducted (Bach et al. 1979, 1980; Climate Research Board 1979, 1980; DOE 1980). Driven by concerns about the magnitude of the risks and the timing involved, a commissioned report by the US President's Council on Environmental Quality (CEQ 1981) reached the following conclusions:

(1) The potential long-term risks of significant social disruption caused by increased atmospheric concentrations of CO_2 are real, although the timing and regional impacts are still uncertain.

(2) If a global response to the CO_2 problem is postponed for a significant time, there may not be time to avoid substantial economic, social and environmental disruptions once a CO_2-induced warming trend is detected.

4.2.3.1 The Greenhouse effect

When modelling the radiation transfer through the atmosphere, it may be seen that carbon dioxide and other trace gases contribute to the greenhouse effect.*

The Earth's surface emits infrared radiation, half of it between 9 and 17 μm, a spectral interval which covers the atmosperic window (8-12 μm) where the atmosphere is, for the most part, quite transparent. As carbon dioxide has a strong absorption band at about 15 μm, and chlorofluoromethanes and nitrous oxide also have absorption bands in the window, increasing the concentrations of these trace gases reduces the transparency of the atmosphere and increases the downward atmospheric infrared radiation. This results in a warming of the surface and

* Although the term greenhouse effect has been critized (Lee 1973) it has become a part of the vernacular (glass in a greenhouse not only absorbs long-wave radiation - as does CO_2 - but also prevents turbulent vertical exchange of heat).

a corresponding cooling of the stratosphere (Newell et al. 1972, Groves and Tuck 1979) with possible destabilization of the troposphere and its consequence upon the hydrological cycle. However, the problem is even more complicated, due to the overlapping water vapour band of absorption.

4.2.3.2 History and projection of CO_2 production from fossil fuel combustion

The history of carbon dioxide procuction from fossil fuel combustion (and cement production, about 2% of the total) has most of the same properties as the history of global energy demand, the rate of growth, 4.3%, at least before the energy crisis (cfr. section 4.2.2.1), being slightly less for CO_2 (Fig. 4.5).

It is also instructive to examine where the fossil fuel is being used today and will likely be used in the future. Figure 4.6 shows that in 1974 nearly half of the CO_2 produced by burning fossil fuels comes from the USA and Western Europe. However, in Figure 4.7, where total carbon dioxide produced in A.D. 2025 is assumed to be 3.8 times that of present, the most striking feature is the very large growth in the developing segments of the world - about 50% of the world total being attributable to China, Latin America, Africa, Asia and the Mid-East.

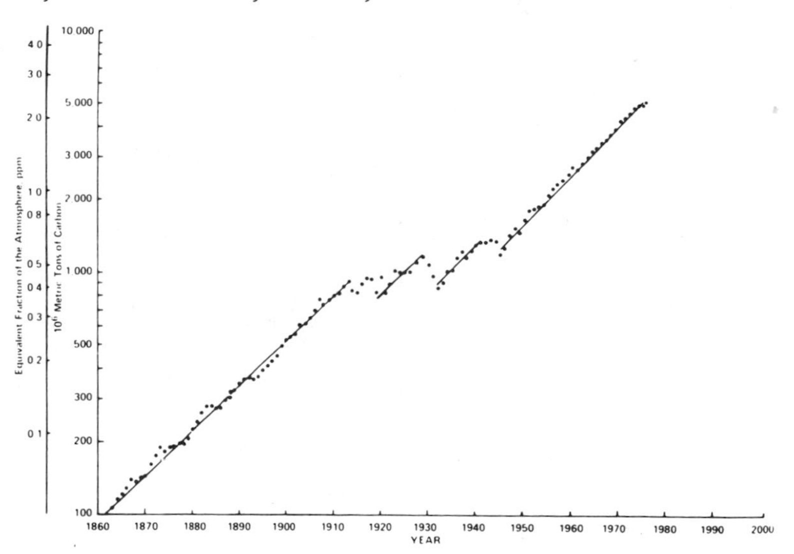

Fig. 4.5 Part of the history of carbon dioxide production from fossil fuel combustion (and cement production - about 2% of the total). After ROTTY 1979).

*4.2.3.3 Atmospheric CO_2 concentrations and the carbon cycle**

The understanding of the CO_2 cycle and estimation of the future

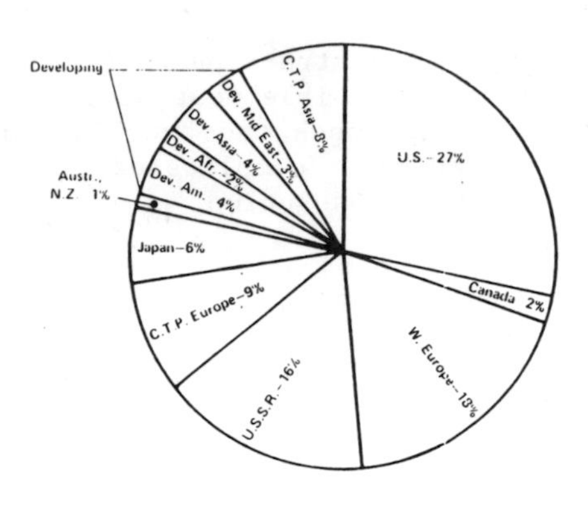

Figure 4.6 Global CO_2 production (1974). Total production is estimated to be 5 Gt of carbon in 1974. (Marland and Rotty 1980).

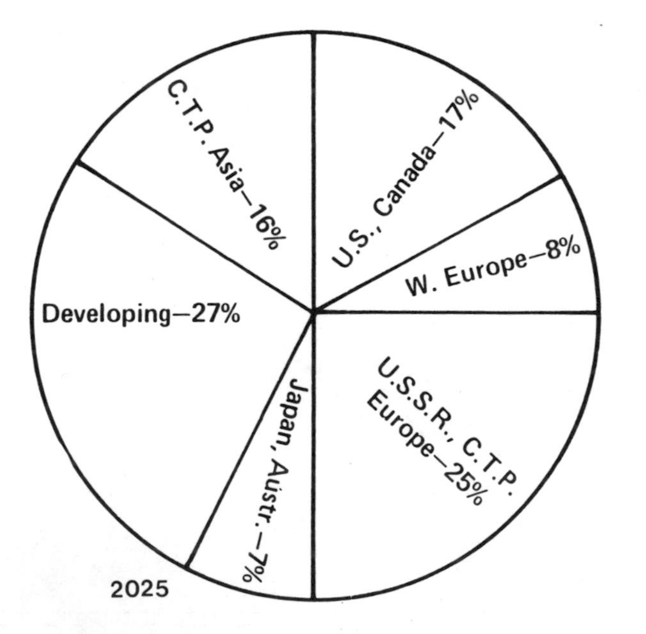

Figure 4.7. Global CO_2 production. The area of the circles representing 1974 (Fig. 4.6) and 2025 (this figure) production are proportional to the respective CO_2 rate of production, estimated to total 19 Gt C in 2025 A.D. (Marland and Rotty 1980).

impact of anthropogenic sources on the CO_2 atmospheric concentration is still limited. Progress in this field may be achieved by, namely, a better knowledge of past changes to supplement insufficient direct measurements. Both CO_2 concentrations and carbon isotope ratios in the different reservoirs (atmosphere, biosphere, ocean and sediments) are of particular interest for such studies.

Atmospheric CO_2 contains the carbon isotope ^{12}C and ^{13}C which are stable (with a respective relative abundance of about 98.9% and 1.1%). It also includes a very small amount (10^{-10} % of the total) of radio-active ^{14}C (half-life of about 5700 years) which has both natural (interaction of cosmic ray neutrons with nitrogen) and anthropogenic (production by nuclear bomb tests and power plant) origins. Carbon isotopes are fractionated during photosynthesis processes which lead to the lighter ^{12}C being preferentially included during the growth of trees. More generally, in nature the isotopic composition can be altered by kinetic processes or isotope exchange reactions which are responsible for the different isotopic composition of the various reservoirs. This isotopic labelling of the different reservoirs may be used to evaluate exchanges which take place. As an example, the addition of fossil fuel CO_2 from organic sediments to the atmosphere reduces both atmospheric $^{13}C/^{12}C$ and $^{14}C/^{12}C$ ratios. Similarly, a decreasing amount of the biosphere reduces atmospheric $^{13}C/^{12}C$ but, to a much smaller extent, $^{14}C/^{12}C$.

Through various calculations related to tree-ring studies, a series of pre-industrial (around 1800) concentrations have been proposed principally in the range of 230-260 ppmv (part per million by volume), but with some values as high as 295 ppmv (Lorius and Raynaud 1983). It must however be emphasized that these conclusions are dependent both upon the hypothesis included in the model calculations and the validity of a significant decrease in the $\delta^{13}C$ tree-ring record reflecting global atmospheric changes. On the other hand, the proxy data available from ice core measurements indicate a pre-1850 atmospheric CO_2 -level of approximately 260 ppmv. Current work in progress will hopefully lead to a firmer estimate of the biosphere non-fossil fuel input (which seems very significant in the late 19th century, the pioneer agriculture effect, while now the fossil fuel source appears to predominate over the last 100-150 years) and of the pre-industrial CO_2 concentrations.

In 1979, the content of carbon as dioxide in the atmosphere was about 709 Gt. Between 90 and 145 Gt of carbon has been added to the atmosphere since 1850 (136.5 between 1860 and 1977 were estimated by Rotty in 1981), the uncertainty arising from the

---

* The accuracy of the carbon cycle is discussed in Section 4.1. The numbers given in this summary are in agreement with the best present estimate and within the overall limits, as given in Section 4.1.

already discussed unknown level of CO_2 in the pre-industrial atmosphere. However, only about half of this carbon introduced in the atmosphere has remained there. In 1978, the carbon content increased by 2.6 Gt even though 5.6 Gt of carbon were introduced by the burning of fossil fuels. A quite similar ratio of airborne CO_2 (2.2 Gt C/yr)* to carbon dioxide introduced by fuel burning (4 to 5 Gt C/yr) has been maintained over the twenty years of direct measurements of the CO_2 concentration in the atmosphere: Rotty (1981) has calculated that the fossil fuel carbon dioxide produced between 1959 and 1978 was 70 Gt C representing 1.75 times the measured increase of carbon in the atmosphere (40 Gt C or 19 ppmv) during those 19 years (Bacastow and Keeling 1981).

The atmospheric concentration of carbon dioxide has in fact been steadily increasing since the Industrial Revolution as a result of the burning of coals, petroleum, natural gas and wood. This concentration was 290 ppmv around 1900, 315 ppmv in 1958 and 331 ppmv in 1975 (JASON 1979), the rate of growth being remarkably constant if we neglect the slumps during World Wars I and II, the Great Depression of 1927-1935 and since the 1973/74 world energy crisis. The best documented period started in 1958 (Figs. 4.2 and 4.8) with the well calibrated continuous measurements made in the Mauna Loa Observatory in Hawaï, the same trend being also measured at the South Pole and from aircraft over the Arctic. As shown in Fig. 4.2 where the annual variation is explained by seasonal changes of the rates of photosynthesis and respiration by plants and the soil, the 340 ppmv value was exceeded for the first time in 1981.

However, if other sources of CO_2 do exist (Table 4.10), as for example the biosphere in the deforestation process and soil management practices, the contribution of fossil CO_2 to the total content of the atmosphere will deminish. Indeed, as during the last decade, the total rate of carbon input into the atmosphere seems to have been between 6 and 11 Gt C/yr (Bolin et al. 1979a), the airborne fraction of man's emission of CO_2 to the atmosphere may have been between 30 and 50% during the last 20 years (Bolin 1979). This deviation, in spite of its uncertainty, is very important for the estimation of future concentrations (Olson et al. 1978).

An understanding of the distribution of carbon among the ocean, biosphere, sediments and atmosphere (Fig. 4.9) is thus of crucial importance (Bolin 1981). The fact that only half of the fuel-based carbon introduced into the atmosphere remains there is conventially explained through ocean absorption, though the potential importance of biospheric interactions has also been recognized (Keeling 1980).

---

*Gt C/yr = 10^9 metric ton carbon per year

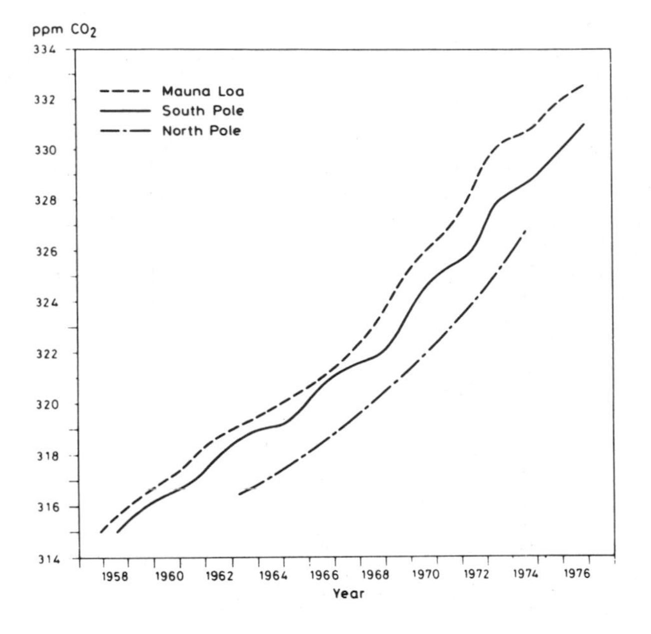

Fig. 4.8. Seasonally adjusted concentrations of atmospheric CO_2 at the Mauna Loa, South Pole (Keeling and Bacastow 1977), and as measured from commercial aircraft over the North Polar region (Bischof 1977). The North Polar curve is a second-degree polynomial deduced from average annual values; the values have not been corrected by intercalibration with the former data series (Bolin et al. 1979a).

Undoubtedly, the oceans are an enormous sink for carbon dioxide, but one with a very slow response time (Fig. 4.9). Most of the carbon dioxide in the oceans is in the form of carbonate and bicarbonate ions, and only a little is dissolved free carbon dioxide. As a rather small change in the amount of free CO_2 dissolved in sea water corresponds to a relatively large change in the pressure of CO_2 at which the oceans and the atmosphere are in equilibirum, the rate at which the ocean can absorb the gas is much lower than had been thought (Revelle 1982). On the other hand, gases are readily exchanged between the oceans and atmosphere only in a well-mixed surface layer some 80 m deep on the average. If absorbed CO_2 is confined to this layer, a given change in the amount of CO_2 in the atmosphere would change the content of the surface water by only about a ninth as much and this buffer factor 9:1 increases with increasing atmospheric CO_2. Downward diffusion, advection and diffusion between the well-mixed layer and the deeper water would increase the amount of CO_2 taken up by the ocean and there by reduce the amount remaining in the atmosphere, but the processes are relatively slow.

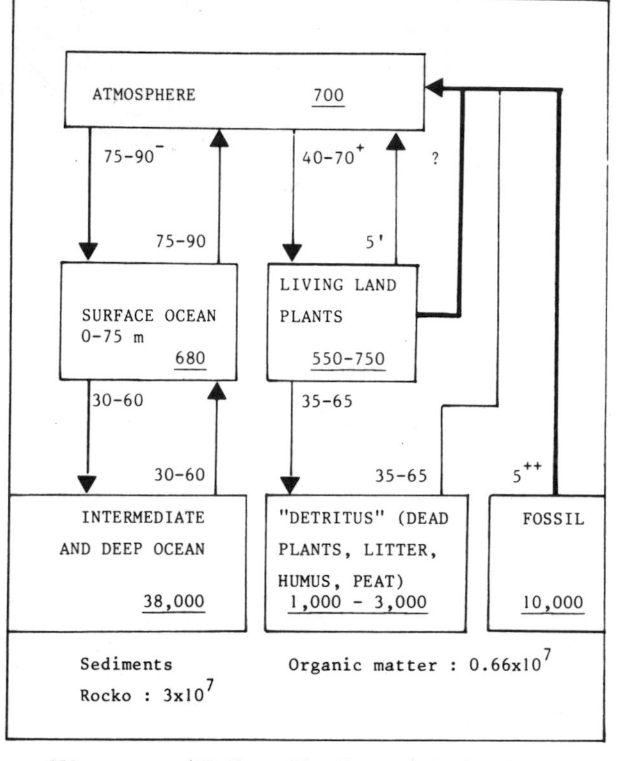

Fig. 4.9. The carbon cycle of the Earth in 1977 (Hampicke 1979b). The data are approximate but within the limits of accuracy given in section 4.1; there are, for example, notable differences in the estimation of C-flux due to deforestation (0.8 for Bolin (1977) to 6 for Woodwell et al. (1978)) and reforestation (0.3 to 5.7).

+ net primary production (estimated to be 63 by Duvigneaud in Bolin et al. (1979a) and 113 by J. Olson (Revelle 1982).

' decomposition via "grazing food chain".

- the gross uptake of atmospheric CO_2 in the ocean may be between 60 and 200 Gt C/year; the net CO_2 uptake of the oceans may be between 2 and 8 Gt C/year (Bolin et al. 1979b).

TABLE 4.10 Man-made transfer of carbon to the atmosphere from the land biota.

Activity	Gt C/year
Forest clearing in the Third World	+ 3.6
Industrial use of wood	+ 0.3
Firewood	+ 0.3
Soil organic matter decomposition	+ 0.6
Reforestation	- 0.3
Regrowth in the tropics	- 1.0
Regrowth in temperate regions	- 0.5
Growth stimulation by CO_2	- 0.3
Growth stimulation by NO_x	- 0.2
Total	+ 2.5

This estimate is in agreement with the estimates of the transfer of carbon to the atmosphere from non-fossil sources as given by Wong (1978), 1.9 Gt C/year, and other authors, the most probable net transfer lying between 1.5 and 4.5 Gt C/year. (after HAMPICKE 1979a).

Hence, the deep water of the ocean (below 1000 m) appears to have played a rather small role as a sink for the emissions to the atmosphere due to mankind's activities: probably less than 5% of the excess CO_2 has gone into the deep sea. However, the intermediate water masses of the ocean (100 - 1000 m depth) have probably served as a considerably more effective sink than the deep water, even if their precise role is still unclear. In any case, it means that, if we were to suddenly stop adding carbon dioxide, it would take from 1000 to 2000 years for the added amount already in the atmosphere to decay to some 15% of its concentration (Junge 1978).

As a consequence, this slow turnover of the oceans (around 1500 years), combined with the relatively low carbonate-ion concentration of surface sea-water, could result in about 80% of the CO_2 added during the next century being retained in the air (Keeling and Bacestow 1977). In fact, the increased resistance of ocean water to absorb dioxide and the probably upper limit of carbon storage reached under human stress in the biosphere are the main forces that cause a larger fraction of the excess carbon dioxide to remain in the atmosphere (Chan et al. 1979).

The other most important natural sink of CO_2 is the biosphere, consisting of all living organic matter. Trees constitute the largest mass and it is logical to ask if the forests of the world will grow progressively faster as a result of the increased atmospheric CO_2 (Goudriaan and Ajtay 1979) or if, on the contrary, they will be being cut down faster than they grow. Unfortunately, these crucial questions are not yet resolved (Elliott and Machta 1977): estimates of the input of CO_2 into the atmosphere due to tropical forests clearings varies between 20% (Bolin 1977) and 100% (Woodwell 1978) of the input due to burning fossil fuels (Table 4.10). Nevertheless, this hypothesis - terrestrial biomass as a net source of CO_2 - is confirmed by observations: in developing countries, mainly due to the need for additional land for agriculture, about 120000 km^2 of natural forest are cleared and burned every year (22 hectares per minute, Bolin 1977), around 20000 km^2 in Africa and 60000 km^2 in America.* If we realize that the terrestrial biomass (at an area of 150 x 10^6 km^2) represents some 830 GT C of which 90% is provided by all forests (49 x 10^6 km^2), 54% being present in the tropical ones (22.5 x 10^6 km^2) alone, it is easy to understand how much the carbon cycle can be sensitive to any important modification of this reservoir.

The ranges of these quantitative uncertainties regarding the carbon cycle (see Section 4.1) lead to several quite different pictures of the distribution of the excess carbon. The still-airborne fraction of the accumulated man-made release varies between 25% and more than 70%, the major causes being the uncertainty regarding the atmospheric CO_2 concentration prevailing before 1860, the potential of the terrestrial biomass to accumulate carbon in organic material on land and the assumptions made on the circulation in the intermediate ocean layer (Björkström 1979).

Considering all the uncertainties in these reservoirs and in the fluxes between them, we may only tentatively deduce different projected CO_2 levels in the atmosphere (Niehaus 1979, Bach 1982b) and, along with other strategies, finally present some reasonable scenarios (not forecast!).

---

* These values lead to a global net man-made transfer of carbon (in the form of carbon dioxide) to the atmosphere from land biota of 1 ± 0.6 Gt C/year. However it must be pointed out that Wong's estimates (1978) of new forest clearings in the tropics range from 8 to 24 x 10^4 km^2. If an average value of 160000 km^2 is considered, the net input of carbon into the atmosphere due to human modifications of land biota and soils would be 1.9 Gt C/year.

4.2.3.4 Temperature changes from doubling atmospheric CO_2

A number of theoretical calculations have been made of the changes in radiative equilibrium and in temperature corresponding to a doubling of the carbon dioxide in the atmosphere, based on a variety of atmospheric models ranging from globally averaged one-dimensional models, to two-dimensional latitude-dependent models with varying degrees of complexity and to three-dimensional models that include a simplified ocean.

Reviews of these simulations have been recently given by CO_2 /Climate Review Panel (1982), Kandel (1983) and Schlesinger (1982). Two types of study have been performed with these models examining the models' response in a state of climatic equilibrium and the transient response (cf. Section 4.2.8) of the climate to the continuous rise of CO_2. In an equilibrium study (Gilchrist 1983), a mathematical model of climate is perturbed with a fixed, time independent doubling of the CO_2 concentration and allowed to reach a new equilibrium. However, the time required to attain that new equilibrium has not been ascertained from the equilibrium simulations and the actual CO_2 increase is a continuous rise, not an instantaneous step increase. In a nonequilibrium study, the CO_2 concentration is made to increase with time and the response of the climate system at any time is compared with the equilibrium response for the CO_2 concentration at the same time. The object of the nonequilibrium study is to determine the lag in the response of the climate system, the knowledge of which is required to estimate when a CO_2-induced climate change may become detectable (Michael et al. 1981).

Although an increase in atmospheric carbon dioxide is expected to lead to a warming of the troposphere and of the Earth's surface, the estimates of this temperature increase are, highly dependent on the physical assumptions made in the model. For example, Augustsson and Ramanathan (1977) have found an increase of 1.98°C with fixed cloud top altitudes, but 3.2°C if cloud top temperatures are held constant; Hansen et al. (1981)observed that, depending on the precise formulation and on the feedbacks included, the resultant change in surface temperature varies between 1.22 and 3.5°C; other examples are given in Table 4.11 where the series of GFDL experiments best illustrates this sensitivity of the predicted global warming, and particularly the warming at high latitudes, to variations in the assumed experimental conditions.

In fact, two major problems arise in climate research, the treatment of the ocean and of clouds. Fully coupled ocean-atmosphere general circulation models exist but they are expensive to run to equilibrium due to the large thermal inertia of the deep ocean. On the other hand, whereas one can predicht with reasonable confidence that the ocean temperatures will rise with increased CO_2, it is still difficult to predict the accompanying change in global cloudiness. Nevertheless, the results, as shown in Table 4.11, turn out to converge on approximately the same conclusion: doubling the concentration (up to about 650 ppmv), which could occur around A.D. 2050, will correspond to a global warming to 2 to 3°C.

Simplified climate models have been used so far as an important tool to give some guidance on the expected change in mean global temperatures due to increases in CO_2. In particular, they are considered as an important addition to GCMs for study the transient response of the climate system to some forcing, certain feedback mechanisms of the global climate system, evaluation of climate sensitivity to individual external forcings of man-made and natural origin as well as interpretation of the causes of the changes of the global climate simulated by GCMs. However, they do not allow an investigation of either the regional changes in temperature which may be very different from the global mean, or the regional changes in precipitation and water balance. Because it is this geographical distribution of a possible CO_2-induced climatic change which is of importance to humanity, GCMs are thus widely used to provide the way this global change is distributed among latitudes. This can be seen on Figure 4.10 which represents the zonal mean temperature difference between the 2 x CO_2 and 1 x CO_2 cases as a function of latitude and height. In particular, at low latitudes the warming is most pronounced at upper tropospheric levels. This is a consequence of the dominance of convective activity at these latitudes, in atmosphere that is usually conditionally unstable. At higher latitudes, the increase of surface temperature is magnified due to the recession of the snow boundary (and the subsequent lowering of the surface albedo in these regions (Lian and Cess 1977)) and the thermal stability of the lower troposphere which hints convective heating to the lowest layer.

Near the pole, the temperature difference is nearly 8°C. In the stratosphere, enhanced radiation to space is not balanced by increased absorption from lower levels, and therefore there is a net cooling.

Because of its direct potential impact on the world environment (cf. Section 4.2.7.3), considerable attention has been directed to the large rise of surface temperature near the pole. In the early experiments (Manabe and Wetherald 1975, 1980), there were reasons for suspecting that the warming at the surface in high latitudes (about 8 - 10°C) might have been exagerated. The use of averaged solar radiation, the lack of ocean currents and heat storage and an idealized topography might have all contributed to such an effect; more recent experiments seem to confirm this view.

TABLE 4.11. Computed surface air temperature changes for a doubling of CO_2 to 600 ppm. Adapted from BACH (1978) and SCHLESINGER (1982) where the references ares listed.

Temperature sensitivity to a doubling of CO_2 (°C)	Specification	References
A. 1-D radiative-convective models		
	1-D R-C model (globally averaged)	Manabe and Wetherald (1967)
2.9	with fixed RH and clear skies	
2.4	with fixed RH and average cloudiness	
1.4	with fixed AH and clear skies	
1.3	with fixed AH and average cloudiness	
1.9	Radiative transfer scheme by Rodgers and Walshaw (1966) with radiative-convective model by Manabe and Wetherald (1967)	Manage (1971)
1.5	1-D R-C model with fixed RH and cloudiness	Ramanathan (1974)
	1-D R-C model	Wang et al. (1976)
≃3.2	with a fixed cloud-top temperature	
≃2.1	with a fixed cloud-top height	
	1-D R-C Model for CO_2 bands 12 to 18 μm and constant RH and CTA	Augustsson and Ramanathan (1977)
1.98	for CO_2 bands 12 to 18 μm plus 10 and 7.6 μm and for	
3.2	same but for constant RH and CTT	
	R-C model (surface temperature 290 K.)	Rowntree and Walker (1978)
	fixed AH	
0.8	PC	
1.3	CA	
	fixed RH, no clouds	
1.7	PC	
1.9	CA	

Temperature sensitivity to a doubling of CO_2 (°C)	Specification	References
	fixed RH (average cloudiness)	
1.4	PC	
1.9	CA	
2.2	humidity generated by model's convection scheme (no cloud)	
2.8	1-D Gray radiating homogeneous	JASON (1979)
	1-D band model for atmospheric IR emission	JASON (1979)
3.8	RH = 50%	
4.3	RH = 60%	
2.0	1-D R-C equilibrium	Wetherald and Manabe (1979)
2.2	1-D R-C model fixed RH, fixed CTA, annual insolation,ocean-atmosphere feedbacks	Ramanathan (1981)
1.2 - 3.5	1-D R-C equilibirum experiments with fixed RH, AH, CTA, CTT snow ice and vegetation albedo feedbacks	Hansen et al. (1981)
transient:	for one simulation (ΔT=2.8), CO_2 signal emerges from climatic noise(2σ =0.2°C) about 1995	
2.4	1-D radiative R-C model fixed RH, no cloud, annual insolation	Hall, Cacuci, Schlesinger (1982)
	2-D annual zonally averaged steady-state hemispherical model with diffuse thin clouds as cloudiness feedback	Temkin and Snell (1976)
1.7	global annual average	
1.2	0 to 30° latitude	
1.2 to 2.0	30 to 60° latitude	
2.0 to 6.0	60 to 90° latitude	

Temperature sensitivity to a doubling of CO_2 (°C)	Specification	References
	CTA is cloud-top altitude; CTT is cloud-top temperature; RH is relative humidity; AH is absolute humidity; PC is penetrative convection; CA is convective adjustment	
B. General Circulation Models		
3.5	3-D GCM realistic topography, predicted cloud, mixed oceanic layer, seasonal insolation	Hansen (1978) GISS
3.9	idem but swamp and annual insolation	Hansen (1979)
	3-D GCM with interactive lapse rate and ocean "swamp" and hydrologic cycle; but fixed cloudiness, no CO_2-ocean buffering, and no coupling of deep oceans to atmosphere; heat transport by large-scale eddies computed explicitly	Manabe and Wetherald (1975) GFDL
2.9	global average	
7	north of 70°N	
	3-D simplified GCM + swamp ocean,idealized geography, predicted SST and cloud, annual insolation	Manabe and Wetherald (1980) GFDL
3.0	global average lower tropospheric warming	
7.0	high latitude	
3.0	3-D GCM + interactive ocean idealized geography, predicted SST, prescribed cloud, annual insolation (inferred from 4 x CO_2)	Wetherald and Manabe (1981) GFDL
2.4	idem but seasonal insolation	Manabe, Wetherald and Stouffer (1981)

Temperature sensitivity to a doubling of CO_2 (°C)	Specification	References
2.0	3-D GCM + interactive ocean realistic geography, predicted SST, predicted humidity, fixed CTA seasonal insolation	Manabe and Stouffer (1979-1980) GFDL
2.0	3-D GCM realistic geography, predicted SST + cloud annual insolation	Schlesinger (1982) OSU
1.3	3-D GCM realistic geography, swamp ocean, predicted humidity and cloud, annual insolation	Washington and Meehl (1982) NCAR
1.4	3-D GCM realistic topography, swamp ocean and sea-ice, predicted humidity and cloud, annual insolation	Alexandrov et al. (1982)
---	3-D + ocean general circulation, idealized topography, predicted humidity, fixed CTA, annual insolation	Bryan et al. (1982)
transient:	10 to 25 yr lag time	
3.0	3-D GCM SST increased by 2 K as a first approach to consider the ocean response to the increase in the atmospheric temperature due to CO_2 doubling, climate cloud cover and sea ice extent, seasonal insolation	Mitchell (1983) UK Met. Office

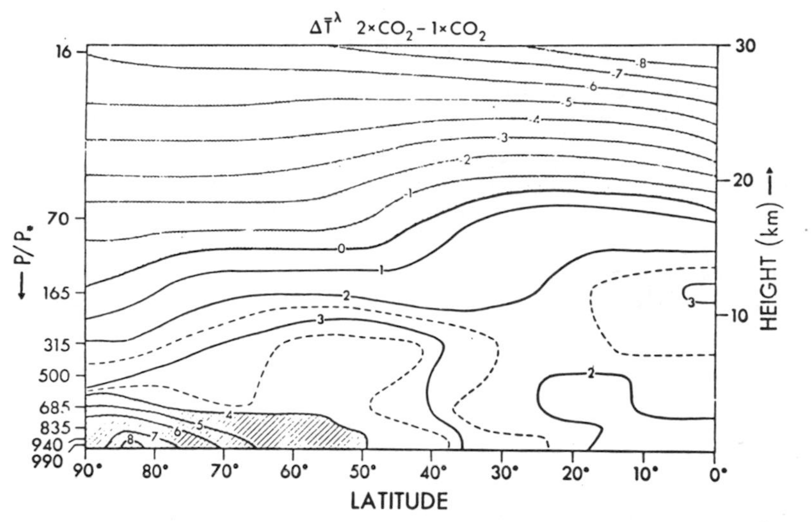

Fig. 4.10. Zonal mean temperature difference between the 2 x CO_2 and 1 x CO_2 cases as a function of latitude and height obtained from a 3-D GCM. (Wetherald and Manabe 1981).

The importance of including the response of the ocean surface is demonstrated in a sensitivity integration by Mitchell (1983) with doubled carbon dioxide concentration and sea surface temperatures increased by 2°C (taken as a first approach to consider the ocean response to the increase in the atmospheric temperature due to CO_2 doubling): tropospheric temperatures increase by 3°C, but 2.9°C are attributed to increasing SSTs. With a mixed layer ocean whose depth (68 m) was chosen to give the heat storage associated with the annual cycle, the surface warming at high latitude obtained by Wetherald and Manabe (1981) was not more than 7.5°C. The use of this model with a seasonally varying solar radiation shows further that the warming near the pole, as a result of doubling CO_2 (inferred from 4 x CO_2 assuming that the effect on temperature should be twice that obtained from 2 x CO_2, although a logarithmic variation is more appropriate) was even reduced to 5.5°C (Fig. 4.13(a)); the warming associated to the (4 x CO_2) experiment amounts, respectively to 15 and 11°C at 85° latitude and 5 and 4°C at 20°C latitude for the annual and seasonal models. The smaller sensitivity of the seasonal model is attributed to the absence of strongly reflective snow cover (or sea ice) during the summer when the insolation has a near-maximum intensity. The Manabe and Stouffer (1980) results are even more significant (Fig. 4.11(a, b)): near the north pole a warming of 4.5°C is observed for doubling, near the south pole it reaches only 3°C because, the Antarctic remaining snow covered, the snow albedo feedback is not very effective, and averaged over all latitudes the near-surface warming is about 2°C. So, as the assumptions were made more realistic the sensitivity of the global surface temperature rise to a doubling of carbon dioxide tends to fall and to reach a value of about 2°C.

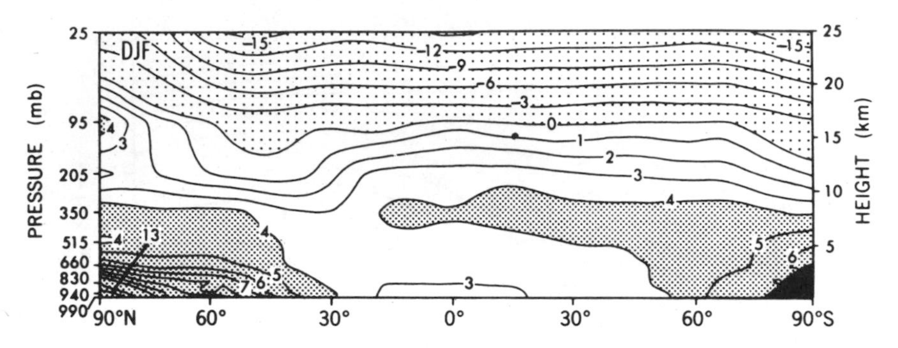

Fig. 4.11(a). Latitude-height distribution of the difference in winter zonal mean temperature between a 4 x CO_2 and a 1 x CO_2 experiment, by Manabe and Stouffer (1980) using a seasonal model.

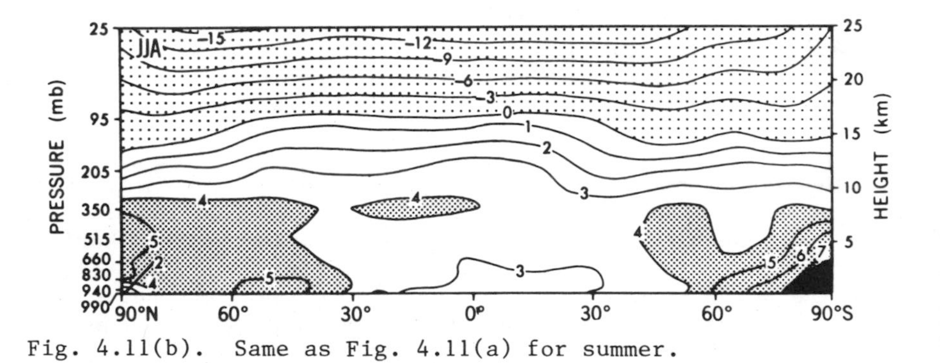

Fig. 4.11(b). Same as Fig. 4.11(a) for summer.

4.2.3.5. *Precipitation changes from doubling CO_2*

With the use of realistic geography in the models, examination of changes in other climatic variables than temperature become possible, particularly in precipitation and availability of moisture in the soil, measured by precipitation minus evaporation (P-E).

As the water vapour content of the air could be expected to be higher in a warmer atmosphere, increased precipitation and a balancing increased evaporation would also be expected. Figure 4.12 shows such an induced general increase in the intensity of the hydrologic cycle (a 6.7% increase in the 4 x CO_2 experiment by Manabe and Stouffer (1980) which does not necessarily imply an increase of wetness everywhere. For example, the following changes are particularly interesting:

- a large increase of the rates of precipitation and runoff in high latitudes.
- a general poleward drift of the climatic zones in middle and

high latitudes; in particular, aridity of the latitudes 34-35° would increase due to the poleward shift of the middle latitude rainbelt.

- a general reduction of the aridity of the eastern half of the subtropical continent.
- a large reduction of the meridional temperature gradient in the lower model troposphere because of the poleward retreat of highly reflective snow cover and large increase in the poleward transport of latent heat.

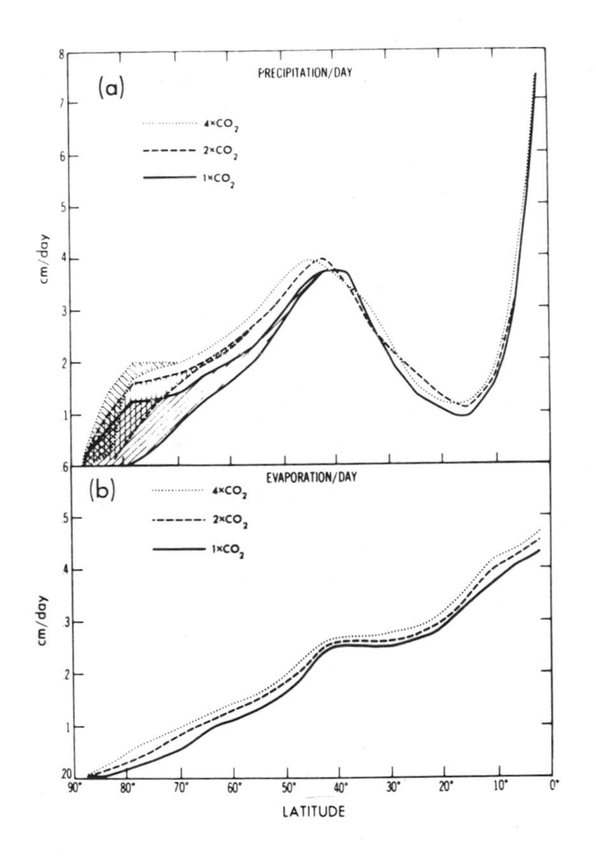

Fig. 4.12. Precipitation and evaporation due to a doubling of CO_2 concentration. The portion of the precipitation rate which is attributable to snow fall is indicated by various types of cross-hatching. After Manabe and Wetherald (1980).

A comparison between the hydrologic response of the annual seasonal models indicates also considerable seasonal variations. The latitude-time distribution of the zonal mean difference of (P-E) between the 4 x CO_2 and 1 x CO_2 experiments over the continent, shows a poleward displacement of the area of negative (P-E) from approximately 35° to 55° latitudes as seasons progress from winter to summer (Fig. 4.13(b)). This seems to be directly attributable to the poleward shift of the mid-latitude rainbelt and therefore reduced precipitation there with season. In high latitudes, there is in general an increase of (P-E) and nearer the equator, there is a tendency for an increase in winter and a decrease in summer, at least in the Manabe and Stouffer experiment (Manabe et al. 1981). Similarly, the zonal mean soil wetness is reduced extensively in summer over two separate zones of middle and high latitudes in response to CO_2 increase (Fig. 4.13(c)).

4.2.3.6 Limitations of the models

However, there are enough uncertainties in the theoretical treatments of all models to cause en "error bar" with a factor of about 2, even for the temperature field. For example, a serious weakness of climate models is that since the type and amount of clouds are difficult to compute, important interactions between the clouds and the radiation field are not accurately represented although some effects of the observed average cloud cover on the radiation budgets are allowed for. Clouds account for nearly two-thirds of the planetary albedo and participate in several important feedback mechanisms that regulate the temperature. Sensitivity analyses have shown that changes of about 5% in cloud amounts and about 500 m in cloud height may produce changes of 3-5°C in surface air temperature, i.e. that a change of only 1% (which could, at present, not even be detected) in the total cloud cover could mask the effects of a 25% increase in carbon dioxide (Mason 1979). However, it has also been suggested that cloud amount is not a significant climate feedback mechanism in the interpretation of the latitudinal change of seasonal or annual mean climate (Cess 1976). Moreover, it has been shown, for an intermediate numerical model, that cloudiness and relative humidity decrease with increasing temperature (Roads 1978). From a comparison between their 1975 model with prescribed cloudiness and their 1980 model with a simple cloud prediction scheme, Manabe and Wetherald found also that overall changes in cloudiness due to increased CO_2 were relatively unimportant though they warn that it would be premature to reach conclusions concerning the sensitivity of climate to cloud cover. Obviously, the possible role of clouds in man-made climatic change should not be overemphasized.

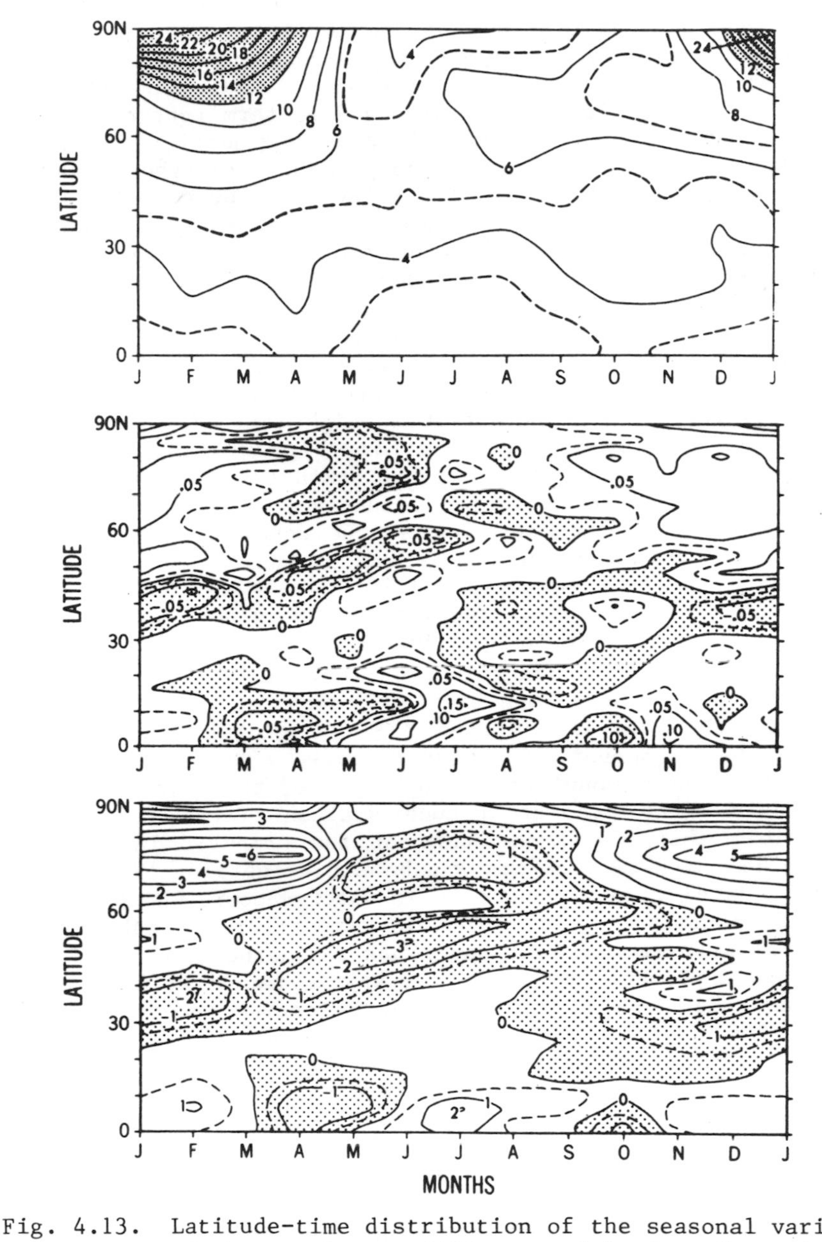

Fig. 4.13. Latitude-time distribution of the seasonal variation of the zonal mean difference (a) of surface air temperature, (b) of (P-E) over the continent, (c) in the soil moisture, between the (4 x CO_2) and (1 x CO_2) experiments with seasonal model. The distributions in two hemispheres of the model are averaged after shifting the phase of the Southern hemisphere variation by 6 months (After Wetherald and Manabe 1981).

A second major deficiency of the models is that they fail to represent interactions between the atmosphere and the oceans which, because they store and transport great quantities of heat, almost certainly exert a strong long-term control on the climate. Although some simplified oceanic general circulation models have been provisionally coupled to atmospheric GCMs, none has taken full account of the effects of the oceanic surface mixed layer, the presence of mesoscale eddies and the formation of sea-ice. Moreover the study of the possible impact of CO_2 on snow and water in high latitudes has also been recommended for future investigation. Some evidence has been presented that the spectral absorption of CO_2 in the near infrared portion of incoming shortwave radiation (0.75 - 5 μm) and the role of this near-infrared in the heating of the uppermost active layer of snow may have a cooling effect in high latitudes (Choudhury and Kukla 1979). However, calculations by Newell and Dopplick (1979) indicate that the thermal radiation enhencement is much larger than the solar radiation reduction, basically negating this possibility. As a consequence, the development of coupled atmosphere-ocean-cryosphere models and extensive and continuous measurements of the parameters of the climate system are indispensable for a full understanding of the global climate and ultimately for a prediction of climatic changes.

Even if empirical determination (Hoyt 1979) of the heating of the Earth by the carbon dioxide greenhouse effect leads us to support the generally agreed 2 - 3°C global warming for a doubling of the amount of carbon dioxide, with larger changes at high latitudes amounting up to 7 - 10°C, there are, however, some models which predict a much smaller effect. Indeed, it is possible that the major part of the computed increase in infrared transfer (3 W m^{-2} in Manabe and Wetherald, 1975) actually originates from an increase in water vapour, which is predicted to accompany an increase in temperature. Breaking down the actual infrared fluxes into bands dominated by CO_2 and H_2O, Newell and Dopplick (1979) using a static radiative flux model have found the expected increase in the energy received at the surface (0.8 - 1.5 W m^{-2} for cloudy conditions and 1.1 - 2.6 W m^{-2} for clear skies) but with a corresponding influence on surface air temperature in low latitudes less than 0.25 K, smaller by a factor of 8 than the finding generally accepted. No results are yet available for higher latitudes. A similar result has been obtained independently by Idso(1982).

However, approaches centered on the surface, limited regions or time-limited observations necessarily become more complex because many fluxes and reservoirs of energy must be explicitly and quantitatively taken into account. Moreover, the sensitivity of climate to a perturbation in some radiative processes can best be assessed by considering the entire global Earth-atmosphere system and its feedback mechanisms. In both Newell and Idso cases, the increase in the downward flux at the surface is underestimated because direct surface heating is considered only. As clearly demonstrated by Ramanathan (1981), the primary contribution to the (2 x CO_2) surface warming (2.2°C) is indeed from the enhanced tropospheric infrared emission (1.7°C) which is an order of magnitude greater than the direct CO_2 radiative heating at the surface (0.17°C). The source for this enhancement is the increased evaporation from the warmer ocean in the CO_2-rich atmosphere. The third process involved is the direct CO_2 tropospheric radiative heating which enchances the infrared emission by the radiatively-active constituents of the troposphere, the downward component of it amplifying the surface warming (0.33°C); this process underlines the importance of the overlap of CO_2 and H_2O bands in the 12-18 μm region.

4.2.3.7 *Scenarios of temperature increase from CO_2 projections in the future*

Having all these limitations in mind, it is nevertheless useful to estimate future changes of global temperature which would result from different scenarios based upon different energy consumption strategies. To illustrate practial impacts, 6 energy projections and their effects on CO_2 and temperature will be documented and compared.

First, from the pre-1980's 30-TW scenarios, a comparison will be made between the respective impacts of fossil fuel versus solar-nuclear strategies. Four projections of energy consumption, decreasing from 36 TW to 5.2 TW, will then be presented (cf. Section 4.2.2).

Figure 4.14(a) shows the scenario based on fossil fuels. Total consumption of coal, oil and gas are, respectively, 2800, 280 and 170 Gt CE (carbon equivalent) and maximum rate of CO_2 emission will be 24 Gt C/year. Atmospheric CO_2 concentration would then increase by the end of the next century to about four times the concentration of today and the global average temperature change would be at least 4°. In Figure 4.14(b), a 30 TW strategy is analysed where most of the energy is of nuclear and solar origin, coal consumption remaining at its present level until the year 2030 and oil and gas being reduced to 240 and 150 Gt CE, respectively. In this case, the CO_2-emissions would reach a maximum of about 8 Gt C/year and the atmospheric CO_2-concentration would increase to about 400 ppmv causing an average global temperature change of only 0.5°C above the present level.

In Figs. 4.15 (cf. Bach 1982a, for further details), the historic development of the CO_2 emission is taken from UN statistics given by Rotty (1980), the historic atmospheric CO_2 concentrations are either calculated from these emission data on the basis of a carbon cycle model or observed (1957-1975) at Mauna Loa, Hawaï (Fig. 4.2). The surface air temperatures

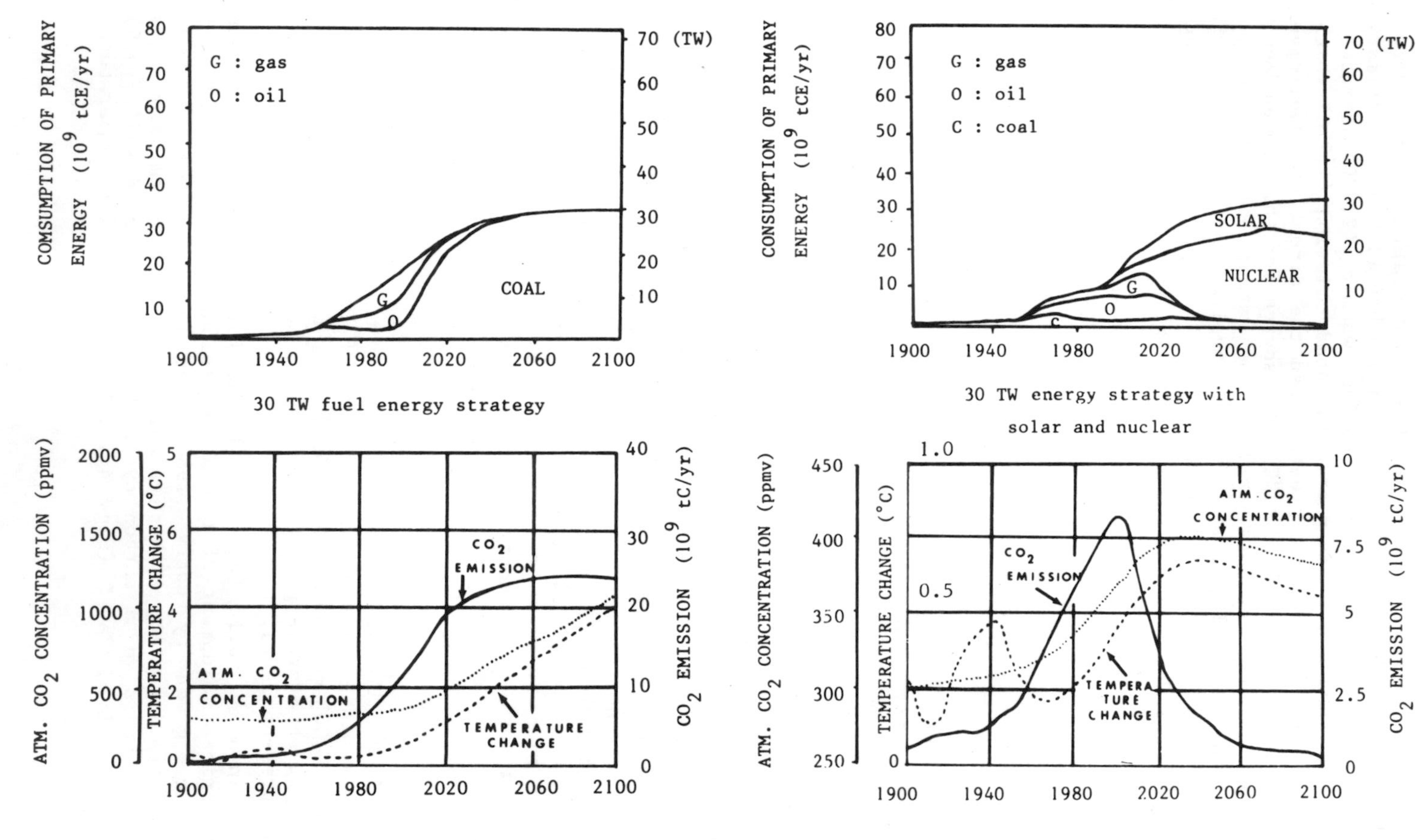

Fig. 4.14. (a) CO_2-impact of 30 TW-fossil fuel strategy.
(b) CO_2-impact of 30 TW-solar and nuclear strategy. After Niehaus 1979.

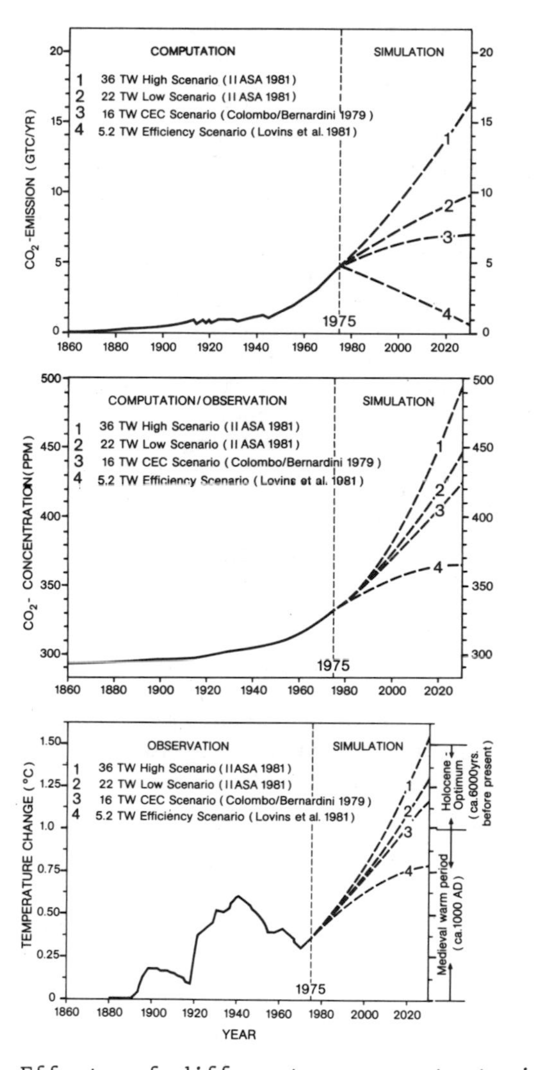

Fig. 4.15. Effects of different energy strategies on CO_2 and temperature (Bach 1982a). (a) CO_2 emission for a variety of energy scenarios. (b) CO_2 concentration for a variety of energy scenarios. They are based on a carbon cycle model consisting of 5 CO_2 reservoirs, namely the atmosphere, the long- and short-lived biosphere, a mixing layer of the ocean and the deep sea. (c) Temperature change for a variety of energy scenarios. There are obtained by employing a time-dependent coupled balance model.

1880-1975 are from Mitchell (1975). The CO_2 emission from 1975 to 2030 are based on energy projections given in Table 4.8(a). The simulation of the atmospheric CO_2 concentrations is from a carbon cycle model consisting of 5 CO_2 reservoirs. The simulated temperature values to the year 2030 are obtained from the Cess and Goldenberg (1981) time-dependent coupled atmosphere-ocean energy balance model. The results are rather revealing. Compared with the base year 1975, the IIASA high scenario shows a 3.4-fold increase in CO_2 emission in 2030 and 50% increase in CO_2 concentrations leading to an increase of 1.6°C. In contrast, the efficiency scenario shows an eight-fold reduction in CO_2 emission and only a 10% increase in CO_2 concentration resulting in a temperature increase of 0.8°C only. Besides the magnitudes involved, it is important to realise that, although a more efficient energy use results in a steady decrease in the CO_2 emission starting shortly after 1975, it is not until the year 2030 that the atmospheric CO_2 concentration also responds with a slight decrease, this being due to the oceanic lag of atmospheric CO_2 uptake.

Even if these models have a large number of uncertainties related to the projected emission of CO_2, to the carbon cycle and to the impact of increase of CO_2 atmospheric content on climate, the order of magnitude of the results can be regarded as reliable enough to make reasonable simulations of different scenarios of man's impact on climate: this may give a conditional prediction under the assumption that we can predict the changing external influence (carbon dioxide) and that our models respond like the real atmosphere.

4.2.3.8 Other infrared absorbing gases

There are many other trace gases that have properties similar to CO_2 for which either climatologically, or anthropogenically induced changes, can also be of significance. Among them, the most important are water vapour, chlorofluoromethanes (sometimes referred to as freons and released from home refrigeration and aerosol sprays (NAS (1976)), nitrous oxide which is produced by denitrification of nitrate fertilizers in the soil (Bolin and Arrhenius 1977), and methane, the major source of which involves the anaerobic fermentation of organic material due to microbial action and the enteric fermentation in mammals.

The purpose here is only to review the climatic impact of these additional minor atmospheric constituants; the following references being recommended for more global consequences: for water vapour, Lvovitch(1979), and for stratospheric ozone, NAS (1976).

Although the strongest absorber of the infrared radiation is water vapour, the direct man-made evaporation over land ($\sim$ 2500 km^3/yr), being equivalent to only 0.5% of the global value, can be neglected in global considerations. About other trace gases, recent studies have shown that, although an estimate of their climatic influence is possible, their sources and sinks are not well known, making estimates of man's impact in this field rather difficult. As a consequence, only a discussion of the impact on climate of an estimated increase of these trace gases is reasonable.

There are several ways in which the amount of climatologically important atmospheric trace gas might be altered:

(1) as a consequence of the direct anthropogenic emission of the gas into the atmosphere, as in the case for CFM's;

(2) as a result of the anthropogenic emission of gases that, through interactive atmospheric chemistry, alter their amounts, e.g. increasing anthropogenic emissions of CO are expected to increase the amounts of tropospheric CH_4 and O_3, both of which being "greenhouse" gases. Moreover, it is also important to note that many hydrocarbons are not only being emitted at increasing rates, but their natural sinks are becoming saturated, the principal sink for methane being hydroxyl ($CH + OH \rightarrow CH_3 + H_2O$). Indeed, OH plays a crucial role in scavenging a number of the less soluble trace gases, such as CO, H_2S, SO_2 and the partially halogenated methanes $CH_x Cl_y F_z$ and $CH_x Br_y$.

(3) Owing to biospheric changes caused by CO_2 warming: e.g. increased CH_4 production by warmer wetland areas or the release of CH_4 now trapped as a hydrate in permafrost regions.

(4) Owing to altered atmospheric temperature and water vapour concentration resulting from increased atmospheric CO_2 which would produce changes in atmospheric chemistry and therefore in trace gas abundances. For example, a depletion of OH may thus aggravate a build-up of infrared absorbing pollutants and their impact on climate. At the present, there is an additional danger of such a depletion occuring because of the increasing production of CO, which is removed mainly by: $OH + CO \rightarrow CO_2 + H$ (Jason 1979).

Employing a coupled climate-chemical model, Hameed et al. (1980) have found that the global warming produced by a 70% increase in atmospheric CO_2 and its subsequent increased H_2O would result in 15% and 10% reductions, respectively, in tropospheric CH_4 and O_3 which, in turn, will reduce the CO_2 warming by a quite minor percentage (10 per cent). Reciprocally, an increase of CO_2 concentration should be accompanied by a cooling of the upper atmosphere (Figs. 4.10 and 4.16), where most of the ozone is formed, and the ozone-production reaction, $O + O_2 + M \rightarrow O_3 + M$, proceeds rather faster at lower temperatures whilst the ozone dissociation reactions proceeds more slowly. The CO_2 cooling of the stratosphere is therefore partly compensated by absorption of ultraviolet and of visible radiation by the increased ozone (Isaksen et al. 1980).

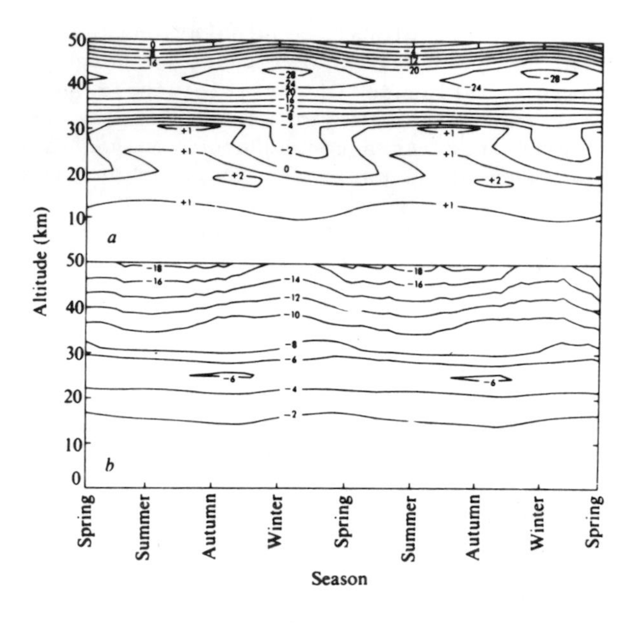

Fig. 4.16. Percentage change in ozone and temperature change given by model when CO_2 increased from 275 to 600 ppmv.
(a) - Seasonal variation of % change in O_3 number density
(b) - Seasonal variation of change in temperature.
After Groves and Tuck (1979). Reprinted by permission from Nature, 280 (5718), pp. 127-129, Copyright © , 1979, Macmillan Journals Ltd. and from the Controller of Her Majesty Stationery Office.

Recent results are summarized in Table 4.12 where the most striking effect comes undoubtedly from CO_2.

As far as the thermal and circulation (Schoeberl and Strobel 1978) effects of decreasing stratopheric ozone are concerned, no definite answer can be given, as the inclusion of airborne particles may change radically the sign of the results. (Ramanathan and Dickinson 1979, Reck 1976).

Although, regionally, Man's perturbation of the nitrogen cycle appears to be significant already, we will probably have to face some serious large-scale environment problem in the next century if the prediction about fertilizer use and combustion of fossil fuels come true, while present-day agricultural and industrial practices remain unchanges (Hahn 1979). Among the environmental consequences to be expected, is a depletion of stratopheric ozone, due to the catalytic effect of man-made NO on the ozone destruction by atomic oxygen, and an increase

TABLE 4.12. Estimated change in global average surface temperature arising from changes of atmospheric constituents based on a one-dimensional radiative-convective model assuming: (a), a fixed cloud-top height; and (b) a fixed cloud-top temperature in the model.

	Initial concentration	Increase in per-cent	Greenhouse effect (°C) fixed cloud-top ...	
			... height	Temperature
CO_2 (carbon dioxide)	320 ppm	+ 100	2.0	3.2
CO_2 (carbon dioxide)	320 ppm	+ 25	0.5	0.8
O_3 (ozone)	0.34cm*	- 25	- 0.3	- 0.5
H_2O (water vapour in stratosphere)	3 μg/g**	+ 100	0.6	1.0
N_2O (nitrous oxide)	0.28 ppm	+ 100	0.4	0.7
CH_4 (Methane)	1.6 ppm	+ 100	0.2	0.3
$CFCl_3 + CF_2Cl_2$ F-11 + F-12 (chlorofluromethanes)	0.2 ppb	factor of 20	0.4	0.5
$CCl_4 + CH_3Cl$ (carbon tetrachloride + monochloromethane)	0.6 ppb	+ 100	0.01	0.02
NH_3 (ammonia)	6 ppb	+ 100	0.09	0.12
C_2H_4 (ethylene)	0.2 ppb	+ 100	0.01	0.01
SO_2 (sulphur dioxide)	2 ppb	+ 100	0.02	0.03

In Section 5.1, the concept of a "virtual" greenhouse effect will be introduced with the aim of combining the thermal impact of all these trace gases.

* cm = depth that would be occupied by the total ozone in the atmosphere if it were at sea level temperature and pressure.

** μg/g = microgram of water vapour per gram of air.

Results are based on models by Augustsson and Ramanathan (1977), Wang et al. (1976), Lacis et al. (1981) and are coherent.

After Flohn (1978) MacDonald (1982) and CO_2 Climate Review Panel (1982) where the references are listed.

of temperature of the atmosphere at the Earth's surface due to the atmospheric greenhouse effect of certain nitrogenous gases. (Apart from climatic effects, the possible decrease in stratosphericozone by some per cent seems to be the major environmental hazard which would demand regulatory action, Whitten et al. 1980).

Main man-made sources of fixed nitrogen are the fertilizer industry (man can alter the rates of release of N_2O by soils and water bodies, by the use of ammonia-based fertilizer on agricultural lands and the addition of sewage to water bodies) and the combustion of fossil fuels. In 1975, about 40 Mt of fertilizer nitrogen were applied to the soil and the average growth in fertilizer use will probably be about 5 per cent per year until the year 2000. The direct production of N_2O during combustion may possible increase from the present 2.2 Mt N_2O nitrogen per year to reach a value of about 30 Mt N_2O nitrogen per year by the year 2050, if the combustion of fossil fuels is to increase by a factor 15 during the same period. Besides, combustion of fossil fuels results in an inadvertent large scale fertilization because of the release of NH_3 and NO_X to the atmosphere (20 Mt fixed nitrogen per year in 1974). From a range of uncertainty for some N_2O atmospheric parameters, Figure 4.17(a) shows that even cautious assumptions yield a doubling of tropospheric N_2O between the years 2024 and 2040.

Taking into account the uncertainties of the assesment of the future increase in tropospheric N_2O and of the one-dimensional radiative-convective model assumed for the thermal structure of the Earth's atmosphere, the range of uncertainty in surface temperature is indicated in Fig. 4.17(b) by the upper and lower curve. Using the same assumptions about the future increase in the combustion of fossil fuels, and using the same atmospheric model, lead to the prediction that by the year 2000 the greenhouse effect due to rising atmospheric N_2O may be about 50% of that due of rising atmospheric CO_2, while by 2050 it is only about 25%. Although its importance seems to decrease with time, as compared to the effect due to rising CO_2, the greenhouse effect of perturbed atmospheric N_2O is still of such a magnitude that one has to take it into account for an evaluation of the total impact of an uncontrolled combustion of fossil fuels on climate. To summarize, doubling the N_2O atmospheric content not only leads to an average warming of 0.7°C, but also its dissociation in the stratosphere produces NO, which in turn adds to the decomposition of ozone (Thrush 1978, Bolin 1979).

Methane (CH_4) has a strong absorption band at 7.66 μm. At present, the rate at which methane is being added to the atmosphere is estimated to be 500-1000 Mt per year. Besides the natural sources, direct man-made sources, as mining and industrial processes, add about 15 - 50 Mt per year. Man, however, is also indirectly responsible for changing methane through the cultivation of rice paddies, the raising of livestock and the filling in of wetlands.

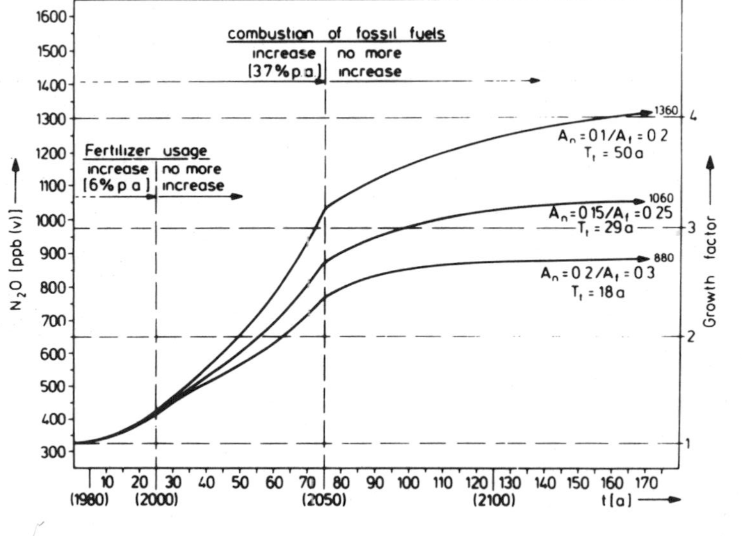

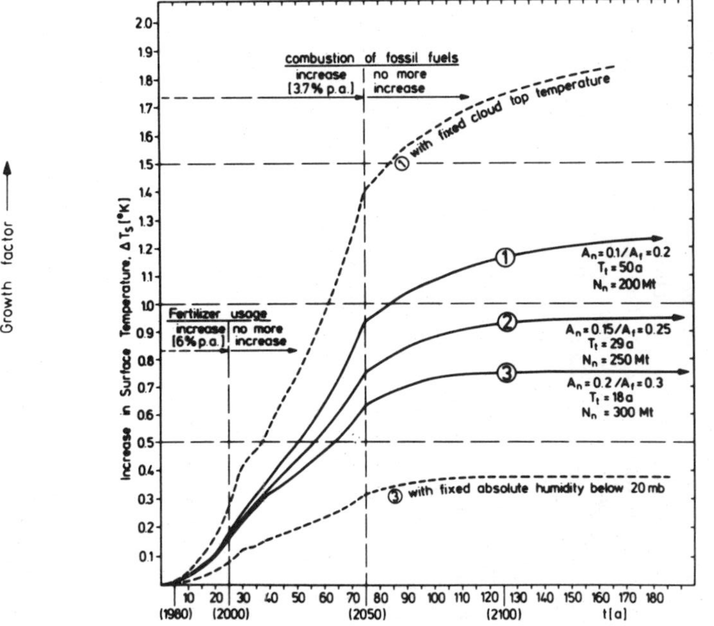

Fig. 4.17 (a) Increase in troposphere N_2O due to increasing use of industrial fertilizers and fossil fuels as a function of time (most likely range).
(b) Increase in mean surface temperature due to increasing use of industrial fertilizers and fossil fuels. The range of uncertainty is given by the upper and lower dashed curve. A_n is the fraction of the fixed nitrogen returned by the biosphere to the atmosphere in the form of N_2O, T_t is the tropospheric turnover time of N_2O and N_m is the natural fixation rate of nitrogen with an atmospheric mass of 1350 Mt of N_2O nitrogen. After Hahn (1979).

Recent observations of atmospheric methane indicate a current concentration of about 1.7×10^{-6} by volume (MacDonald 1982). Predictions of future levels of CH_4 in the atmosphere is difficult because of the many unknowns. If the methane concentration were to double, the surface temperature would increase by about 0.3°C (Wang et al. 1976) to 1°C (Jason 1979), for NH_3 it would be 0.1°C and for N_2O - CH_4 - NH_3 - HNO_3 globally this increase could be as much as 1 - 2°C.

To date, the only known significant change in trace-gas abundance, other than that of CO_2, has been in the CFM's, which have increased from a essentially zero abundance a few decades ago to 0.3 ppb of CCl_2F_2 and 0.2 ppb of CCl_3F (Fig. 4.18). Assuming the atmospheric content of freons ($CFCl_3$ and CF_2Cl_2) reach the 3 ppbv level (Figure 4.18), the expected temperature rise would be about 1°C (Fig. 4.19), this effect being enhanced

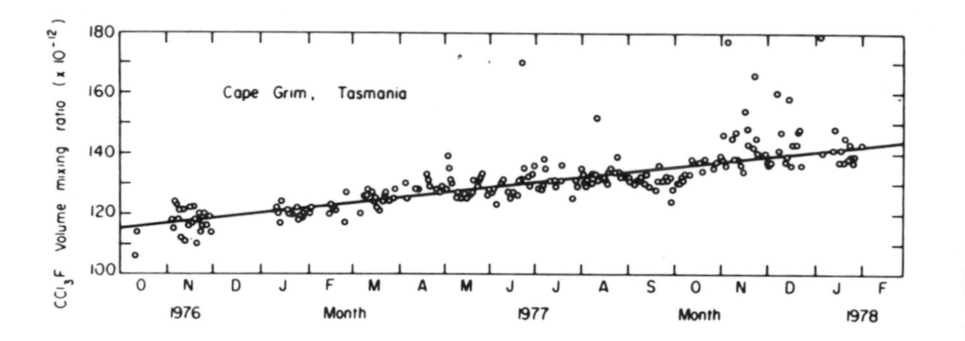

Fig. 4.18. Daily mean atmospheric concentration of $CFCl_3$ (trade name Freon 11) at Cape Grim, Tasmania. After CSIRO (1978).

by their very long residence time in the atmosphere (several decades (Thrush 1978)). We cannot say exactly when this should be (since the production rate will depend very much on the passage of legislation), but it is not inconceivable that it could occur as early as 2000.

4.2.4 *Aerosols*

Another product of human activity is the particles that are produced by industry, space heating (soot, sulphate, hydrocarbons) and from agricultural practices. There is little doubt that since the turn of the century there has been an increase in the rate at which aerosols have been produced by mankind,

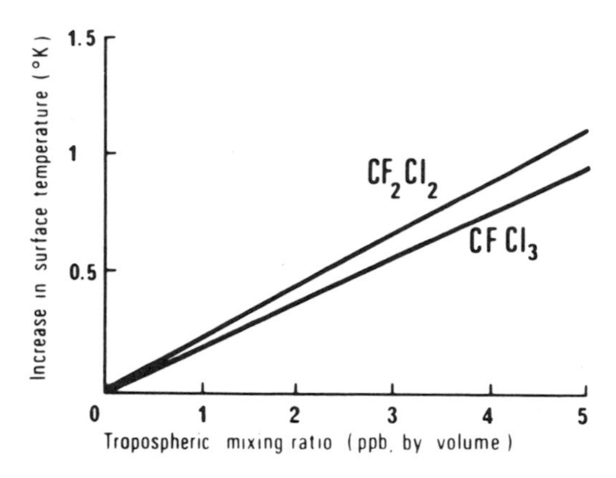

Fig. 4.19. Increase in global surface temperature as a function of the tropospheric concentration of CF_2 Cl_2 and $CFCl_3$. After Ramanathan (1975).

particularly in the more industrialized countries and in regions where desertification is most intense (Table 4.13).

Many of the aerosols produced by industries (steel mills and lead compounds) and burning coal and fuel oil (sulphates) have the property of acting as condensation nuclei and play an important role on the precipitation process, at least at the regional scale (Detwiller and Changnon 1976). Following the same idea, Krypton-85 released from nuclear power plants may play a similar role. Its radioactive disintegration (half-live of 10.7yr) increases the ionization of the lower atmosphere; it may also have an effect on the fair weather electric field and thus, on the thunderstorm activity and on precipitation.

The radiation effect on aerosols is far more complex than that involving trace-gas changes, since this depends on their composition, size and vertical and global distributions. Stratospheric aerosols, which persist for a few years following major volcanic eruptions, can produce a substantial reduction in global surface temperature and even explain much of the observed natural climatic variability of this century (Hansen et al. 1981; Gilliland 1982). Although there are indications of an anthropogenic source of stratospheric aerosols, it is not clear if or when this will become significant in comparison with volcanic sources.

The climatic effect of tropospheric aerosols is much less certain. These aerosol particles can both scatter and absorb sunlight and also absorb and re-emit infrared radiation to a more limited extent (Dalrymple and Unsworth 1978). In order to decide whether lower atmosphere aerosols cause an increase in the net albedo (cooling) or a decrease (warming), we must

TABLE 4.13. Particulate production estimates (diameter < 5 μm)

	10 tonnes yr^{-1}	Per cent
Anthropogenic input		
Fossil fuels		
Gases (subsequently converted to particles in the atmosphere)	311	45
Particulates	54	8
Wind-blown dust	180	26
Agricultural burning		
Hydrocarbons	72	10
Particulates	62	9
Nitrates	7	1
Fuel wood		
Particulates	4	1
Forest fires		
Particulates	2	0
	692	100
Natural input (excluding volcanoes)		
Sea salt	500	46
Sulphates (natural decay)	335	31
Wind-blown dust (natural decay)	120	11
Hydrocarbons (natural decay)	75	7
Nitrates (natural decay)	60	5
Forest fires (lightning)	3	0
	1093	100
Volcanic input*		
Dust	25-150	37
Sulphates	42-255	63
	67-405	100
Grand total	1852-2190	

It can be seen that, for the moment, natural emissions still predominate (Dittberner 1978a). For a global inventory of natural and anthropogenic emissions of trace metals, see Nriagu (1979).

* Mass and Schneider (1977) and Bryson and Goodman (1980) hypothesized that historical volcanic eruptions have had a measurable short-term effect on climate (cf. also Rampino and Self (1982), Robock and Mass (1982) for the 1980 Mount St. Helens Eruption and Mitchell (1982) for El Chicon (1982).

take into account the ratio of the particle absorption to its backscatter and also the albedo of the underlying surface. When aerosols are over a dark surface, such as the ocean, they are more likely to increase the net albedo than when they are over a light surface, such as snowfield, a low cloud or over land generally (Chylek and Coakley 1974). As most of these anthropogenic aerosols exist over the land, near where they are formed, and they are sufficiently absorbing to reduce the albedo rather than increase it, they have a warming effect (Kellog et al. 1975) which is maximum for small zenith angles of the sun (Böhlen 1978), instead of a cooling one as is widely believed. However, although aerosols are shown to have a pronounced influence on the heating rate of the lower clear and hazy atmosphere, they have not in cloudy conditions (Liou et al. 1978).

Because of vigorous attempts to control industrial aerosols the number of large particles has definitely decreased in many cities. Although, no worldwide long-term trend in these anthropogenic aerosols has yet been established, it is not true, in general, for the total aerosol content in Europe and in the eastern US because smaller sub-micron particles produced by burning high sulphur fuels have become a dominant factor in regional air pollution (Weiss et al. 1977).

Recent measurements indicate that man-made increase of both sulphates and windblown dust lead to cooling on a global scale, but perhaps with important regional exceptions (CO_2 Climate Review Panel 1982). Wind-blown mineral dust of soil particles, originating from agricultural practices and overgrazing, generally absorb less solar radiation than industrial or slash-and-burn particles. (Grams et al. 1974). Nevertheless, these fine particles (notably loess) will play an increasingly important role due to the large areas under cultivation (about 35 x 10^6 km^2) or subject to overgrazing ($\sim$ 5 x 10^6 km^2) (Flohn 1975).

Finally, industrial soot, a highly absorbing anthropogenic aerosol, would lead to global warming. Again, it must be emphasized that insufficient observations have been made to detertime its global trend. But recent chemical and optical analyses of Artic haze indicate that during spring and early summer, the haze particles contain a high concentration of graphite carbon and it has been suggested that this Arctic soot may be influencing the Arctic climate (Porch and MacCraken 1982).

4.2.5 Water Consumption

Continents, which represent 29% of the Earth's surface, contribute about 13% of the total evaporation but receive 29% of the

TABLE 4.14. Long-range approximate forecast of transformations of the world balance by the early 21st century

Resources	Water balance* $P = E + D$ (km^3/year) Present	Future	Transformations
Precipitation on continents(P)	110 300	110 300	
Total river runoff (D)	38 800**	37 500***	Conversion of flood runoff into soil moisture (700 km^3) and increase of evaporation from forests and from reservoirs (600 km^3)
Evaporation on continents (E)	71 500	72 800	An increase from higher crop yields and greater evaporation from reservoirs (1300 km^3)
Stable runoff including:	14 000	22 500	Increase of stable runoff through:
1. Groundwater runoff into rivers and renewable groundwater stores	12 000	17 000	Storage of groundwater
2. Lake- and reservoir-controlled runoff	2 000	5 500	Regulation of floodwater by reservoirs
Surface (flood) runoff	26 800	20 500	Use of surface runoff, as moisture in the soil (1300 km^3) and stored as groundwater (5000 km^3)
total infiltration	83 500	89 800	An increase consisting of additional moistening of non-irrigated land and the storing of groundwater

World water use for water supply (km^3/year)

Forecast period and alternative	Water withdrawal	Consumptive use	Wastewaters	Volume of river runoff polluted by wastewaters
Present (1970 (A.D.)	600	140	460	5 600
Forecast for the year 2000 A.D.				
- Present principles of use	7 170	1 080	6 090	38 000
- More rational principles of use	1 500	1 500	0	0

* Round figures.
** This does not include the runoff of water (ice) from polar glaciers.
*** Exclusive of the runoff of water (ice)from polar glaciers and consumptive use of water for economic needs.

After Lvovitch (1977).

total precipitation amount*. Increase of world population and large-scale industrialization create a use of continental waters at an increasing rate, especially for irrigation and cooling which, in turn, increases evaporation (man-made evaporation is estimated to amount to around 3% of the natural continental evaporation). For example, artificial lakes (Obeng 1977) already cover at least 300000 km^2 and globally there are some 1500 litres per day and per capita that are evaporated from consumed waters (Lvovitch 1977).

Table 4.14 provides a long-range approximate forecast of transformations of the world water balance by the early 21st century.

Related to this problem of water use, it must be noted that there exist some very ambitious plans for water deversion which most certainly would modify the world climate (Schneider and Mesirow 1976, Hollis 1978). For example, to increase the supply of water available for irrigation, Soviet engineers are considering diverting southward a significant fraction of the flow of several major rivers that normally empty into the Arctic Ocean. Doing this would increase the salinity of the Arctic. Since saline water does not freeze as rapidly as fresh water, this would conceivably reduce the present stability of the Artic Ocean and lead to a reduction of the fraction of the Arctic Ocean covered by ice and a modification of the climate of the surrounding regions (an overview is available in Sellers (1977) and Lamb (1971)).

4.2.6 *Land Use*

4.2.6.1 *Urbanization*

Increasing urbanization is responsible for modification of the climate at the local and regional scale (Table 4.15). Urban-produced modification of most climate conditions had been clearly recognized for many years (e.g. Landsberg 1970 and 1981, Oke 1974, and in Section 3.1.3). Moreover, the clouds and fogs induced by large power plants are obvious (Hanna 1977) and large-scale solar energy conversion systems could alter climatic boundary conditions (Williams et al. 1977).

* It must be mentioned that the accuracy of evaporation and precipitation data is very poor over sea and even over land; a critical review of the world water balance has been given by Baumgartner and Reichel (1975) and more recently the variations in the global water budget are analysed in Street-Perrot et al.(1983).

TABLE 4.15. Climate of a city as compared with that of the surrounding countryside

Radiation	
Total insolation	15 - 20% less
Ultraviolet (winter)	30% less
Ultraviolet (summer)	5% less
Sunshine duration	5 - 15% less
Temperature	
Annual mean	0.5 - 1.0°C higher
Winter minimum	1.0 - 2.0°C higher
Relative humidity	2 - 3% less
Cloudiness	
Cloud cover	5 - 10% more
Fog in winter	100% more
Fog in summer	30% more
Precipitation	
Total quantity	5 - 10% more
Snowfall	5% less
Particulate matter	10 times more
Gaseous pollutants	5 - 25 times more
Wind speed	
Annual mean	20 - 30% lower
Extreme gusts	10 - 20% lower
Calms	5 - 20% more frequent

After Landsberg (1970).

Cities are great converters of energy and matter; the by-products of conversion directly affect the heat and moisture budgets, and the exchange of materials in the atmosphere. Obviously, the degree of change of any of these elements at any time depends on the areal extent of the urban-industrial complex, its types of industry, its juxtaposition to major water bodies and topographic features, time of the day, season of the year, existing weather conditions and regional climate. The effects of urban areas on weather and microclimate that are easy to measure include decreased visibility, reduced winds, changed humidity and increased temperature.

The temperature within a moderate (100000 population) to large-size urban area is generally higher than it is in rural areas because the most radical change produced by cities is in the heat balance (Böhm 1979). METROMEX (the Metropolitan Meteorological Experiment; Changnon and Semonin 1979) measurements at St. Louis, in the central U.S.A., showed that the daytime urban heat excess often extended vertically for 2000 m, while a thermal and pollutant plume might extend 50 km downwind. Urban areas act to decrease the wind speed and to alter its direction near the surface, to increase turbulence and vertical motions in the immediate area, and to create localized rural-urban circulation patterns at the surface (Dreiseitl and Reiter 1978). Humidity over cities generally decreases because much of the rain runs off and there is less vegetation to provide moisture to the atmosphere. Investigations of data from larger cities showed that amounts and frequencies of rain, thunder and hail increased (Dettwiller and Changnon 1976). METROMEX findings revealed local increases of 30% in total rainfall, 40% in heavy rainfall rates and rain storms, 45% in thunderstorms, 100% in strong surface winds and hailfall intensities.

Another principal concern is major power plants and the release of latent heat and moisture through cooling water or through large cooling towers. The magnitude of the latent heat (in the form of water vapour) released from cooling towers in a 0.6 km^2 area for a single large power plant (2200 MW) is 1/7 of that released by St. Louis over an area of 400 km^2. Coal-fired power plants, petrochemical industries and wood-processing plants are prolific sources of cloud condensation nuclei which have now been noted to increase fogs, clouds, snowfall and rainfall under certain weather conditions.

4.2.6.2. Deforestation

Alteration of the pattern of vegetation is certainly the way that mankind has been effective longest, at least since the Neolithic revolution of 8-10000 years ago, influencing the heat balance of the Earth. Table 4.16 give us the opportunity to estimate the effect of such new forest clearings in the tropics.

TABLE 4.16. Estimates of new forest clearings (in 10^4 km^2 year^{-1}) in the tropics.

	Africa	Latin America	Asia	Tropics
Manshard (1974)	1.5	10	10.5	22
Bolin (1977)	2	6	4	12
Wong (1978)				
Based on FAO (1966)	8	7.5	8.5	24
Based on rural population increase	1.2	1.3	5.0	7.5

After Wong (1978) where all references are listed.

When a forest is cleared for a pasture or wheat field, the result is an area that generally reflects more sunlight since crops and grassland are usually less absorbing than trees; for example, in the tropics, albedo goes grom 0.13 to 0.23 (Baumgartner et al. 1978). Such changes in the solar radiation absorbed by the surface must certainly have an effect on the heat balance and climate of a region (Table 4.17), and influence its precipitation (see Section 4.2.7.2) as well its mean temperature. Potter et al. (1975) have shown that the removal of tropical rain forests would lead to a global cooling and a decrease in precipitation* between 45 and 85°C and 40 and 60°S. The sequence might be as follows: deforestation 5°N-5°S → increased surface albedo → reduced surface absorption of solar energy → surface cooling → reduced evaporation and sensible heat flux from the surface → reduced convective activity and rainfall → reduced release of latent heat, weakened Hadley circulation and cooling in the middle and upper tropical tropo-

* It is, however, possible that this removal of forests will have secondary effects related to non-linear feedback mechanisms characterizing the real atmosphere. For example, it is not impossible that a decrease of the latent heat flux, heat which is available in the middle troposphere, will be accompanied by a slight increase of the sensible heat available to warm the air at the surface (Idso 1977). In this case, the subsequent vertical destabilization of the tropical atmosphere will tend to increase convective activity and maybe to reduce precipitation deficit.

TABLE 4.17. Climatic and hydrologic properties of forest in relation to other Earth surfaces.

	Forest	Grass	Crops	Bare	
Absorption (A_s)	85-90	75	70	65	%
Overheating (ΔT)	2-4	4-6	6-10	10-20	K
Net Radiation (Q)	60-65	45-55	45-55	35-40	W/m^2
Potential Evapotranspiration (Ep)	850	550-750	550-750	400-500	mm/a
Evaporation-ratio (E/P)	70	65	40-50	30	%
Runoff-Ratio (D/P)	30	35	50-60	70	%
Roughness Parameter (z_o)	100-300	5	30	0.1-1	cm

After Baumgartner and Kirchner (1980).

sphere → increase in tropical lapse rates → increased precipitation in 5 - 25°N and 5 - 25°S and a decrease in the equator-pole temperature gradient → reduced meridional transport of heat and moisture out of equatorial regions → global cooling and a decrease in precipitation between 45 and 85°N and 40 and 60°S.

Baumgartner and Kirchner (1980) have also calculated the effects of total deforestation of the globe. The changes in albedo are largest in the tropics and in the northern mid-latitudes where they amount up to 6% in winter; the global average surface albedo would shift from 16.7% to 17.4%. The average roughness of the Earth will decrease from 15 to 3 cm. But the greatest impact, however, should be on the carbon dioxide cycle:

(1) A remarkable rise of atmosphere CO_2 content should follow, caused by the reduction of assimilation and by the release of carbon from biomass and within the soil.

(2) The atmospheric CO_2 stock would be depleted totally by forest production within 26 years, if no CO_2 return from that metabolism existed.

4.2.6.3 Overgrazing, bush fires and desertification

Desertification, estimated to increase by around 60000 km^2/yr and affecting some 50 million people, characterizes the arid zone at the desert borders where average annual precipitation ranges between 100 and 400 mm (Le Houerou 1979). This phenomenon is far from being completely understood (Hare 1977) and so far there is quantitative information only for some localities;

for example, the desert in Sudan has expanded by some 100 km between 1958 and 1973, affecting an area equivalent to 63% of France.

Principal regions of desertification are located north and south of the Sahara, in the arid zones of the Near and Middle East and in Eastern Africa. These man-made deserts extend now over more than 5×10^6 km^2 in Africa and Middle-East (30% of all arid zones of Africa and Asia together), which represent an area almost as important as the natural climatic deserts of this zone.

Causes of desertification are the periodical occurrences of persistent drought and an alteration of the ecological equilibrium by man and domestic animals. It is particularly effective when nomadic tribesmen allow their cattle and especially goats to overgraze on marginal land, the destruction of vegetation markedly increasing the surface reflectivity (U.N. 1977). This increase of albedo, with or without a change of evaporation, contributes to a net decrease of radiative flux into the ground, causes sinking motion and a net decrease of convective cloud and precipitation with consequent additional drying and, therefore, perpetuates and intensifies the arid conditions (Charney et al. 1977). It should be noted that while the albedo feedback effect seems to be positive for tropical desert precipitation, it is negative both locally and globally in its effect on temperature, at least if deserts are associated with high temperature. This negative feedback is, in fact, enchanced through the reduction in precipitable water, reducing the greenhouse effect of water vapour, an effect amplified locally by the increased subsidence (Ellsaesser et al. 1976).

More recently, Potter et al. (1980) conducted a numerical experiment combining desertification of the Sahara and deforestation of the tropical rain forest. Over an area of 9 x 10 km^2 at 20°N, the desert albedo was increased by 0.16 to 0.35, and over 7×10^6 km^2 at the equator and 10°S, the rain forest albedo was increased from 0.07 to 0.16. While the most significant direct climatic responses were observed in the modified zones, high northern latitudes exhibited the greatest cooling (up to 3°C near the surface at 70°N) through a reduced poleward transport of sensible and latent heat and an activation of the ice-albedo feedback process. The mean surface temperature cooling was 0.2°C, although the northern hemisphere surface temperatures cooled by 0.6°C (this is related to the temperature increase at 30°S associated with a reduction in stability in the southern hemisphere, producing clouds with more vertical and less horizontal extent, which allow more radiation to reach the Earth's surface).

A very likely secondary effect of this desertification is the increase in windblown soil and sand which also affects the radiation at the surface and the stability of the atmosphere above it.

4.2.7 Global effect on climate

Even if we could anser all of the questions associated with the carbon cycle and the energy policy, the fundamental question underlying the CO_2 issue would remain: What is the effect of increased atmospheric CO_2 concentration on the global climate? The answer to this (Flohn 1979) must ultimately come either from climatic data collected over a long period (Lorius and Raynaud 1983) or from mathematical models (CO_2 Climate Review Panel 1982) that can predict the consequences of a man-made climatic change early enough and convincingly enough to permit intelligent decisions to be made (Glantz 1977).

4.2.7.1 Temperature

A number of anthropogenic causes for climatic change have been mentioned and Table 4.18 summarizes these effects on the mean air surface temperature. To a first approximation they can probably be considered as additive because they represent small fractional changes (for much larger changes some of the effects defenitely become non-linear).

Other scenarios tend to confirm this global view with differences. On the basis that the most likely value of the atmospheric CO_2 concentration in the year 2025 is actually discussed to be 450 ppm, the CO_2 Climate Review Panel (1982)'s scenario of global man-made alterations postpones to the end of the 21st century the changes projected here to occur in the year 2050. From presently observed or assumed rates of trace gases, Volz (1983) projects an increase of temperature due to CFMs, N_2O, CH_4 and O_3 to be 0.1 to 0.5°C in 2000 and 0.1 to 1.2°C in 2030, against, respectively, 0.8 and 1.3°C for the CO_2 IIASA Low scenario.

Among all these effects, that of adding carbon dioxide is the largest; deforestation and some other related land-uses seem to be the only man-made activities which would contribute, by increasing surface albedo, to cool the lower troposphere, depending on their areal extension and time-scale. However, such a cooling can also be generated by natural causes such as a hypothetical decrease of the solar constant (for which little convincing evidence exists) and volcanic dust injection into the upper troposphere and lower stratosphere.

In order to be able to make a valuable "long-range" climate forecast, based upon an extrapolation both of our social and economic activities and upon climatological models, it is first necessary to test the reliability of the models, i.e. their capability to reproduce recent past climatic fluctuations. Although these fluctuations are different in different regions and for different time spans (Kukla et al. 1977), it is gene-

TABLE 4.18. Anticipated warming of the atmosphere in the mid-21st century as a result of man's activities.

	Time period for the effect to occur	Influence on surface temperature (°C)	Rate of change toward the end of the time period (°C/decade)
CO_2	+25% by 2000 +100% by 2050	+0.5 to 2[a] +1.5 to 6[a]	0.2 to 0.8 0.3 to 1.2
Chlorofluorocarbons	0.8 ppbv by 2000[b] 2.5 ppbv by 2050[c] 3.5 ppbv by 2000	0.1 to 0.4[d] 0.25 to 1[d] 0.4 to 1.5[d]	0.04 to 0.2 0.02 to 0.1 0.2 to 0.8
CH_4, NH_3	+2-fold by ?	0.4[h]	?
Nitrous oxide	+100% to 2050	0.25 to 1[e]	0.02 to 0.1
CO[i]		0.6 to 0.9	0.06 to 0.1
H_2O in stratosphere	+25% by ?	+1.0[h]	?
O_3 in stratosphere	-25% by ?	-0.47[h]	?
Adding aerosols to lower troposhere	?	heating[f]	?
Direct addition of heat	50-fold increase by 2100	0.5 to 2	0.05 to 0.2[g]
Patterns of land use	?	?	?
Global surface albedo	+1.4% by 2020	-0.23	
Natural climatic change		+0.25	

The estimated changes in global average surface temperature arising from changes of atmospheric constituants are consistent with Table 4.12.

a See Table 4.12
b Assuming continued FC production at 1973 level (NAS 1976).
c Assuming a 10% per year increase in FC production rate (NAS 1976).
d Estimate by Ramanathan (1975), and reviewed and extended by NAS (1976).
e Estimated by Yung et al. (1976). This now appears to be an upper limit on the possible increase of N_2O on this time periode.
f It is not clear whether the upward trend in anthropogenic aerosols will continue; it will depend to a large extent on control of sulphur-dioxide emissions, which will probably have to be reduced in some areas.
g In this case the total effect (about 1°C) would be more significant than the rate of change because it builds up over a fairly long period.
h from Wang et al. (1976).
i Indireet effect by reducing OH and increasing the greenhouse effect of troposheric O_3.

After Kellogg (1977), Bach (1978) and MacDonald (1982) where more references are listed.

rally admitted that the global surface air temperature rose about 0.5°C between 1885 and 1940, has experienced a maximum in the 1940's with a slight cooling trend of around -0.15°C/decade thereafter and a reversal in the mid and late 1960's, and was, in 1980, almost as high as in 1940 (Jones et al. 1982). Taking into account both volcanic (Miles and Gildersleeves 1978) and solar (Eddy 1977) activities, on the one hand, and man-made variations of the atmospheric carbon-dioxide and dust (Dittberner 1978b), on the other, reconstruction of the climate of the last centuries has been possible. Hansen et al. (1981) took into account past carbon dioxide levels, the absorption of heat by the oceans, variations in the quantity of volcanic aerosols and a hypothetical variability in the Sun's luminosity. The general agreement between modeled and observed trends in temperature strongly suggests that carbon dioxide and volcanic aerosols are responsible for much of the global temperature variation in the past 100 years. However, following Gilliland (1982), the problem is not so straightforward. Indeed, he reported that the Sun's radius and the related solar luminosity vary with periods of 76 years, 11 years and on longer time scales. These measurements have shown that the Sun has been cooling off in the past two decades, counterbalancing the warming expected from CO_2 increase. As the next solar maximum is due in 2010, the projected man-made 1°C warming by then may now be bigger than the greenhouse theory alone can explain.

Moreover, despite the wide disparity of results from various climate models, there is widespread agreement that a general warming of the Earth will occur if atmospheric CO_2 levels rise as predicted. This allows us to admit the best estimates of each of the anthropogenic effect (Tables 4.12 and 4.18) and to conclude that the net influence of mankind is as shown in Table 4.19 and Fig. 4.20.

TABLE 4.19. Estimate of the influence of mankind on long term global mean surface temperature.

	To-day	2000 AD	2050 AD
Absolute change (°C)	0.5	1.0	3
Rate of change (°C/decade)	0.1	0.4	0.5

Assumption: manufacture of chlorofluorocarbons will remain at 1973 level. Direct addition of heat will not be important globally until after 2050 A.D.; effects of aerosols and patterns of land use are not included. Adapted from Kellogg (1978).

When we compare the rate of change of global mean surface

temperature due to anthropogenic factors with the expected natural changes around ± 0.1 - 0.2°C per decade, the lower limit of these estimates (Table 4.19) seems to be sufficient to cause a signal - to - noise ratio greater than 2 by the end of this century, assuming that this warming will not be counteracted by a simultaneous cooling due, for example, to volcanic activity and large-scale deforestation.

Again, it must be emphasized that all this refers to the global mean surface temperature change; the corresponding changes at high latitudes will be much (3 - 5 times) larger as shown in Figure 4.10. One of the consequences will be an important modification of the latitudinal temperature gradient and thus, of the principal characteristics of the general circulation and precipitation patterns.

4.2.7.2 Precipitation

Most of the discussion so far has concerned the surface temperature and the effects that mankind might have on it, but it is even more important to know what might happen to precipitation and its distribution (Lettau et al. 1979). In fact, it is the rainfall that largely determines whether vegetation will thrive and whether a region can grow food. Particularly important is the role of water supply in many semi-arid or semi-humid areas (e.g. Israel, California, Sahel, etc.).

In spite of their limitations, general circulation models have shown that when there is an increase in the total heat supplied to the system, there is an overall warming of the model lower troposphere, but the response of the model climate is far from uniform geographically (Manabe and Wetherald 1980): some regions will warm very much and their should also be cooling in some places. Since the same complex response undoubtedly applies to the patterns of precipitation, the problem is still more complicated, as can be appreciated from the large spectrum of results obtained from theoretical modelling.

The hydrologic cycle has also been modelled in assessing not only induced CO_2 warming (cf. Section 4.2.3.5) but also deforestation effects. Limiting the model treatment to a definite large-scale portion of the Earth's surface has indeed advantages with respect to the testing of parameterization as well as computed results, similar tests being difficult when the climatic response to tropical rainforest removal is investigated with the aid of a global atmospheric model. For example, if the forest cover in central Amazonia is assumed to be reduced to one-half of its original value, a climatonomical experiment (Lettau et al. 1979) shows that, due to recycling, the precipitation increases slightly at all longitudes between 57.5 and 77.5°W. However, the primary augmentation of evaporation outweights the precipitation increase and consequently the total runoff from the entire basin decreases from the original 1075 mm yr^{-1} to 1018 mm yr^{-1}.

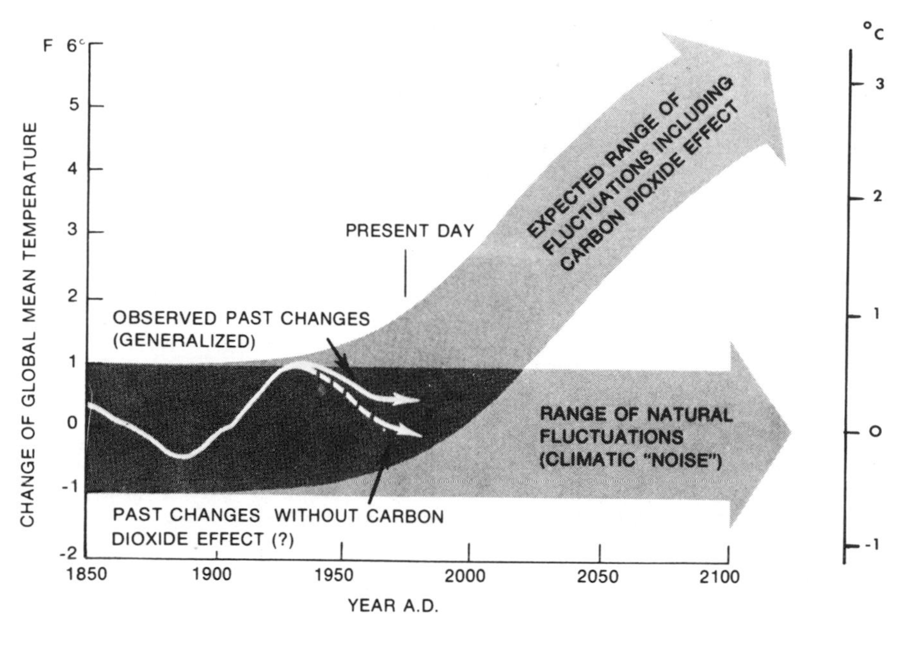

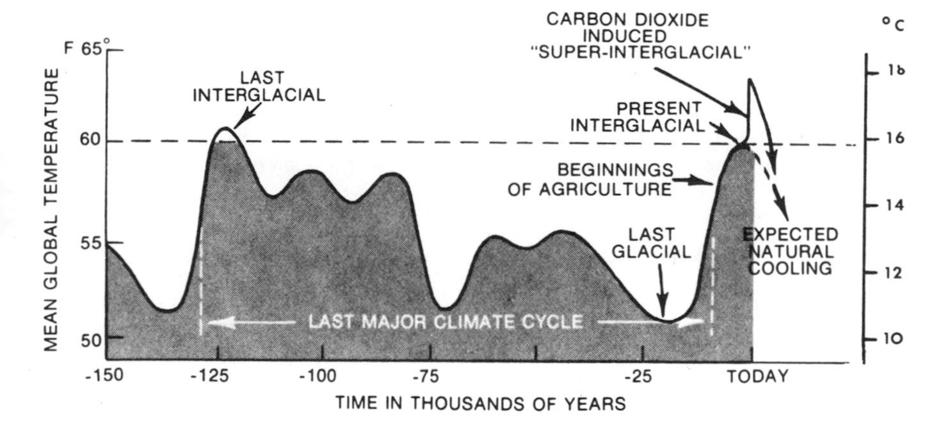

Fig. 4.20. Above: range of global-scale mean temperature, 1950 to 2100, with and without the projected CO_2 effect. Below: projected CO_2 temperature disturbance in the context of ice-age chronology of the past 150000 years.

After Mitchell (1977).

The effects of a warming on snowfall and snowcover will also differ according to latitude (Barry 1978). In low and latitudes, where the occurrence of snow rather than rain is frequently marginal, warming will decrease the frequency of snowfall and duration of snowcover on the ground. In high latitudes, where snowfall is limited by the low vapour content of the air, due to low temperature, warmer winters with maritime air advection are generally more snowy.

4.2.7.3 Cryosphere (ice and snow)

Given the observational record of temperature change with latitude during the present century (Williams and van Loon 1976) and model predictions, especially those envisaging large changes at high latitudes, we have to wonder about the consequences on the polar ice (see Section 5.5). As albedo plays a major role in climatic-variation feedback mechanisms, any changes in floating sea ice and the extend of large ice sheets could have major climatic significance, but the consensus of opinion seems to be that the ice sheets of Greenland and Antarctica will respond only very slowly to a temperature increase (Hollin and Barry 1980; cf. Section 5.4.3).

Because of the large contrast between the seasonal behaviour of the Arctic and Antarctic polar sea ice regions (Untersteiner 1975, Barry 1980), the major question is how much warming would be required to remove the Arctic pack-ice either seasonally or completely.

For an ice-free Arctic Ocean, temperature changes in high latitudes (Fletcher et al. 1973) seem to be of the same order of magnitude as those resulting from the 2 x CO_2 experiment, and it is reasonable to expect a melting of the polar ice pack, at least partially. From a large-scale numerical time-dependent model of sea ice, Parkinson and Kellogg (1979) have shown that as a result of a 5°C atmosphere surface temperature increase, the pack-ice would disappear completely in August and September but would reform in the central Arctic Ocean in mid-autumn (as also quoted in 4.2.3.4 from Manabe and Stouffer (1979)). However, the sea ice boundary is itself determined only by the location of the -2°C isotherm.

Even if a totally adequate combined atmosphere-ocean-sea-ice model does not yet exist to estimate the response of Arctic sea ice due to a global warming, empirical and theoretical estimates suggest that considerable warming would be required to eliminate it. Once removed, there are at least two reasons for believing that the open freely mixing ocean will not freeze over again: (1) more energy would be absorbed by the system in summer, and (2) the thin stable upper layer of the ocean would be removed and would then allow mixing and exchange of heat between the surface layers and the warm waters below.

An open Arctic Ocean would, of course, allow a great deal more evaporation and this would certainly represent a major difference in the pattern of temperature and rainfall that exist now. Several model experiments have been performed to assess the possible atmospheric effect of such an ice-free Arctic Ocean (Fletcher et al. 1973). The results all show a major warming of the lower troposphere over the Arctic Basin (by as much as 10°C in winter, Fig. 5.17) and a cooling over some continental areas in middle latitudes. Since the oceanic advection effects have not been taken into account, these are only preliminary results and changes in precipitations are certainly poorly simulated.

Turning to the large ice sheets (Greenland and Antarctic), the effect of a warming is even more complex, their total volume being determined over a long time period mainly by a balance between the snowfall on their tops and the melting, ablation or calving at their edges. The possible consequences of a global warming for the ice-sheets of Greenland and West Antarctica are discussed in detail in Section 5.4 and of an ice-free Arctic in Section 5.5.

Finally, it has to be pointed out that a warming trend would also be most significant, in terms of effects on soil moisture and vegetation, in the areas of discontinuous permafrost along the southern margins (Fig. 5.8). Here, the observed trend of retreat should be continued or even intensified.

4.2.7.4 Possible feedback from large carbon reservoirs.

A man-made change in climate does not only affect the atmosphere. The biospheric reservoir would also respond to a temperature increase by an increase in net productivity, particularly if the temperature increase is accompanied by an increase in precipitation. Because of its opposite influence on the decay rate and its dependence upon the availability of land and nutrients other than carbon, this increase remains problematical. A one-degree rise in temperature could release as much as 100 - 200 Gt of carbon from the enormous amount of the organic pool in the soils (1600 Gt) but it is probable that the time scale for the response of the soil is long compared with the time scale for fuel additions to the atmosphere. The impact on the chemical composition of the upper layers of the oceans is now better understood, a relatively small quantity of carbon being released into the atmosphere by the oceans, about 20 Gt/degree. The methane hydrate reservoir (terrestrial, deep ocean and permafrost contain some 4.7×10^4 Gt) is the least well understood of the large carbon reservoirs. The temperature sensitivity could be a release of about 70 Gt/degree. In summary, a worldwide warming will tend to produce a positive feedback, releasing carbon from the soils, methane hydrates and oceans, but the rate of release is not known.

4.2.8 *Precursor indicators of a CO_2 warming and detection strategies*

Up to now an atmospheric warming caused by the past increase of atmospheric CO_2 has not been unequivocally identified (DOE 1982), perhaps for the following reason:

(1) oceanic influences might cause the climate response to lag 5-20 yr behind the CO_2 forcing (Schneider and Thompson 1981, Bryan et al. 1982);

(2) climate changes caused by other factors (volcanic aerosols and solar irradiance) may mask this CO_2-induced warming;

(3) available climatic observations do not enable the statistical significance to be established of a CO_2-induced signal above the natural variability of climate (noise).

Recent sea ice observations of Kukla and Gavin (1981) and the model versus observed surface temperature trend comparisons of the GISS group (Lacis et al. 1981) are at least consistent with consensus of modellers' expectations for a climatic signal from increasing "greenhouse" gases. If Lacis et al. are right then, by 1990 and not the often-cited year 2000, a temperature-rise signal from the combined trace greenhouse gases should stand out clearly from the background of natural climatic variability; assuming, of course, that current trends in CO_2 and other trave gas concentrations continue.

However, Thompson and Schneider (1982) believe that the statistical significance of these results is simply too small and the number of unverified modelling assumptions too large to allow one to proclaim detection. Moreover, the change found by Wigley and Jones (1981) in surface temperature between 1901-20 and 1934-53 being similar to that found in the equilibrium CO_2 response resulting from a CO_2 increase of $\sim$14% (which will occur probably by 1990) based on Manabe and Stouffer's model (1980), analysis of spatial and/or seasonal details of future variations in surface temperature will be of little help in distinguishing natural changes from the effects of CO_2. However, if transient response patterns do differ from equilibrium-model response patterns as shown by Schneider and Thompson (1981), this may make it easier to distinguish CO_2 effects from natural changes in climate.

A better way of identifying regional climatic changes with potentially good signal-to-noise ratios would thus be to search for these in the results of detailed time-dependent (instead of steady-state) simulations performed with a coupled atmosphere/ocean model using realistic geography and plausible CO_2 increase scenarios. But if this is true, this would cast some doubt on the value of both equilibrium model results and instrumental analogues (Wigley et al. 1980) in generating spatially detailed scenarios for a high CO_2 warmer world.

The problem of estimating the transient climate response also has a bearing on the selection of potential precursor indicators of a CO_2 warming. Preliminary attemps to identify such indicators have already been made by Madden and Ramanathan (1980) and Wigley and Jones (1981). These studies noted that the zonal-mean surface air temperature in mid-latitudes have relatively large signal-to-noise ratios in summer and for the year as a whole, although the CO_2 signal in the Manabe-Stouffer model is maximum in winter high latitudes (4.6°C for the zonal belt 55-85°N). These conclusions depend strongly on the spatial and seasonal distribution of the natural variability of surface temperatures, relatively low in summer and in low-latitudes. The early detection of the CO_2-climate signal (Schuurmans 1983) requires not only a prediction of the CO_2 -induced climate change, but also a knowledge of the natural climate variability and their non-CO_2 causes. It is, therefore, advisable to monitor not only the usual climatic data but also the temporal variations of other climate parameters, surface albedo, minor atmospheric constituants, atmospheric aerosols and solar irradiance.

Verification of model predictions by the real atmosphere takes time; the recent results strongly suggest that perhaps the time is closer at hand than previously believed.

4.2.9 Summary and conclusions

Anthropogenic effects on climate can be tentatively estimated only if we first assume that some degree of predictability is present in the climate system and that no large natural climate changes will occur in the near future (e.g. no major changes in solar radiation, in worldwide volcanic activity and in the stability of the Antarctic ice sheet). In view of our insufficient knowledge of the cloud-radiation interaction, we must also exclude changes in global cloudiness.

With a demonstrated increase of atmoshperic CO_2 over the last two decades and the implication that this is ultimately likely to alter global climate patterns, this problem is the most pressing at present. As it is inherently difficult to predict future levels of atmospheric CO_2 because of the uncertainties related to the future energy demand (world population and per capita energy growth) and to the complexity of the carbon cycle, only scenarios rather than forecasts can be proposed.

Although the effects of unknown or improperly modelled feedback mechanisms (particularly those related to clouds, ocean and polar ice) could modify the result several-fold, a state-of-the art, order-of-magnitude estimate, of global surface temperature change would be 3°C with a probable error of $\pm$ 1.5°C for a doubling of CO_2, with a general increase in the activity of the hydrologic cycle. Due to the sensitivity of high latitudes, a much larger temperature increase has to be envisaged there, especially in winter, with important consequences for the arctic sea ice. As two very important aspects of climatic change are the implications for agriculture (Chapter 6) and the extent of regional variations, climate models with a much higher regional resolution are fundamental to understanding the full impact of atmospheric CO_2.

The challenge to science is how to narrow the uncertainties and define the prospects more precisely so that society may have sufficient time to reach informed decisions as to whether any action is advisable and, if so, which actions are most advantageous (Kellogg and Schware 1981). Due to the overall complexity of the problem and the different time responses of the atmosphere, the oceans and the biosphere, it is not at all clear how long will be required to obtain sufficient evidence to achieve a consensus of opinion for action, and how long is available for a decision.

In view of worldwide concern, progress is particularly needed through intensified research in multidisciplinary areas like:

(1) collection and analysis of CO_2-related data from the atmosphere and oceans;

(2) climate modelling with due regard to land processes, hydrologic cycle and air-sea-ice interactions;

(3) evaluation of the principal reservoirs and fluxes in the carbon cycle, with special attention to ocean mixing, CO_2 uptake and biospheric carbon fluxes;

(4) evaluation of energy policy issues and their effects on future global carbon flows.

Although the climate system possesses many resilient qualities, man's activities may well alter greatly the future climate and, as a consequence, our society itself. It therefore behoves us not to let this experiment, the greatest inadvertent geophysical experiment ever begun, proceed unobserved and uncontrolled.

Two very important statements were issued since January 1983.

(i) In a recent World Climate Program Newsletter (4, 1983), Dr. Th. D. Potter, director of the World Climate Program department concludes: "If energy consumptions follows current projections, it seems probable, based on present knowledge of the carbon cycle, that atmospheric CO_2 will reach twice the preindustrial level around 2050 A.D.".

(ii) Trough a press release dated October 19, 1983, the U.S. Environmental Protection Agency claims that "the atmospheric CO_2 increase is not more a potential but a real threat to which Man must start to be acquainted with right now. Progressively, starting 1990, and in less than 150 years, the global warming might be potentially catastrophic, reaching 5°C. Following EPA there is no way to escape and even the most drastic actions will postpone this warming by only some 15 years. The foreseen consequences are related to a reduction of food supply at the planetary scale and a partial melting of the ice sheets with its subsequent rising of sealevel all over the world". Following EPA we must learn to live with the major changes which will happen during the next two decades.

Acknowledgments

Figures 4.4, 4.5, 4.8, 4.11, 4.12, 4.13, 4.16 and 4.17 have been reproduced with the kind permission of the following authors: R. Bacastow, W. Bach, B. Bolin, K.S. Groves, J. Hahn, C.D. Keeling, S. Manabe, R.M. Rotty, R.T. Wetherald and of the following publishers: Elsevier Scientific Publishing Co., Macmillan Journals Ltd. (London), Controller of Her Majesty's Stationery Office, American Geophysical Union, Edward Arnold (London) and Consencus (Belgium).

I would like to thank also A. Gilchrist and J.M.B. Mitchell (U.K. Met. Office), M. Schlesinger (Oregon State University) and A. Volz (Jülich, RFA) and the lecturers of the Second International School of Climatology on CO_2 and Climate (held in Erice, July 1982) for sending the preprints of their more recent papers. I especially want to thank Dr. C.D. Keeling for providing me with the up-to-date atmospheric CO_2 concentration at Mauna Loa, Hawaï, W. Bach, R.M. Rotty and R.T. Wetherald for their judicious advices.

I express also my gratitude to the members of the Institute of Astronomy and Geophysics Georges Lemaître (UCL, Louvain-la-Neuve) and especially Chr. Tricot, whose fruitful discussions were very helpful, and Mrs. N. Materne for taking care of the manuscript.

Preliminary drafts of the manuscript of Section 4.2 were reviewed by R.G. Barry, University of Colorado, Boulder, and W.W. Kellogg, NCAR, Boulder, to whom the author expresses sincere thanks.

Chapter 5

SELECTED CLIMATES FROM THE PAST AND THEIR RELEVANCE TO POSSIBLE FUTURE CLIMATE

5.1 Introduction

In Chapter 4, the physical role of the various anthropogenic impacts on climate was outlined. The role of gases which absorb parts of the terrestrial (infrared) radiation-water vapour (H_2O), carbon dioxide (CO_2) and several trace gases such as ozone (O_3), nitrous oxide (N_2O), ammonium (NH_3), methane (CH_4) and the chlorofluormethanes (CF_2Cl_2 . $CFCl_3$) - is enhanced by their long residence time in the atmosphere. Apart from H_2O and O_3, most of them remain on the average many years (5-50 or more) in the atmosphere. The absorption bands of O_3, NH_3 and the chlorofluoromethanes in the so-called "window" (in which the atmosphere is transparent) do not overlap with the powerful bands of H_2O and CO_2 on both flanks of the window, and those of N_2O and CH_4 only marginally (Chapter 4). It is therefore possible to combine the effects of these partly man-made trace gases (Wang 1976) with the effects of an increasing CO_2 content. It has been proposed (Flohn 1978, Machta and Munn 1979) to add 50% to the effect of increase in CO_2 content - above the 320 ppm base used by Wang - and to express this "Combined Greenhouse Effect" (CGE) in units of a virtual CO_2 concentration (in ppm). This assumption is equivalent to a contribution of 33% by the trace gases to a "virtual" CO_2 increase, to be compared with a contribution of 67% from the real CO_2. Table 5.1 shows this virtual CO_2 content together with the real CO_2 content.

The significance of this concept arises from the large uncertainties of the future evolution of the composition of the atmosphere (see Chapter 4). Uncertainties of the future CO_2 input into the atmosphere by fossil fuel, deforestation and soil deterioration and of the fate of this CO_2 input in its three compartments atmosphere, terrestrial biosphere and soil, ocean do not allow a reasonable estimate of the atmospheric CO_2 content with time beyond, say, 30-50 years. On the other hand the presently available climate models allow a fairly realistic estimate of the climatic "greenhouse effect" or, more precisely, the CGE.

Based on these considerations, it is possible to foreshadow the future evolution of temperature in temperate and arctic latitudes as dependent on selected threshold values of the CO_2 concentration in the atmosphere. This scenario takes into account only the lesser uncertainties of the CO_2-climate relation and omits the much larger uncertainties of the future economic and political development and of the carbon cycle. A statement by an independent group of scientists, who were asked by the Science Advisor to the President of the U.S.A. to review critically the present state of the CO_2 climate issue and to look for hidden errors, mis-interpretations and gaps (see Wade 1979 and Section 4.28), came to the conclusion that no overlooked effect could be found. "If the CO_2 concentration of the atmosphere is indeed doubled... our best estimate is that changes in global average temperature of the order of 3°C will occur, and that these will be accompanied by significant changes in regional climatic patterns". This value of 3°C is somewhat higher than hitherto assumed from all available models (1.5-3°C).

Levels of representative temperature changes have been selected (Table 5.1) from characteristic warm climates of the past -they have been derived from the most recent paleoclimatic evidence as well as from available text books (Schwarzbach 1976, Frakes 1979). Using the Augustsson-Ramanathan model, they can

Table 5.1 Past Climates, change of representative temperatures (ΔT) and equivalent CO_2 levels (with probable limits of errors in %) 1) 2) 3)

Warm phases	ΔT	Equivalent virtual CO_2 content	Equivalent real CO_2 content
Early Medieval Optimum (ca. 1000 AD)	+ 1.0	455 ppm ± 8%	408 ppm ± 6%
Holocene Optimum (ca. 6000 BP)	+ 1.5	525 ppm ± 10%	455 ppm ± 8%
Eem Interglacial (120.000 BP)	+ 2.0	600 ppm ± 12%	508 ppm ± 9%
	+ 2.5	675 ppm ± 13%	555 ppm ± 10%
Late Tertiary (12-2.5 Million years BP)	+ 4.0	965 ppm ± 19%	755 ppm ± 17%

AD = Anno Domini; BP = before present; for "virtual" and "real" see text.

1) Equivalent CO_2 levels averaged from two versions of the Augustsson-Ramanathan model; the error limits in % give values of these extreme versions.

2) If the contributing role of trace gases should be neglected, values of the "virtual" CGE are valid for the CO_2 produced greenhouse effect.

3) If the contribution of other trace gases to the greenhouse-effect would increase to about 70% - mainly due to their higher residence time in the atmosphere - the equivalent "real" CO_2 content decreases, e.g. for ΔT = 4°C to 600-650 ppm.

be converted into equivalent levels of virtual and real CO_2 content. This procedure is intended only to provide an estimate of future climate evolution- it should not be misunderstood as indicating that such CO_2 levels existed in the climatic past*. Indeed the probable causes of past climatic variations of purely natural origin were far different from those to be expected in the near future. Among these past causes were different patterns of land-sea distribution, the extent of continental ice-sheets, different values of the elements of the Earth's orbit around the sun, the frequency and intensity of volcanic activity and possibly also variations of the extraterrestrial radiation from the sun (the so-called "solar constant") outside the recent range of less than 0.2%. All these "external" factors may be neglected when concentrating our attention on man-made effects on future climate.

The most interesting of the past warm climates as listed in Table 5.1 will be examined in Sections 5.2, 5.3 and 5.5 together with a critical discussion of the role of different boundary condition (see Chapter 1 and of some possible impacts on man's activity. They can be used to provide analogues of possible future climatic evolution, bearing in mind their limitations (as well as the limitations even of well-advanced interactive models which are not yet available). Our climatic description of these analogues is centred on Europe, but since climatic change is a world-wide event affecting strongly other climatic zones which are connected with Europe in many ways, physically and socially, large-scale changes in other climatic belts (especially in the tropics) can never be neglected.

Apart from these warm phases, Section 5.4 considers the possibility of a strong cooling leading to a climate similar to that of the Little Ice Age (Chapter 2) or, eventually, to a new Ice Age. It is quite unlikely that such an event will be produced by Man's activity: dust in the lower troposphere whether produced by combustion, mostly but not exclusively in industrialized areas, or by release of mineral particles from the soil by agricultural operations, scatters sunlight and reduces the radiation energy available at the Earth's surface. But most of it also absorbs solar radiation and emits terrestrial infrared radiation (Chapter 4) and tends therefore (at least in continental regions) also to warm the atmosphere slightly. An increase of surface albedo through the destruction of vegetation and through

*In recent years, however, convincing evidence has been found that, in the geological past, under purely natural conditions, the CO_2 content of the atmosphere underwent significant fluctuations. Carefully scrutinized measurements from bubbles enclosed in Greenland and Antarctic ice cores (Delmas et al. 1980, Neftel et al. 1982) have shown that during the last glacial, about 20-15.000 years ago, the CO_2 content was 180-200 ppm and rose, about 6-10.000 years ago, to about 300 ppm or even more. The paleoclimatic consequences of these facts are not yet understood.

land use conversion is quite a slow process which (together with decreasing evapotranspiration) might cool the lower atmosphere, through mostly on a local scale; it will most probably be counteracted by the CO_2-induced warming. The same is true for a possible reduction of the O_3 content of the atmosphere; this would be accompanied by an increase of stratospheric water-vapor with an opposite (warming) effect.

However, natural processes such as a cluster of heavy volcanic eruptions can occur (as e.g. in the period 1811-1840) which inject large amounts of particles (in the submicron range) into the Junge layer (20-22 km) of the stratosphere, where they accumulate mostly in the polar regions. They scatter sunlight, absorb infrared radiation and warm the adjacent layers of the stratosphere, but this has only a small cooling effect on the surface temperatures. Since volcanism and large-scale earthquakes are both caused by the same geophysical mechanism - the slow motion of tectonic plates - the frequency of severe earthquakes in recent years may perhaps be taken as an indication of a period of enhanced volcanic activity following the lull between 1912 and 1948. But no convincing evidence has yet been presented, and any forecasting of volcanic activity is at present impossible.

In Section 5.4 the decay of the existing large ice-sheets under the impact of a man-made global warming will also be considered. Due to a different shape of the solid bedrock, the ice-sheets on Greenland and both parts of Antarctica may behave quite differently. While the greater part of the Antarctic ice has covered in East Antarctica an extended continental shield well above sea-level for many million years past (Section 5.5), much of the smaller ice-sheet in West Antarctica (extending from the Antarctic peninsula towards the Ross shelf ice near longitude 150°W) is potentially less stable, since its bedrock is situated below sea-level. The possibility of its disintegration, perhaps accompanied by a sea-level rise of 5-7 m. as during the Eemian (Section 5.3), has recently led to some discussion. This problem is serious enough to deserve detailed investigation in order that is may be raised above the level of more or less speculative hypotheses (Section 5.4.3).

5.2 The Holocene warm phase

After the peak of the last ice age (about 18000 years BP) the main climatic amelioration did not start immediately. The transition between the end of the Glacial Period and the present Interglacial (Holocene) began first in the Southern Hemisphere, where the thin subantarctic drift ice (Hays 1978) retreated in a few centuries, from its northernmost edge about 600 km nearer to the equator, about 14000 years ago, just at the start of the melting of the northern continental ice-sheets. The warmest period occured here centred around 9500 BP (Hays 1978, Lorius 1978). High temperatures in the Biscaya have been verified (Duplessy 1981) around 6000-8000 BP. All dates in this chapter are given in "radiocarbon years"; Their deviation from calendar years may reach as much as 8-12%, but this is still a matter of some controversy.

The retreat of the northern continental ice sheets and the gradual warming was interrupted by one(or two) abrupt coolings and glacier advances between about 13000 and 10500 BP. At least one of these coolings extended far into the tropics, accompanied by a return to a dry climate with nearly glacial temperatures.

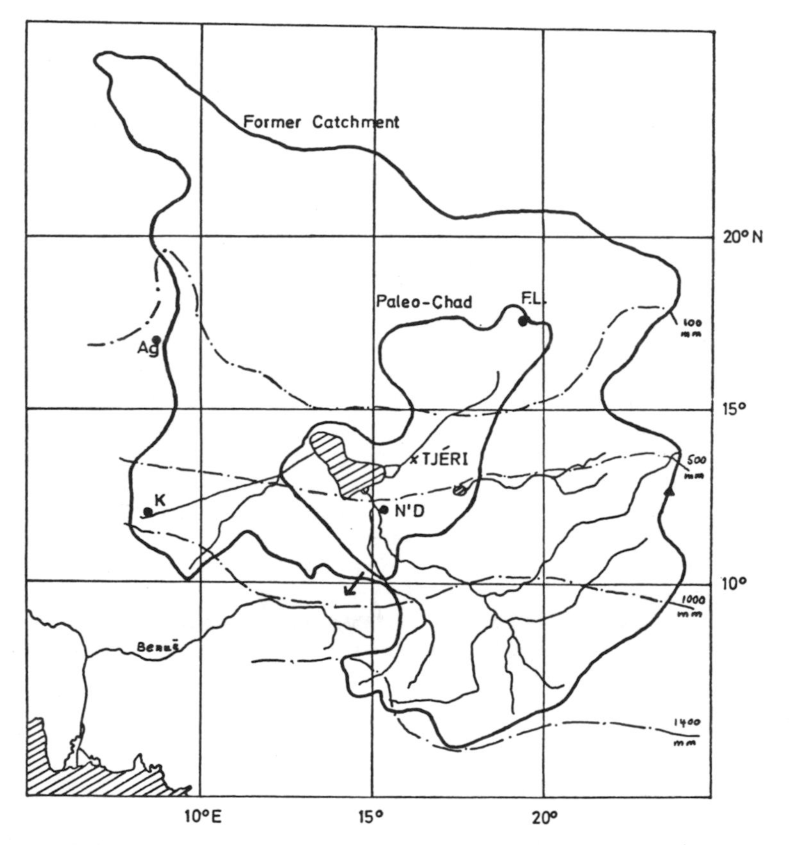

Fig. 5.1. Lake Chad now (hatched) and during its maximum level 6-7000 years BP (Paleo-Chad), catchment at this level, arrow = overflow to Benuë-Niger. Tjéri = core site (Chronology, pollen data).
Places: K = Kano, Ag = Agadés, N'D = N'Djamena (former Ft. Lamy), F.L. = Faya Largeau. Dash-dotted lines: lines of average annual rainfall (actual): 100, 500, 1000, 1400 mm/a.

Thereafter, a general warming occurred (usually defined as the beginning of the Holocene period), together with the occurrence of a moist period especially in tropical and subtropical latitudes. The equatorial rain-forest, which was drastically reduced during and after the peak of the last ice age, reappeared around 14000 BP (Shackleton 1977) and spread northward into the southern fringes of the Sahara. The arid belt of the Old World, extending over about 9000 km in longitude from the West African coast to the deserts of Pakistan and northwestern India (Rajasthan), enjoyed between about 11000 and 8000 BP a period with much higher rain frequency, and was still relatively cool. The valleys, which are now dry wadis, were covered by grassland with many trees (Maley 1977). Among a multitude of lakes, Lake Chad should especially be mentioned because its area increased to nearly 340000 km^2 (about the size of the present Caspian Sea), with an overflow to the Niger-Benuë catchment (Fig. 5.1) of which there is evidence in a sudden enormous increase of the sedimentation rate off the Niger delta (Pastouret et al. 1978).
Its maximum depth reach more than 300 m as compared with 2-3 m now; its present area is variable, with a maximum of about 20000 km^2. Similar evidence (e.g. Rognon and Williams 1977, Street and Grove 1976) has been found in East Africa, where Lake Turkana (formerly Lake Rudolf) stood 70 m higher and had an overflow to the Nile catchment, in Arabia (McClure 1976) and Rajasthan (Singh et al. 1974). Even in the arid centre of the Sahara - around Kufra, now with an annual rainfall below 5 mm - the rainfall has been estimated 300 mm or more, sufficient to nourish a grassland with a dense population of big animals (buffalo, elephant, giraffe) which were hunted by late paleolithic man (Gabriel 1977).

This periode ended rather abruptly around 8000 BP, simultaneously with two important climatic events (Flohn and Nicholson 1980):
a) Disappearance of the Scandinavian Ice Sheet, which had still covered up until 10000 BP most of Scandinavia and Finland north of the line Oslo-Stockholm-Helsinki.

b) Incursion of the sea right into the center of the Laurentide (=North American) Ice Sheet, the Hudson Bay area, which took place catastrophically over a period of not more than 200 years (Andrews et al. 1972), simultaneous with a rise of the world's sea level of at least 7 m, as evidenced at the coast of Lancashire (Tooley 1974) and at the Kattegatt (Mörner 1976). During this event, about 3 million km^3 of ice (about 140% of the present volume of the Greenland ice cap) calved as enormous icebergs through the Hudson Strait into the Atlantic, where the glacial debris at the bottom spread towards the southeast. This event deserves special attention, since a similar event might perhaps happen in West Antarctica (see Section 5.4.3) after any prolonged climatic warming. The disappearance of the Scandinavian ice, together with the still existing areas of the Laurentide ice together with the still existing areas of the Laurentide ice in Labrador and territories west and north of Hudson Bay, produced quite a peculiar climate during the period between 8000 and about 5000 BP. During the warm season, the occurrence of an extended area of snow and ice around Hudson Bay created a type of atmospheric circulation which dominates at present the winter/spring season: a deep cold trough above Eastern North America, a flow of warm air ahead of it with southwesterly winds across the Atlantic and a quasi-stationary second trough reaching from Central Europe towards northwestern Africa accompanied by frequent rains. In many parts of Europe this circulation type was responsible for summers about 1-2°C warmer than now (Table 5.2) and with 10-15% greater rainfall (Lamb 1977). During the cold season, the distribution of snow/ice on the North American continent did not differ very much from that of recent winters. Since the Arctic sea ice had retreated to about its present position, the winter circulation should have been rather similar to the present one. Due to the well-known North Atlantic temperatures see-saw (van Loon and Rogers 1978, 1979) a cold winter in the Greenland-Labrador-Maine region is usually accompanied by a mild and rainy winter season in Western and Central Europe. Such a circulation pattern (cf. Lamb 1977, Fig. 16.8-9) was responsible for the occurrence of a rather long warm and moist period over Europe between about 7000 and 5500 BP. the so-called Atlantic period (Table 5.2).

TABLE 5.2. Climate estimates, England and Wales (Lamb 1977)

	Temperature (°C)			Rainfall(1)		Evaporation(2)	Runoff(1)
	Year	Jul./Aug.	Dec.-Feb.	Year	Jul./Aug.	Year	Year
"Atlantic",6000 B.P.	10.7	17.8	5.2	110-115	?	108-114	112-116
Little Optimum 1150-1300 A.D.	10.2	16.3	4.2	103	85	104	102
Little Ice Age, 1550-1700	8.8	15.3	3.2	93	103	94	92
Recent Warm Period, 1916-50	9.4	15.8	4.2	932 mm		497 mm	435 mm

(1) percent of 1916-50 average.
(2) Turc's formula

An important change of the climatic boundary conditions during the Holocene warm phase is given by the shift of the perihelion - the shortest distance between the Sun and the Earth at its elliptical orbit. While the present perihelion (early January) favors the Southern Hemisphere during its summer, its position was 9-10000 years ago during July. Then the extraterrestrial summer radiation at the Northern Hemisphere was up to 7% higher than now:

this caused higher summer temperatures (and slightly lower winter temperatures) at the northern continents, while at the northern oceans the temperatures could not be much different from to-day due to the oceanic heat storage. A low-resolution circulation model (Kutzbach and Otto-Bliesner 1982) simulated a distinct intensification of the monsoon circulation, which coincides well with all available paleoclimatic data.

In other parts of the globe this "Holocene climatic optimum" did not everywhere occur at the same time, depending in each case on the timing of the disappearance of glaciers and sea ice. Evidence from northern Greenland and in the northernmost parts of the Canadian Archipelago shows driftwood coming from Siberia during that time: this seems to indicate that the Arctic sea ice had retreated during summertime farther than in the warmest period of the 20th century. The climatic optimum around 4500 BP (Barry et al. 1977) improved Man's living conditions in the Arctic to their highest ever level. Along the coast of the Arctic Ocean, as well as in continental Canada, subarctic forests spread 200-300 km further north than now, equivalent to a warming of the summer temperatures by 2-3°C (Nichols 1975). Permafrost in Siberia retreated to several 100 km north of its present position.

After a drier climatic phase in the Sahara belt around 7500 BP, perhaps related to a cooling of the Atlantic after the above-mentioned Hudson Bay event, the moist period recovered between about 7000 and 5000 BP, enabling a rather dense population (Gabriel 1977) of neolithic cattle-raising nomads to live even in the now driest areas of the Sahara (Fig. 5.2).

A review of the existing literature (Sarnthein 1978) has demonstrated that during the Holocene optimum most continental areas enjoyed higher rainfall. Only two exceptions should be mentioned here: the "Prairie triangle" expanded at this time from 95°W to 85°W into Wisconsin and Illinois indicating a temporary desiccation and, simultaneously, some areas in South-West Siberia became drier, and the level of Lake Van in Eastern Turkey fell about 300 m (Kempe 1977).

This climatic optimum came to an end rather suddenly shortly before 4800 BP, when a sudden retreat of the boreal forest in Canada equivalent to a summer cooling of 2-3°C was observed(Nichols 1975). Simultaneously, new glaciations began to develop in several high mountainous areas; another, even more pronounced "neoglaciation" occurred after about 3400 BP. The available evidence in the Swiss and Austrian Alps indicates a sequence of at least six warm phases after the final warming at 10000 BP, separated by local glacier advances and lowerings of the upper forest boundary. In contrast to conditions in the subarctic, the climatic fluctuations between the warm and cold phases seem to have been of rather constant amplitude (Patzelt 1973), so that here the Atlantic period differs only in length and not in amplitude from the other warm episodes.

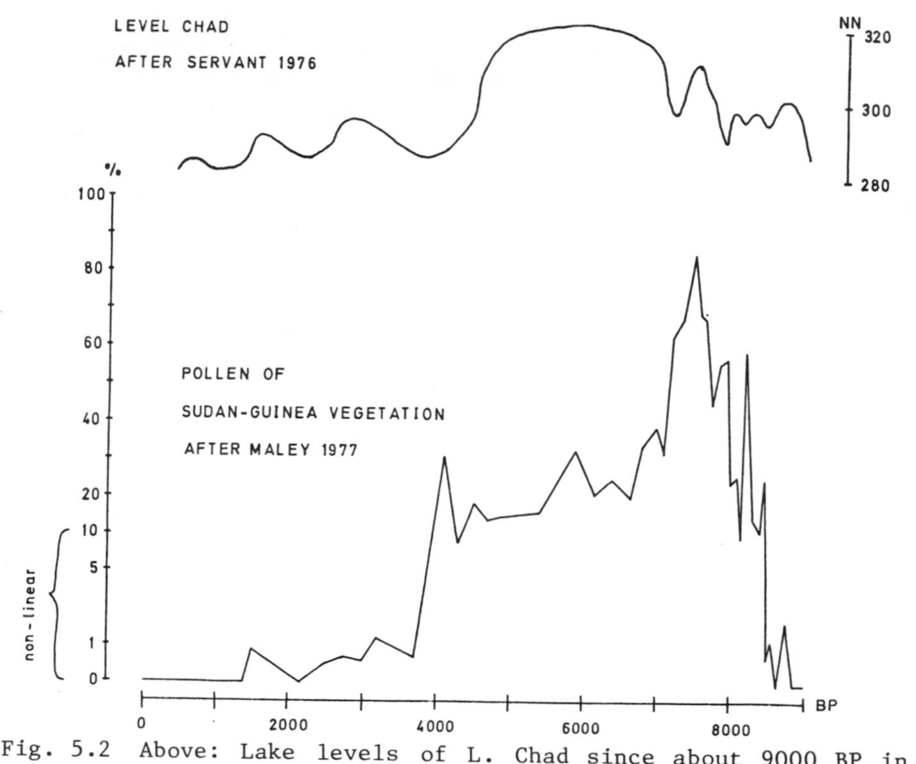

Fig. 5.2 Above: Lake levels of L. Chad since about 9000 BP in m above sea-level (NN). Below: percentage of pollen from Sudan-Guinea vegetation (humid Tropics, equivalent to an annual rainfall of 1200 mm and above). For data see Maley (1977); chronology before 8000 BF assumed to be partly uncertain.

The large-scale boundary conditions of the Holocene climatic optimum are the following:

a) disappearance of continental ice-sheets, except in Greenland, Labrador and Baffin Island;

b) retreat of the Arctic sea ice to its northermost position reaching near 81-83°N in the Atlantic section around 5000 BF;

c) reduction of the meridional gradient of temperature, leading to a weaker atmospheric circulation above the Northern Hemisphere and the tropics. The latter is apparently correlated with a greater frequency fo equatorial oceanic downwelling (El Niño, see Chapter 1) and a reduction of equatorial upwelling in frequency and intensity, leading to a substantial increase of evaporation in both the tropical Pacific and Atlantic Oceans.

Taking a global view, the climate at that time was in many regions more beneficial than the present one with, however, some serious exceptions. Is it really possible to use it as an analogue for future climatic patterns during a CO_2 -induced global warming? Can we indeed expect a substantial increase of rainfall in the arid belt extending from West Africa to Rajasthan? Here we have to check the boundary conditions, including the present distribution of vegetation and some relevant physical parameters (albedo, soil moisture). In the case of a future global warming, due to increasing CO_2 , boundary condition (a) will not apply: Labrador and the greater part of Baffin Island is now ice-free and snowfree during summer; even in the case of a new "Little Ice Age" which cannot be entirely excluded (see Section 5.4.25) a largescale shallow glaciation of these areas cannot be expected during the next 100 years or so. Its absence will prevent the formation during summer of a deep trough at the eastern coast of North America and consequently of a deep secondary trough over Europe; only weak troughs are observed over both areas during the melting period of the Arctic ice from mid-June to the end of August.

An argument against a strong amelioration of conditions in the arid zone is presented by several recent model computations on the role of surface albedo on the aridity of the Sahara (Charney 1975, Ellsaesser et al. 1976, Berkofsky 1977), with rather different physical assumptions and mathematical treatment. All these models describe the role of an increase of the albedo due to destruction of the vegetation by Man, which has been going on since the end of the neolithic moist phase 5000 years ago and is now expanding into marginal areas in causing an intensification of the average downward motions of the air and thus prohibiting effective rainfall. It should be mentioned that in several of these regions no evidence for a downward trend of rainfall since the beginning of this century could be found. This is true, e.g. for southern Tunisia, the Sahel belt, Rajasthan and interior Australia. But the process of desertification (Hare 1977), including soil erosion, deflation and salinization, is obviously rather slow and the effect of rainfall may be detectable only over very large regions after many centuries.

Regarding the possible future evolution of climate in the case of a CO_2 -induced global warming, one has to expect, for this state, a weak displacement northward of the rainfall belts, i.e. of the subtropical belt of winter rains at the northern flank as well as of the belt of tropical summer rains at the southern flank of the arid zone. Given an increase of evaporation from both tropical Pacific and (especially) Atlantic of near 5 - 10%, this shift should gradually become effective only in the Sahel belt. Here, rainfall in the vicinity of the northern edge of the moist equatorial air-mass arriving with winds from SW and W is, under present conditions, very scanty because of the role of large-scale forced subsidence above a shallow layer of this moist and relatively cool air-mass (Flohn 1966). A gradual northward shift of the whole system could eventually lead to increased rainfall in the Sahel belt, i.e. to a northward expansion into the present desert and a extension of the rainy season. But if the above-mentioned models are correct in substance, any perceptible increase of rainfall can occur only if the reestablished vegetation can be protected, under the stress of a still increasing population, against the positive feedback effect: man-made desertification-increased albedo-increasing downward motion-less rainfall-extended desertification. The reestablishment of a sufficiently dense vegetation during a period of increasing and expanding rainfall is then a socio-economic problem, which would call for a major concerted effort.

The future evolution of the climate in southern Arabia, Pakistan and Rajasthan is even more difficult to predict. Here, the summer monsoon system, controlled by the powerful effect of the elevated and heated Tibetan plateau, is more pronounced than in Africa. The physical reasons for the large interannual fluctuations of monsoon rainfall are far from being understood; at least part of the mechanism may lie in the air-sea interaction above the Arabian Sea and in the variations of the low-level jet-stream along the East African coast (Findlater 1977). In view of this situation it si difficult to see how weak global warming could modify these processes and significantly increase average rainfall.

In the European area, however, a weak warming should lead to an extension of the length of the vegetation period and probably to a slight increase in rainfall, with one notable exception namely the strongly variable Mediterranean winter rains (see Section 6.4). If this belt during winter is displaced only, say, 100-200 km poleward, the intensity and duration of the winter rains should be significantly reduced. Such a reduction follows from boundary condition (c): if the temperature gradient Equator-Pole is decreased, the intensity of the rain-producing disturbances should diminish and the length of the rainless summer season should increase. The already critical hydrologic balance can be expected to change in a negative sense: decreasing rainfall together with increasing evaporation. This would be mostly felt in the area south of about Latitude 38°N.

5.3 The Eemian interglacial and its termination

5.3.1 Introduction

In the last 25 years, a great deal of information about long-term climatic changes in the last million years or more has been collected by studies of various kinds of proxy data, obtained from marine or continental sediments, moraines, sea level terraces, etc,... Dating problems and lack of continuity of the records have sometimes caused confusion, but a reasonabl consistent picture is beginning to emerge, not least due to stable isotope studies on the fossil fauna found in the deep-sea floors.

In this section, we shall first outline a few important methods hitherto applied in long-term climatology (Section 5.3.2) and we shall then see how to estimate global and some regional climatic conditions during the last warm period prior to the present one, the Eemian (Section 5.3.3). Finally, we shall try to describe how the Eemian ended (Section 5.3.4), approximately 115,000 years ago, or 115 ka BP (1 ka = 1 kiloannum = 1000 years; BP = before present).

5.3.2 Methods

The study of the marine fauna deposited on the ocean floor has led to the most spectacular results to data in long-term methods of palaeoclimatology. However, like any other indirect recording of a phenomenon as complex as climatic change, it involves several sources of error that have to be taken into account. It should be borne in mind that the parameters accessible to observation (the proxy data such as the isotopic composition of fossils or ice, sea level, pollen distribution, etc, ...) are, at best, a few consequences, rather than direct measures of the climatic changes themselves. For the same reason, the interpretation of such data involves several potential pitfalls; after all, each of the observable parameters describes no more than a boundary condition (volume of continental ice, regional sea surface temperature, etc,...) and not the climatic changes themselves, much less their causes. Therefore, interpretation of any time series of proxy data in terms of climate implies some degree of speculation that can be reduced only by a never-ceasing effort to cross-check for inconsistencies, internal ones by applying a given method on material collected under widely different conditions, and external inconsistencies by unprejudiced comparison with other methods.

5.3.2.1 Deep-sea cores

Cores from the ocean floort contain material deposited at a fairly constant rate of the order of a few centimeters per thousand years. Detailed analyses allow the establishment of very long times series ($\sim$ 1000 ka) of various parameters, but the resolution (1-3 ka) is lower than in many other sedimentary time series. The stable oxygen isotope composition δ_c of carbonate*, formed by micro-organisms (e.g. foraminifera) and deposited in marine sediments, depends on the temperature and the isotopic composition, δ_o, of the ocean water, when and where the micro-organisms lived (Emiliani, 1955, 1966).

At any given time, the mean δ_o of all ocean water must be the greater, the more isotopically light (low δ) water in removed from the oceans and deposited as glacier ice on the continents.

* δ is the relative deviation of the $^{18}O/^{16}O$ ratio in a sample from that in a standard. The small δ-values dealt with in this work may be considered as ^{18}O concentration deviations.

The maximum change of δ_O, i.e. from glacial to interglacial conditions, has been estimated at between 1.2 and 1.6‰. This is close to the highest variation in δ_C observed in Caribbean and equatorial deep-sea cores, showing that δ_C may be indicative of the degree of glaciation rather than ocean temperature (Olausson 1965, Shackleton 1967, Dansgaard and Tauber 1969, Duplessy et al. 1970, Yapp and Epstein 1977, Duplessy 1978).

The δ_C data are therefore often interpreted in terms of continental ice volume in excess of the present 31 x 10^6 km^3. The ice volume scale is calibrated by assuming a 50 x 10^6 km^3 excess of ice some 20 ka ago, cf. the scale to the right in Fig. 5.3.

However, there are many sources of chemical and mechanical disturbances, that have the effect of partially blurring and smooth-

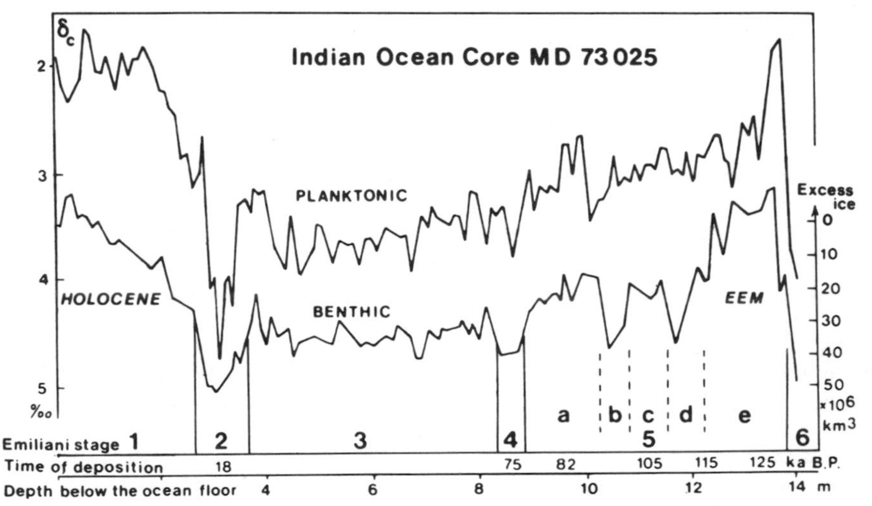

Fig. 5.3 Oxygen isotope δ_C records of planktonic foraminifera, Globigerina pachyderma (upper curve), and of the benthic foramifera (lower curve) in the same core from the South Indian Ocean. The benthic record is assumed to reveal past continental ice volume in excess of the present, as indicated on the scale to the right (valid for the lower curve, only). The time scale at the very bottom is established by reconciling characteristic features (e.g. the Emiliani-stages 2, 4 and 5b) with the corresponding features in other cores that have been dated independently. The figure suggests (1) that it is more than 100,000 years since the continental ice volume was as low as it has been in our own warm period, and (2) that at the end of the last warm period (Eem), the continental ice sheets and glaciers grew by some 40 x 10^6 km^3, or 15 times the present volume of the Greenland ice sheet, within some 10,000 years.

ing a δ_c profile along a deep-sea core. First of all, one might ask if a δ_c record could be biased by a biological element, since δ_c is the isotopic composition of shells made by organisms. And indeed a biological bias exists, but it seems to have remained constant through the last 150 ka (Duplessy et al. 1970, Shackleton 1977a), as long as one and the same species is considered.

The choice of species is very important. For example, planktonic foraminifera live in surface waters, and the δ_c of their remains in a core therefore depends on the then δ_o of the surface water, which is not only determined by the volume of continental ice, but also by the precipitation/evaporation ratio and the influx of isotopically light fresh water. Furthermore, δ_c of the planktonic foraminifera is significantly influenced by the surface water temperatue changes, and finally, during the long fall from surface to bottom, progressive dissolution of calcite causes a selective removal of thin walled and highly porous variants within the same species. This enriches the fossil population in the sediment with thick walled and terminal forms that built up their shells in deeper and cooler water, thus giving the fossil planktonic shells higher δ_c than that of the original population as a whole (Berger 1971, 1977, Savin and Douglas 1973). Although all of these effects have existed at all times, and thus partly cancel in a record of δ_c changes, they to represent sources of error, which contribute to the fact that planktonic foraminifera δ_c records from various parts of the ocean agree only as far as the main features are concerned.

In contrast, δ_c records based on benthic foraminifera living on the bottom (greek benthos = bottom) are much less influenced by the dissolution process and by the other sources of error mentioned above, since the dead organisms remain where they lived, and since the shell precipitation has always occurred essentially in the same environment. Therefore, benthic foraminifera should always be the object of any investigation with the aim of estimating the degree of glaciation in the past. Unfortunately, the benthic foraminifera are relatively rare, which sometimes creates continuity problems in the δ_c record. Hence, only few continuous, high-resolution benthic records have been published. Figure 5.4 shows smoothed versions of two such records from cores recovered at locations no less than 13,600 km apart: the heavy curve is the sub-Polar Indian Ocean (43°49'S, 51°19'E) record from Fig. 5.3 plotted on a linear time scale, and the thin curve is from the Eastern Equatorial Pacific (03°35'S, 83°13'W: core V 1929: Shackleton 1977b). The synchronism is fictitious, because the Pacific record has been slightly adjusted to fit the Indian Ocean record along the time axis, but it is remarkable that δ_c values of all of the extremes defined by Emiliani (1966) are essentially equal in the two records. This supports the view that the main features of the benthic records are broadly independent of local effects, and that they do represent changes of a general physical characteristic of ocean water, its isotopic composition, and hence the amount of continental ice, assuming that the temperature of the ocean deep water has remained essentially constant in the past.

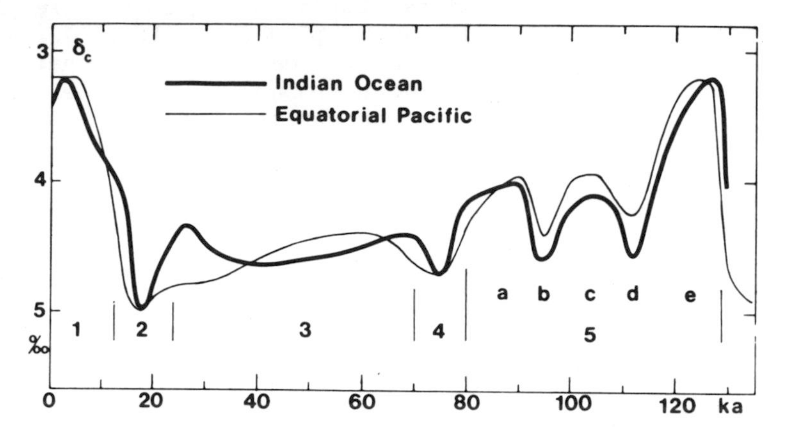

Fig. 5.4 Smoothed versions of two δ_c records spanning the last 130,000 years and measured on benthic foraminifera from two deep sea locations 13,600 km apart: Indian Ocean, 1800 km south of Madagascar (heavy curve, from Fig. 5.3), and Equatorial Pacific, 400 km west of Ecuador (light curve, from Shackleton 1977b). The synchronism is ficticious, because the curves have been slightly adjusted for optimum fit along the linear time scale. But, the δ_c oscillations having nearly the same amplitudes indicates that these records describe the slowly changing isotopic composition of the oceans, independent of many of the sources of error that influence planktonic records.

All kinds of sedimentary records may be disturbed by mechanical mixing of the upper few centimeters of the sediment due to the activity of benthic micro-organisms living in the oxygen-rich bottom water (Berger and Heath 1968, Peng et al. 1977). The consequent temporal smoothing of the δ_c records is of course most serious in areas of low sedimentation rate, which on the other hand offer the longest time series for a given core length. In practice, the most detailed information is obtained in areas with sedimentation rates higher than 5 cm/ka, which gives for a 20-m core a time span of less than 400 ka. Since this core length is just about the longest one that is technically feasible to obtain at present, records spanning 1 or 2 million years have to be established in low sedimentation rate areas, but such δ_c records reveal only the main trends. The only way to estimate the influence of benthic mixing is to compare various δ_c records as is done in Fig. 5.4.

The composition of the fauna in ocean surface water depends mainly on the temperature and, hence, if the composition of the fossil fauna in a deep sea core is determined by counting, it may be transformed into an estimate of the surface palaeotemperature by statistical analyses using empirical transfer functions (Imbrie and Kipp 1971). The standard error of such estimates is rather high (1.5°C), which makes the method more useful at higher latitudes, because the amplitudes of the climatic temperature oscillations increase with latitude.

Dating the cores is done either by radioactive methods ^{14}C, ^{230}Th, ^{231}Fa) or by simply (1) interpreting the last geomagnetic reversal in the cores as being identical with the Brunhes-Matuyama boundary that has been potassium/argon dated at 700 ka BP by Dalrymple (1972), and (2) assuming a constant sedimentation rate over the entire core length, cf. Fig 5.12. In other cases, a time scale along the core is established by just identifying characteristic features in the record with their analogues in independently dated records, cf. Fig. 5.3 .

5.3.2.2 Pollen

Just as the fossil fauna pattern in deep-sea cores is indicative of the palaeotemperature of the surface ocean water, the fossil pollen pattern in continental sediments is indicative of the palaeoclimate in the region, in as much as most climatic belts have a characteristic flora. Caution should be applied, however, when ascribing a change in the pollen pattern to a climatic change, because it could also be due to changing conditions in the soil (Andersen 1969). Pollen diagrams often appear as time series of the occurrences of several individual pollen species, which of course gives a differentiated picture of the floral evolution in the region. The simpler diagram in Fig. 5.5(b) from an estimated 140 ka long profile in a peat bog in northeastern France (Woillard 1975, 1978) gives only the percentage number of forest pollen relative to the number of forests plus herbs pollen. An immediate way of interpretation is to ascribe low percentages to cold an/or dry conditions.

A common drawback of all pollen records is that they cannot be independently dated beyond the range of the carbon-14 method, which is about 35 ka or, with an improved enrichment method, about 75 ka.

5.3.2.3 Sea level and coast lines

The eustatic falls and rises of sea level due to Pleistocene glaciations and deglaciations are recorded along shorelines in the form of marine erosional benches and depositional terraces. Radioactive dating of the coralline material found at the terraces shows when the high sea stands and the corresponding glacia-

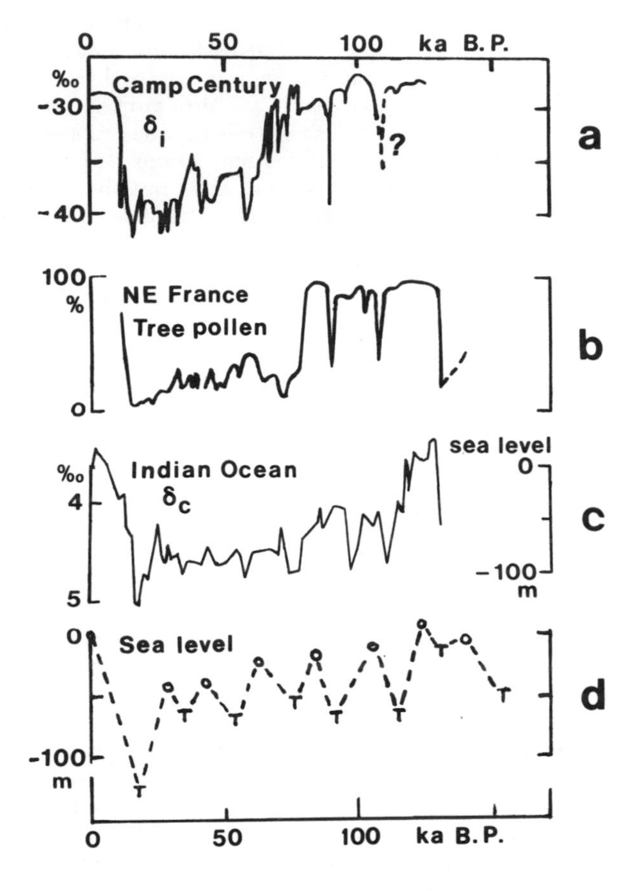

Fig. 5.5 The last glacial cycle (0-130 ka B.P.) revealed in four records of different climate-sensitive parameters.
(a) The isotopic composition of Greenland ice sheet precipitation. Low values correspond to cold surface temperatures. The record is dated by ice flow calculations that are inaccurate for the old part of the record. (A new tentative time scale has been suggested by Dansgaard et al. 1982).
(b) The ratio between forest and tundra type pollen deposited in northeastern France. The record is undated beyond 35 ka B.P.
(c) Continental ice volume (converted into sea level deviations) estimated from the isotopic composition of fossil foraminifera that lived close to the ocean floor. The record is undated, cf. Fig. 5.3 and its text.
(d) Sea level deviations from the present. The sea level sinks when glacier ice piles up on the continents. The sea level maxima are dated by radioactive isotopes.

tion minima occurred. However, the absolute positions of sea stands during these times are very difficult to establish. One major complication is the difficulty of identifying tectonically stable land areas and, in fact, it is questionable if the palaeo-geoid is adequately defined. But, the Eemian deposits are widely exposed and many investigators claim that enough relatively stable coasts, or areas with uniform tectonic uplift rate, have been found around all the oceans to allow a reasonably accurate estimate of the Eemian sea level, which is generally agreed to have been 6 to 8 meters higher than the present one (Bloom et al. 1974, Ku et al. 1974, Steinen et al. 1973, Lalou et al. 1971 and bibliography therein), cf. Fig. 5.5(d).

5.3.2.4 Ice cores

The oxygen-18 concentration δ_i in polar ice is an index of the ice sheet surface temperature at the time and site of deposition of the snow that formed the ice (Dansgaard 1964, Dansgaard et al. 1973). Like the δ_c index in the deep-sea cores, δ_i contains at least two components, because not only atmospheric temperature changes, but also surface elevation changes influence δ_i. For example, at the end of a glaciation δ_i increases, not only because the atmosphere warms up, but also because of the consequent shrinkage of the ice sheet, which leads to lower surface elevation and therefore higher surface temperature. As in the case of δ_c profiles along deep-sea cores, it is difficult to separate the two components in an isotope profile or, in other words, to correct the δ_i record for the effect of changing surface elevation in order to obtain the pure atmosphere temperature signal.

As an example, Fig. 5.5(a) shows a continuous δ_i profile along a 1400 m long surface-to-bottom ice core from Camp Century, Northwest Greenland (Johnsen et al. 1972). The Holocene part of the record has been dated with an accuracy of a few percent by measuring annual layer thicknesses from the top (Hammer et al. 1978), but the rest of the record has been dated by less accurate ice flow modelling with later corrections based on the assumption of persistence of a ca. 2.5 ka oscillation in the data (Dansgaard et al. 1971).*

Unfortunately, Camp Century was a poor drill site, because this area may have been exposed to greater surface elevation

*In a recent paper Dansgaard et al. (1982) suggested a new tentative time scale for the deep part of the Camp Century record. It is based on reconceliation of δ_i and δ_c trends and therefore, it brings nothing new into the discussion of the duration of the Eemian. The final dating of the deep ice must await the development of ^{14}C dating based on the new accelerator-mass spectrometer technique.

changes than most other areas in Greenland.** This is one of the reasons why the Camp Century δ_i record has never been provided with a temperature scale.

A more suitable drilling site in Greenland would be the central part that has been exposed to only small elevation changes. Furthermore, the ice flow is extremely simple, and no bedrock or surface melting occurs, which makes it possible to establish very long records of several environmental parameters, including surface temperature and atmospheric carbondioxide concentration. The recovery of a Central Greenland deep ice core is the main objective of the American-Danish-Swiss Greenland Ice Sheet Program. This ice core is expected to reach more than 500 ka back in time (Dansgaard et al. 1973), and new dating techniques (Muller 1977, Hammer 1977) open wide perspectives for establishing a reliable time scale along the core.

5.3.3 The Eemian interglacial

In this section, we adopt the interpretation that δ_c profiles measured on benthic foraminifera along deep sea cores reveal the continental ice volume in excess of the present 31 x 10^6 km^3, as discussed in Section 5.3.2.1, while δ_c profiles measured on planktonic foraminifera reveal the climatic history of surface water superimposed on the effect of the major changes of continental ice volume. It should be stressed that a δ_c curve cannot be directly interpreted in terms of any climatic parameter, such as temperature, although for example an increasing continental ice volume indicates cold and, probably, moist conditions at high latitudes.

The simple dating and interpretation of the δ_c records, combined with their nearly perfect correlation with astronomically governed insolation changes (see Section 5.4) has left the impression in wide circles that the problem of long-term climatic changes in the past has been solved, in priciple. Perhaps this is the case - the evidence seems abundant - and yet, the issue is still open to discussion, not least regarding the details and interpretation of the basic time series, and in the present context, the very details are of special interest. In this section, we shall therefore study the fine structure of the last interglacial and its termination, not because they were representative for interglacial to glacial transitions in general, but rather because it is the only one that is enlightened by time series of

** Recent ice flow modelling studies (Reeh, 1984) show that the surface elevation changes in the Camp Century area during the glacial-to-postglacial transition was probably much too small to explain, why the corresponding shift in δ_c at Camp Century (11-12‰) is considerably higher than in other parts of Greenland and in Antarctica (6 $\pm$ 1‰). It may therefore reflect a more drastic climatic change than elsewhere, due to the disappearence of the North American ice sheet at the end of the glaciation (Dansgaard et al. 1983).

several independent proxy data. For the same reason, what we can hope to deduce will, at best, be just a suggestion of how the present interglacial might end.

Five detailed δ_C records of the last 500,000 years indicate that the ice volume was almost similar to the present during the short interglacial periods (Emiliani 1972, Shackleton and Opdyke 1973, Hays et al. 1976, Bé and Duplessy 1976). Figure 5.3 shows that during the last climatic cycle, the only period with a sea level having remained continuously close to the present one but slightly higher, is isotopic substage 5(e). We thus consider it as the last interglacial. Isotopic substage 5(e) has been correlated with the Eemian interglacial on the continent (Mangerud et al. 1979, 1981). The Eemian has further been recognized by isotopic and faunal studies of deep sea cores from the North Atlantic Ocean and the Norwegian Sea (Sancetta et al. 1972), Duplessy et al. 1975, Kellogg et al. 1978) and in many other proxy data series. It lasted only about 10,000 years, from 125 to 115 ka B.P.

Most low and mid-latitude oceanic areas experienced climatic conditions similar to the present ones (see Cline and Hays 1976, and most literature cited in this section). Departures from the modern conditions have been recorded only at high latitudes, and only as barely significant deviations (cf. the standard error or 1.5° mentioned in Section 5.3.2.1.): in the central and eastern parts of the North Atlantic Ocean, sea surface temperature may have been higher by 2°C in Summer, and 1°C in Winter (Sancetta et al. 1972, Ruddiman and McIntyre 1977). In the western part, the Polar front was closer to Iceland, and the warm Irminger current did not exist. Accordingly, the mean sea surface temperature between Iceland and Greenland was at least 3°C colder than that experienced today. Oceanographic considerations suggest that the influx of warm water into the Norwegian sea was brought about solely by the Norwegian current, which was probably stronger than now (Kellogg et al. 1978). The Norwegian sea was also 1-2°C warmer and the Polar front was located some 200 km north of its present position (Kellogg 1975).

In the Southern Ocean, all the palaeoclimatic records show a strong warm peak (at least 1-4°C above the present temperature) followed by a sharp cooling, up to 6°C. Preliminary estimates suggest that the major cooling occured there <u>during</u> the Eemian interglacial. Let us once more turn to the <u>oceanic</u> record in Fig. 5.3, which compares the δ_C records of <u>benthic</u> foraminifera with that of <u>planktonic</u> foraminifera species <u>Globigerina pachyderma</u> <u>(left coiling)</u> in core MD 73025 from the <u>Southern Indian Ocean.</u> G. pachyderma is a species well adapted to cold water, generally secreting its shell at intermediate depths around 70-80 m (Kellogg et al. 1978). It is evident in Fig. 5.3 that the planktonic and benthic records correlate fairly well only as far as the major trends are concerned. If we define the Eemian by the peak in the <u>benthic</u> record, which is less sensitive to local variations, the <u>planktonic</u> record peaked early in the Eemian and fell drastically long before the benthic record began its descent. On the face of it, this would seem to suggest that the surface water cooled off long before the ice began to pile up on the continents. However, it could also be that the early fall of the planktonic curve reveals a regional change in the isotopic composition of the surface water due to decreasing precipitation to evaporation ratio.

Anyway, we may conclude that a major climatic shift occurred during the Eemian in the southern Indian Ocean, prior to the onset of the major deposition of continental ice.

As mentioned in Section 5.3.2.3, the extremely low degree of glaciation made the sea level 6-8 m higher than at present. As a consequence, the coastal lines were markedly different from those of today in the lower flat areas. In Europe, Baltic Sea water spread 200 km southeast from the Gulf of Finland and extended northeastward, converting most of Fennoscandia into an island (Fig. 5.6).

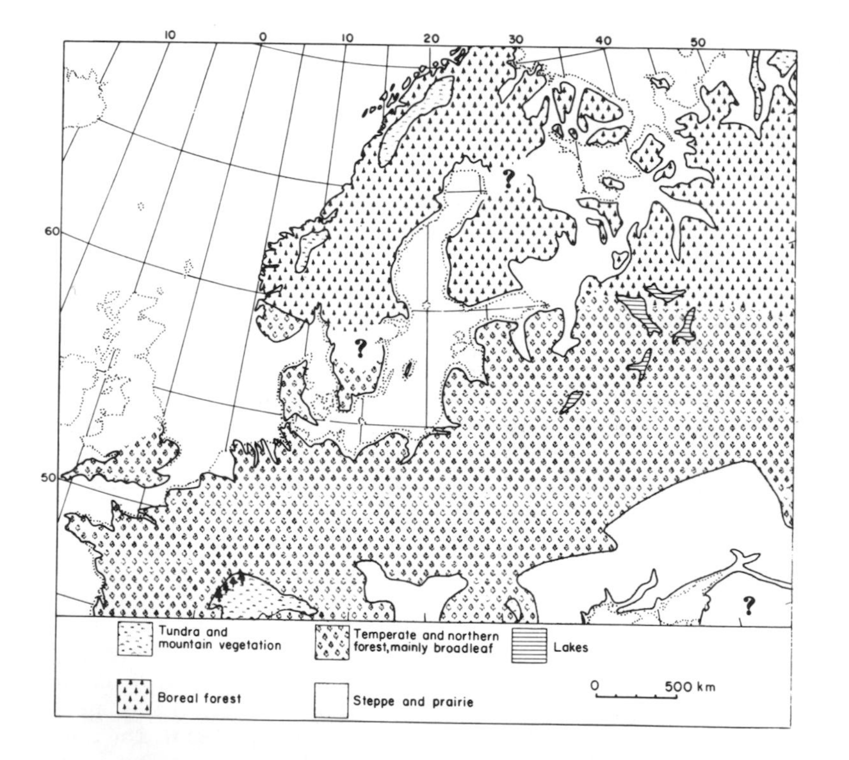

Fig. 5.6 Sketch map showing approximate coast line and generalized distribution of vegetaion in part of Europe in the middle part of the Eemian Interglacial (Flint 1971).

The sea penetrated deeply also into western Siberia, along the Ob and Yenissey flood plains, up to 62°N. On the East coast of America, the lowest terrace forms a fairly continuous broad stretch of flat terrain, which was submerged like southern Florida (Fig. 5.7).

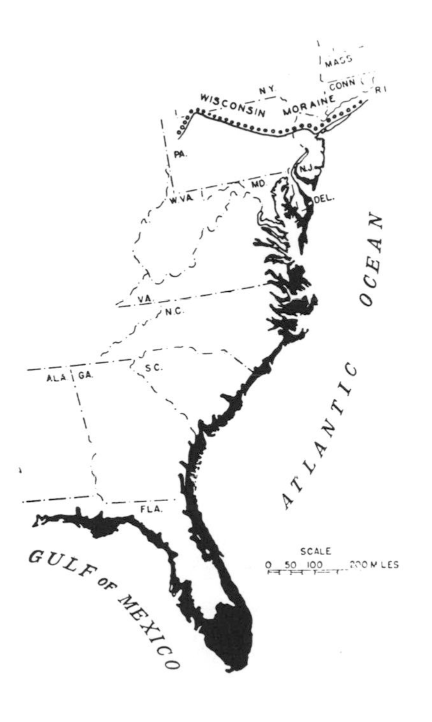

Fig. 5.7 The Eemian shoreline along the US Atlantic coastal plain. The areas set out in black were submerged. (Richards and Judson 1965).

Quaternary deposits are rare on the continents, and it is generally impossible to find pollen or other fossils from a priori interesting locations. During the Last Glaciation, large glaciers or ice sheets eroded previous deposits, and many remains from the Eemian Interglacial disappeared. Eemian sediments are thus highly discontinuous and the correlations between deposits from different geographic areas are difficult. On a world-wide scale, only the broad features of the Eemian climate can be reconstructed: the Eurasian record is good, the American one is fair and the other continents are mostly represented by no more than one or two deposits of which the stratigraphy is dubious.

Nevertheless, the changes in western Europe during the Eemian are revealed by several pollen profiles (Frenzel 1967, Zagwijn 1977, Woillard 1978): after glacial steppe conditions, a subarctic tundra appeared and marked the beginning of the Eemian. A boreal forest took over and was succeeded by a warm temperate forest at the end of the Eemian. Unfortunately, this secenario does not indicate the time when continental ice began to accumulate.

TABLE 5.3. Climatic differences: Eem-Actual (Frenzel 1967).

Area	Temperatures (°C)			Rainfall
	January	July	Annual	Annual (mm)
Denmark	+ 2	+ 1-2	+ 1-5	0
N.+ C. Germany	+ 1-2	+ 3	+ 2-5	0
Central Poland	+ 3-4	+ 3	+ 3	+ 50
N.E. Poland	+ 3-5	+ 3-5	+ 5	+ 50
Byelo-Russia	+ 5-6	+ 5	(+ 6)	0
Central Russia	+ 9-10	+ 2	+ 5	+ 100
N.W. Ukraina	+ 2-3	± 0	+ 1	+ 50
W. Siberia	+ 4	∓ 3	+ 3	+ 100?
Central Siberia	?	?	+ 6	?
Toronto	+ 3-4	+ 2	+ 2-4	+ 250
S.E. Alaska	?	+ 4-5	?	?
Banks Isl.(72°N)	?	+ 4-5	?	?

All the pollen data indicate that the climate was warmer and moister than today (Table 5.3) in northern Eurasia. A similar trend is indicated also in the Mediterranean areas: in Italy, Marchesoni et al. (1960) report January temperatures as 3-4°C higher, July temperatures about 2°C higher and a somewhat higher total rainfall. However, it is not clear whether the rain fall mainly during the Summer months or was spread throughout the year Similarly, in southern France, the Eemian period represents one of the few periods within the Upper Quaternary, when the rainfall was sufficient to allow cave-concretion growth.

Other European proxy data also suggest a climate slightly warmer and more humid than at present. From the Rhine to Poland, the buried Eemian soils are nearly everywhere of the brown-earth type, characteristic of a temperate forest. In the Ukraine, black earth (chernozem), characteristic of a temperate steppe, is dominant. This pattern is quite comparable to that current today.

Higher temperatures are indicated by marine mollusks from the Baltic and White seas. Since most of the Arctic ocean was still ice covered (Herman 1969, Clark 1971), the Arctic drift ice was probably displaced poleward; the area of open sea was thus a potential source of moisture for the neighbouring continent.

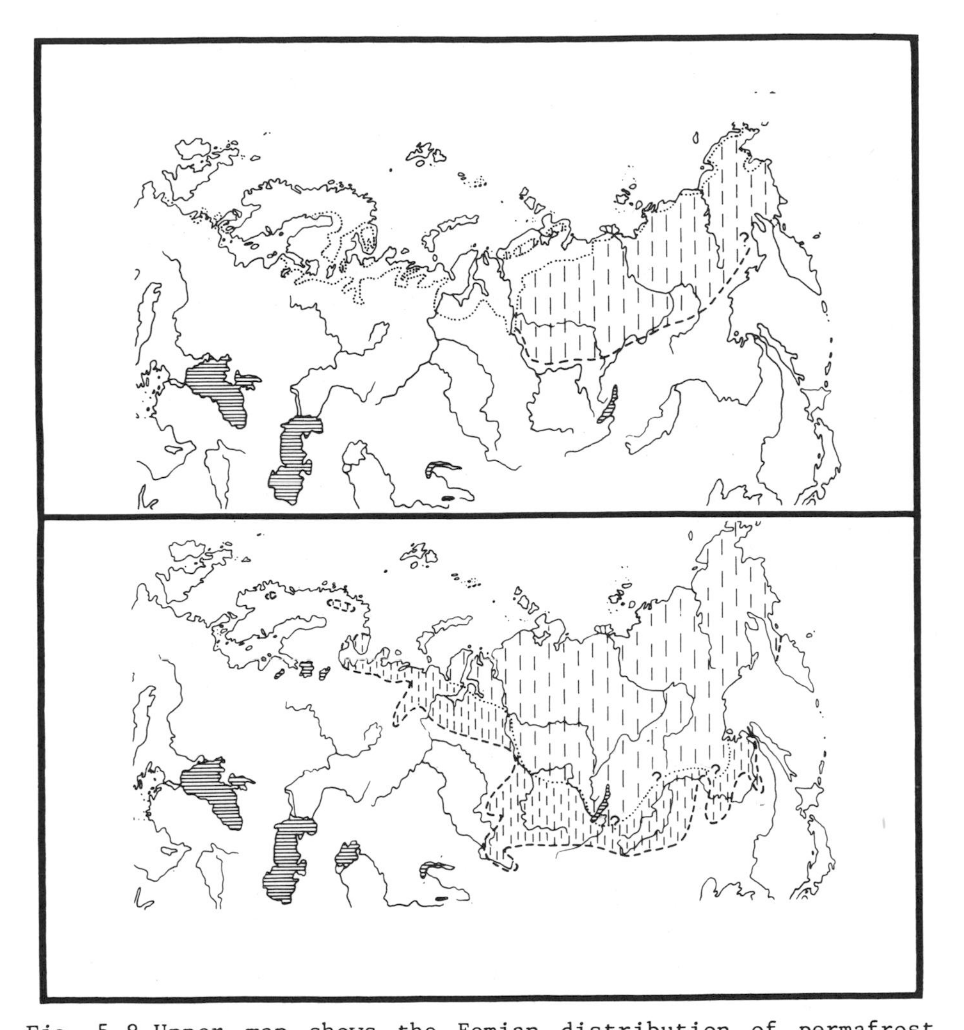

Fig. 5.8 Upper map shows the Eemian distribution of permafrost in northern Eurasia (area hatched by broken lines). Lower map shows the present distribution of permafrost (dotted line). The widely spaced broken lines indicate the permafrost area some 6000 years ago, i.e. during the warmest period of our own interglacial (Flint 1971).

A similar warming is suggested by the southern limit of the permafrost zone (Fig. 5.8), indicating mean annual temperatures lower than -2 to -3°C. This line was displaced far to the North. As a result, the permafrost area in northern Eurasia was not only smaller than it is now, but it was even smaller than during the Holocene climatic optimum (about 6 ka B.P., see Section 5.2). These data suggest a complete deglaciation of Scandinavia during the Eemian.

In North America the Eemian has been given the stratigraphic name of Sangamon, but it is now clear that the deposits represent the same climatic period. A climate similar to or warmer than today is generally recorded. The warm trend was enhanced in high latitudes: on Baffin Island, macrofossils and pollen indicate that the July temperatures were 1-4°C warmer than at present, precipitation more abundant and, as a consequence, the growing season was 20-25 days longer than at present (Andrews et al. 1972). Similarly in southern Canada, the January temperatures were 3-4°C higher than today, the July temperatures about 2°C higher and the mean annual rainfall about 250 mm higher than now (Goldthwait et al. 1965).

At mid-latitudes in the United States, going from east to west, the oceanic climate was progressively replaced by dry conditions, often drier than today (Wright and Frey 1965). Along the west coast, climatic conditions were slightly cooler and drier than now. In Southern California, deposits on the coastal terraces contain a cold water fauna on coasts exposed to the ocean, but a warm water fauna in protected shallow bays, suggesting that upwelling and coastal currents may have had a strong effect on the local climate.

In other parts of the world, fossil remains from South America, Africa, and Australia confirm that the continents were warmer than today.

All of these data suggest that the Eemian Interglacial (isotope stage 5e) was a warm interglacial, with a climate comparable to that of the present interglacial (Holocene), which has now lasted for 10,000 years, like the Eemian did until it came to an apparently abrupt end.

5.3.4 *The termination of the Eemian interglacial*

The Eemian interglacial was succeeded by a drastic deterioration of the climate at mid and high latitudes. The climatic shift is revealed in many different proxy data series, oceanic as well as continental. But, once again it should be stressed, that many of these data series are undated (e.g. the δ_c records in Fig. 5.4; and all pollen records beyond some 35 ka B.P.). Hence, they can only be provided with a time scale by interpretation and comparison with dated series, and most of the statements below therefore rest on the assumption that the interpretations are correct.

It has already been pointed out (Section 5.2.3) that the cooling of the southern oceans seems to have started already in the middle of the Eemian, i.e. long before the ice began to pile up on the continents. In contrast, the presence of warm planktonic foraminifera in post-Eemian sediments suggests that the subpolar North Atlantic from 40°N to 60°N maintained warm surface temperatures during the first half of the ice-growth phase (Ruddiman and McIntyre 1979), thus providing enough moisture to rebuild the Scandinavian glaciers, once the continental temperatures fell. In western Norway, a complete interglacial sequence in coastal marine sediments has been correlated with the Eemian (Mangerud et al. 1979). The study of continental pollen and marine molluscs suggests a drastic fall in temperature, simultaneously in the sea and on land, and possibly related to weakening of the North Atlantic Current into the Norwegian Sea.

Anyhow, the forest vegetation disappeared in northwestern Europe (Andersen 1969, Wijmstra 1969, Woillard 1978, 1979, cf. Fig. 5.5(b), and the Scandinavian glaciers advanced considerably, merging into an ice sheet at some later stage after at least one temporary retreat (Mangerud et al. 1979).

Once again, let us turn to the benthic foraminifera δ_c record that has been transformed into a glacio-eustatic sea level curve in Fig. 5.5(c), assuming 10^6 km^3 of ice deposition on the continents being equivalent to a 2.67 m decrease in sea level. It seems that the sea level sank drastically within 10,000 years, from 120 (substage 5e) to 110 ka B.P. (substage 5d), corresponding to rapidly advancing glaciers. The same situation seems to have re-occurred around 95 ka B.P. and 75 ka B.F. (stages 5b and 4), of which the latter event marks the onset of a long period (stages 3 and 2), conventionally called the Wisconsin or, in Europe, the Weichselian glaciation. the period 115 - 75 ka B.P. is now considered to have been pre-stages of the "real" glaciation, because most of the time (substages 5c and 5a) the ice volume was intermediate, yet still 20 x 10^6 km^3, or 60% larger than today. The Norwegian Sea remained at or close to the freezing point and was almost continuously covered by ice (Duplessy et al. 1975, Kellogg et al. 1978), but northwestern France (Fig. 5.5(b) and Denmark (Andersen 1960) were re-invaded by forest vegetation, particularly in stage 5a, assuming this stage to be identical with the St. Germain II interstadial in France and the Brørup stadial in Denmark. The southern ocean temperatures were about 3°C cooler than today (Hays et al. 1976).

Many features of the (undated) δ_c record can be reconciled with the sea level deviations from the present (Fig. 5.3.3(d)), estimated from geological evidence at Barbados (Steinen et al. 1973) and New Guinea (Veeh and Chappell 1970, Bloom et al. 1974) after correction for calculated tectonic uplift rates. Most of the high sea level stages are dated by ^{230}Th/^{234}U, whereas the low sea levels are undated. The record suggests maximum sea level close to 120 ka B.F. and fast build-up of continental ice around 115, 95 and 75 ka B.P. with a final sea level minimum at 20 ka B.P.

Sea level estimates around 110 ka B.F. are -60 to -70 m in the geological record and -90 m in the δ_c record. If the geological estimate is correct, the discrepancy can be explained only if the deep water cooled off to near the pressure melting point.

However, the disagreement between the rest of the data series cannot be accounted for by deep sea cooling, nor by possible errors in the dating or in the isotopic analysis: the geological sea level record indicates two sea level rises culminating at around - 15 m close to 85 and 105 ka B.P., while in the δ_c record, there is no evidence of a sea level as high as that until the beginning of the Holocene. Furthermore, between 20 and 70 ka B.P., the geological record suggests several sea level rises close to or above -40 m, whereas the isotopic record shows that the sea level remained below -50 m throughout this period. Hence, it seems inevitable to conclude that the interpretation of one or both of the data series has been pushed too far.

As pointed out in Section 5.3.2.3, estimation of past sea levels is hampered by the lack of reliable reference markers, because the continents are floating on a semi-elastic earth. This difficulty is evident, if the above mentioned + 8 m Eemian sea level estimate at Barbados is compared with - 8 m in the Netherlands, which are recognized as a subsiding area however. The example also poses the question if the tectonic correction made for the Barbados and new Guinea estimates could be wrong.

No help is to be found in the pollen record and the Greenland isotope record (Fig. 5.5(a) and (b)). Both of them are climatic rather than ice volume indicators. The pollen, and probably also the Greenland isotope record, suggest short lasting, and drastic cooling episodes close to 110 and 90 ka B.P., with minima of full glacial severity. But, if the time scales are even approximately correct, these episodes did not last nearly long enough to allow the substantial build-up of continental ice recorded in both the δ_c and sea level records. Independent dating of the pollen and ice records is thus urgently needed.*

Moreover, the pollen and the Greenland isotope records suggest warm conditions in stages 5c and 6a (again: if the interpretation is correct), which is difficult to reconcile with the great amount of continental ice in the same stages suggested by the two lower curves in Fig. 5.5.**

5.3.5 *Summary and conclusion*

The development and particularly the termination of the Eemian Interglacial are important, because it may serve as a model show-

* Recent papers by Woillard and Mook (1982) and by Hollin (1980) confirm the synchroneity between continental and oceanic climatic shifts and present some evidence on the abruptness of the vegetation shift during the cooling episodes.

** The new dating of the Camp Century record (Dansgaard et al. 1982) suggests that the short-lasting δ drop shown at 90 ka B.P. in Fig. 5.5(a) may be 30 ka older. If correct, it marks the termination of the Eemian, and the succeeding intermadiate δ_i values may reflect the period of warm North Atlantic surface water that allowed the build-up of large ice sheets.

ing how the present warm period might end in case of no anthropogenic impact.

The most important methods for studying the Eemian (isotopic and faunal analyses of deep sea cores, pollen in continental sediments, sea level and coast line variations, isotopic composition of ice cores) have been outlined and critically evaluated.

In spite of the scatter of the available data and the difficulties of interpretation, it seems justified to conclude the following about the Last Interglacial (Eemian, or Sangamon) and its termination:

(a) the duration was no more than 10-15 ka;

(b) it was warmer than the Holocene;

(c) the sea level was 5-10 m above the present, corresponding to 2-4 x $10^6 km^3$ less continental ice than today;

(d) approximately 120 ka ago the climate deteriorated drastically; in 5-10 ka the sea level dropped considerably, perhaps as much as 70 m, corresponding to deposition of 26 x $10^6 km^3$ of continental ice;

(e) in the new stage, the sea level was some 65 - 90 m below the present, corresponding to twice the present continental ice volume.

However, major details of the benthic isotopic record cannot be reconciled with the past sea level variations estimated from geological evidence, and many aspects of the Interglacial to Glacial transition are still unclear.

5.4 The future degree of glaciation

5.4.1. Introduction

It is now generally agreed that a smooth development of the human society is dependent on climatic stability. Any substantial deviation from the present climatic conditions is tantamount to an economic and sociological upheaval. This is why increasing efforts are being devoted to climatic studies, but we are still far from reaching their main objective which is a physically well founded prediction of the future trend of climatic development - we do not even know, if climate is predictable at all.

Under these circumstances it is useful to look into the past to learn what kind of variability, including extremes, has been imposed on the climate by Nature herself. It is absolutely possible that such studies may lead to statistically significant probabilities for any given change to happen within, say, the next hundred years, still assuming Nature to be left alone. Unfortunately, this assumption does not hold, in so far as the increasing pollution of the atmosphere with carbondioxide, etc. disturbs the heat balance (see Section 4.2). Estimates of the consequent climatic warming range from negligible to catastrophic. However, this uncertainty detracts nothing from the importance of palaeo-climatic studies, because the natural and the anthropogenic effects may be considerd additives, at least to a first approximation.

In this chapter, we shall discuss the risk for a drastic climatic change in the next few hundred years and the consequent change of the cryosphere, i.e. land and sea ice. The latter point is important, because one cannot exclude the possibility that a considerable change of the extent of the cryosphere may trigger long-term, irreversible changes, whether it be an increase (glaciation) or a decrease (melting away of the Greenland Ice Sheet, for example). In view of the above-mentioned uncertainty, it seems most rational to proceed by assuming the future climatic changes be dominated by, first, natural causes (Section 5.4.2), then anthropogenic (Section 5.4.3) impact.

5.4.2 *Natural changes*

The climate, and thereby the extent of the cryosphere, has been changing at all times in the past, and on time scales from a few years and up to many millions of years. The greatest changes are known as shifts between glacial and interglacial stages, and vice versa. In the glacial stages that have prevailed in most of the past two million years, northern Europe was covered by an ice sheet several thousand meters thick (Fig. 5.9), and the greater part of the rest was governed by a tundra climate. As to the next few centuries from today, it is obvious that piling up of such huge amounts of continental ice is out of the question - it takes much longer time, of the order of ten thousand years. But, how fast and how severe was the climatic deterioration that made the growth of ice sheets possible on previous occasions? We have already seen in Section 5.3 that the last interglacial, the Eem, apparently terminated rather abruptly 120,000 years ago; for example, European pollen records suggest that the forest vegetation disappeared within a period of time that may very well have been as short as a few centuries or even less. For this reason, and also because the present interglacial has lasted approximately as long as its predecessors, it is absolutely relevant to study the glacial-interglacial sequence in the past, which will be the main topic of this section.

On the other hand, much less than a full glaciation could force a substantial change of the living conditions in Europe and other regions at mid and high latitudes. In fact, we only have to go a few hundred years back in time to find a well documented example of this in the so-called "Little Ice Age", culminating in the 17th century, cf. Chapter 2. Before we turn to the main topic, the great glaciations, let us briefly discuss the little one and its possible causes, hoping to find a hint as to the risk of a new short-lasting advance of land and sea ice.

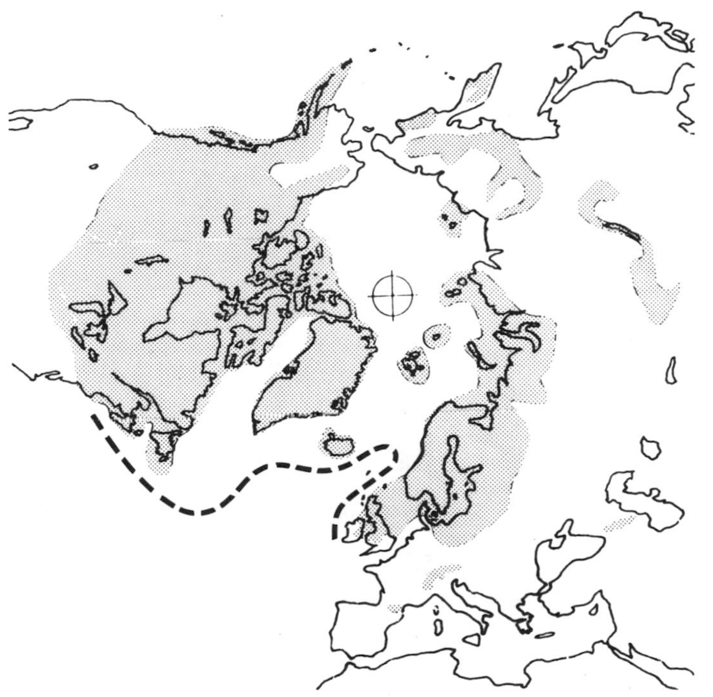

Fig. 5.9. Maximum extension (shaded areas) of continental ice during the last glaciation. South of the great ice sheet in Europe the climate was tundra-like, except for the Mediterranean countries. The heavy dashed curve shows the approximate extension of permanent pack ice in the North Atlantic at the same time, ca. 20,000 years ago. After Flint (1971).

5.4.2.1. *The Little Ice Age (see also Section 2.3)*

There is plenty of direct and indirect evidence that cooler conditions prevailed in most of the Northern Hemisphere in the centuries prior to our own century, but only few detailed and well documented records extend far enough back to allow a comparison of the Little Ice Age with the preceeding period. The "Northern Hemisphere temperature index" shown in Fig. 5.10 (from Hammer et al. 1980) is based on three records from Central England (Lamb 1968, cf. Fig. 2.5a); and Greenland (isotopic composition of ice sheet precipitation). It shows the "medieval warmth" A.D. 1000-1300, and the Little Ice Age (colder than the long-term average), A.D. 1350-1900.

In Scandinavia and in the Alps all glaciers grew considerably; and some of them continued to grow till after A.D. 1900, when the great retreat began around 1920.

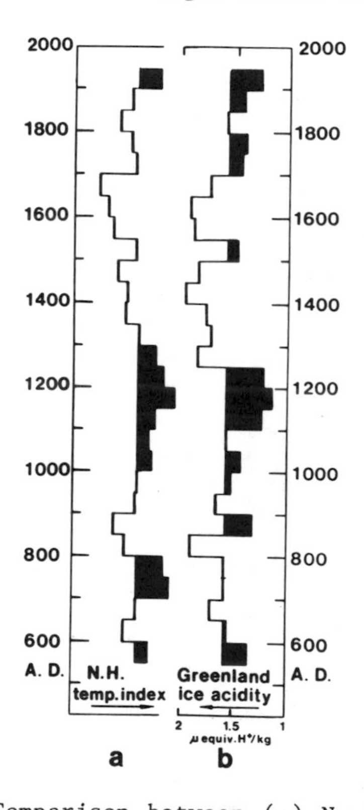

Fig. 5.10. Comparison between (a) Northern Hemisphere temperature index since A.D. 550 (shaded valued denote warmer than average conditions), and (b) the acidity of Greenland ice sheet precipitation (shaded values denote lower than average acidity in periods of less than normal volcanic activity). The correlation suggests that climatic cooling of up to several centuries duration is caused, in part, by series of violent volcanic eruptions that inject acid aerosols into the stratosphere, thus increasing the back-scattering of solar radiation. From Hammer et al. (1980).

According to Koch (1945) drift ice reached the coasts of Iceland only on the average of a few weeks per year since A.D. 1200, but after A.D. 1600 it appeared more frequently, reaching a maximum average of some 17 weeks per year in the 19th century, cf. Fig. 5.11.

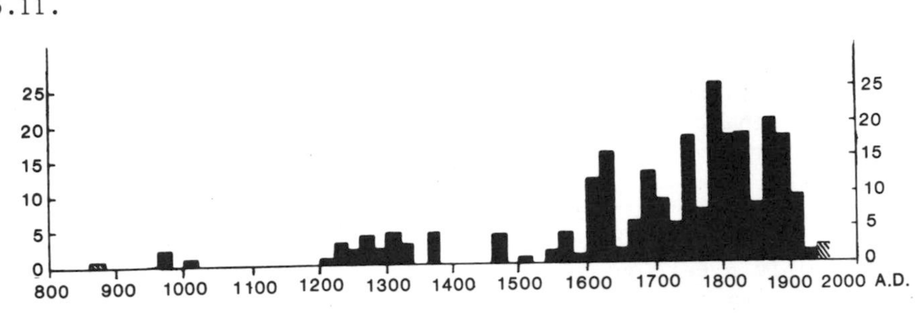

Fig. 5.11. The occurrence of pack ice on Icelandic coasts (number of weeks per year) since A.D. 800. From Koch (1945), cf. Fig. 5.10.

What would be the risk of another Little Ice Age in the centuries to come, if no anthropogenic effect were to influence the climate? Answering the question is impossible without knowing the cause(s) for the onset of the Litte Ice Age, and since the question is important there has been a great deal of speculation on the basis of most of the suggestions listed below in Section 5.4.2.2 as possible causes for the great glaciations. One of the suggestions will be considered in some detail here, not because it is new, but because it has been supported quite recently by strong evidence by Hammer et al. (1980).

Violent volcanic eruptions inject silicate micro-particles and sulphuric gases into the stratophere. The main effect of the silicates is probably to heat up the stratosphere by absorption of thermal radiation, but they disappear by settling during the first few months. At the same time a sulphate aerosol, mainly sulphuric acid, is building up from the sulphuric gases to become the dominant aerosol, the climatic effect of which seems to be cooling the lower troposphere by back-scattering of solar radiation, as previously suggested by various authors (e.g. Humphrey 1940, Lamb 1970). The volcanic fallout is stored on the Greenland ice sheet in a continuous and easily datable sequence. The ice layers deposited up to a few years after a violent eruption are therefore marked by elevated acidity, the bulk of the volcanic fallout being sulphuric acid, sometimes also chloric acid.

Figure 5.10(b) shows a continuous acidity profile measured along a 404m-long ice core from Central Greenland (71°N, 37°W), reaching back to A.D. 533, according to year-by-year stratigraphic dating. 50-yr-mean acidity values (in micro-equivalents of hydrogen ions per kg ice) are plotted on the reversed scale at the bottom, and the values *below* the overall mean are shaded to indicate half centuries of *less* then normal volcanic activity in the northern hemisphere (in fact, down to 20°S). The two records in Fig. 5.10 have a correlation coefficient of 0.52, significant at the 95% confidence level for an estimated 12 degrees of freedom.

The significant correlation, and the obvious cause-and effect mechanism behind it, makes it natural to conclude that the stratospheric load of volcanic aerosols explains a significant part (0.52^2=27%) climatic fluctuations of up to at least several hundred years duration. Thus, the generally increasing severity of the Little Ice Age throughout the four centuries A.D. 1300-1700 was probably due to generally heavy volcanic activity, i.e. long series of violent eruptions. Conversely, in the first half of our own century, there was only a few great eruptions, and none between Katmai 1912 and Agung 1963, which might be the cause for the drastic climatic warming that began around 1920 and made the 1920-1960 period the warmest one in the last six centuries or more.

And now back to the question: What is the probability of a new Little Ice Age? The above paragraph suggests that the question implies an answer to another one: What is the probability of heavy volcanic activity in the coming centuries? So far, nobody has tried to answer that one, because it depends on the future movements of the continental plates, which cannot be predicted at present.

Although *natural* short-term cryospheric changes might prove to be unpredictable, those caused by *anthropogenic* effects might very well be predictable with specific assumptions regarding the forcing function, in case the pollution, cf. Section 5.4.3.

5.4.2.2 Possible causes of long-term glaciations

Few if any, geophysical problems have been discussed as extensively as that of the causes of glaciations. Some of the numerous hypotheses are briefly listed below, and as to details reference is made to more comprehensive reviews by Mitchell (1965) and Flint (1971) and Berger (1980a). Some investigators have ascribed the fluctuations of the ice cover to internal parameters in the atmosphere-ocean-cryosphere system, e.g. periodic surging of the Antarctic ice sheet (Wilson 1964); fluctuating carbondioxide distribution between the oceans and the atmosphere (Plass 1956); oscillating circulation pattern in the oceans (Newell 1974); or spontaneous alterations of the climatic system between two different, relatively stable states (Lorenz 1968). Other investigators are in favour of external forcing functions, e.g. volcanic aerosols in the atmosphere (Humphrey 1940, Lamb 1970, Kennett and Thunell 1975, 1977, Bray 1976,

1977), which may conceivably trigger a sudden onset of a glaciation (Flohn 1974), if the aerosol load becomes considerably higher than during the Little Ice Age A.D. 1350-1900, cf. Section 5.4.2.1 and Fig. 5.10; changing solar constant (Dennison and Mansfield 1976); changing seasonal and latitudinal distribution of the insolation, due to varying Earth-orbit parameters (Milankovitch 1930, 1941); or changing geomagnetic field (Wollin et al. 1971), which may in fact also be linked to the varying Earth-orbit parameters (Wollin et al. 1978).

The true explanation may include elements of several of these hypotheses, because the fluctuating ice cover influences all of the internal characteristics of the climate system. But only the Milankovitch hypothesis offers the possibility of exactly calculating the forcing function through a long time interval. Of course, this does not make it more plausible, a priori, but it does open the door for tests by comparison with proxy data, leaving the other hypotheses tinged with speculation which does not necessarily mean that they are wrong. We shall return to the astronomical hypothesis in Section 5.4.2.4.

5.4.2.3 The last million years

In the last 25 years, a great deal of information about long-term climatic changes in the past million years and more has been collected by studies of various kinds of proxy data, obtained from marine or continental sediments, moraines, sea level terraces, etc. Dating problems and lack of continuity of the records have sometimes caused confusion, but the reasonably consistent picture is beginning to emerge. In the present context, the greatest interest is associated to the problem of the duration of interglacials in the sense of periods of climatic conditions similar to those that have prevailed in the present interglacial (Holocene).

Many investigators have argued in favour of the opinion that previous interglacials, and at least the Eem that perceeded the last glaciation, lasted only 10-15 ka (kiloannum = 1000 years), and this view is indeed supported by a great variety of evidence, e.g. sea level changes; pollen records; loess varves; and, in particular, deep-sea sediment records, cf. Section 5.3. It has further been argued that, since the Holocene has now lasted 10 ka, the world may be faced with a imminent glaciation. This has caused concern, which would hardly be justified, if the prediction were based only upon simple considerations of analogy. However, there is more to it than that, as will be outlined in Sections 5.4.2.4 and 5.4.2.5.

But first, let us discuss the upper part of Shackleton and Opdyke's (1973) famous oxygen-isotope (δ_c) record in Fig. 5.12 (c), from deposits in the Equatorial Pacific. The time scale is established by (i) interpreting a magnetic reversal found at a depth of 12 m as being identical with the Brunhes-Matuyama boundary that has been potassium-argon dated at 700 ka by Dalrymple (1972), and (ii) assuming a constant sedimentation rate of 1.7 cm per ka over the entire core length. The figures above the curve refer to Emiliani's (1966) numbering of the glacial/interglacial stages. The δ_c curve in Fig. 5.12(b) is combined of two from nearly subantarctic Indian Ocean cores (Hays et al. 1976).

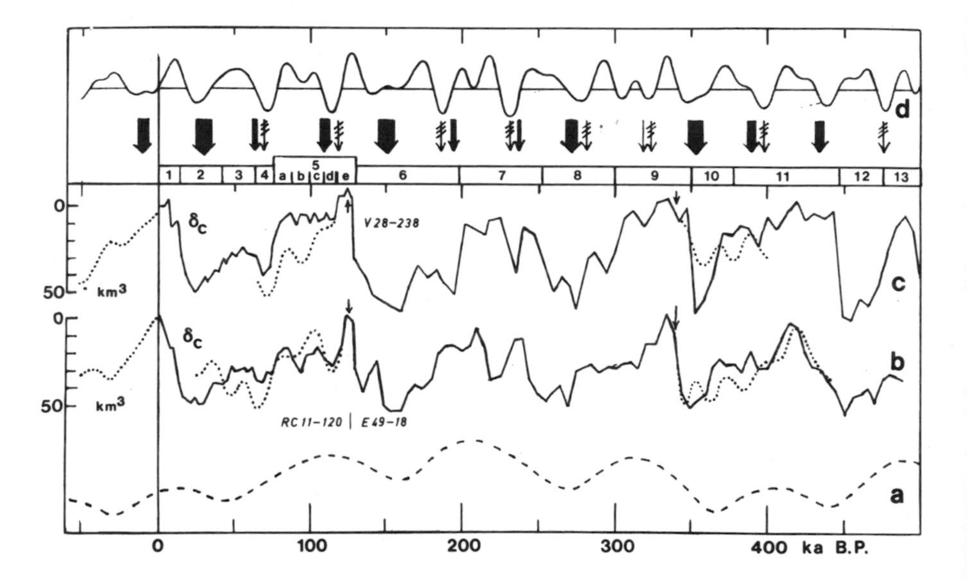

Fig. 5.12 (a) Variations in the eccentricity of the Earth's orbit from 500 ka B.P. (kiloannum = 1000 years before present) to 50 ka A.P. (after present). From Berger (1978).
(b) Full curve: The oxygen-isotopic composition δ_c of planktonic foraminifera found in two deep-sea sediment cores from subantarctic Indian Ocean (from Hays et al. 1976). If the δ_c variations are assumed to reveal fluctuations in the amount of continental ice, the scale to the left indicates excess ice relative to the present. Dotted curves: Checks on the predictability and, to the outer left, prediction of the δ_c curve by the maximum entropy filter technique, cf. Section 5.4.2.5. (c) Same as (b), but related to the upper part of a deep-sea sediment core from the equatorial Pacific Ocean (Shackleton and Opdyke, 1973). The figures above the curve refer to Emiliani's (1966) numbering of past glacial/interglacial stages. (d) Summer insolation at 65°N in excess of the present (from Berger, 1978). The arrows indicate periods of favourable conditions for glacier advance, see Section 5.4.2.5.

As explained in Section 5.3.2.1, interpretation of δ_c data in terms of continental ice volume in excess of the present 31 x 10^6 km^3 is most straightforward when the δ_c data are measured on *benthic* foraminifera. But such records are not long enough to allow studies of several glaciation cycles. This is only possible on *planktonic* foraminifera δ_c records that are influenced by more sources of error, cf. Section 5.3.2.1. If, nevertheless, we interprete the δ_c records in Fig. 5.12 (b) and (c) as above (the scale to the left is calibrated by assuming a 50 x 10^6 km^3 excess of continental ice 20 ka ago), the great glaciations seem to have occurred at intervals of approximately 100 ka. But the interglacials in between did not have the same duration, they did not elapse the same way, nor did they reach the same degree of deglaciation, and none of them came close to full deglaciation. Only on three occasions in the last 8-900 ka, in stages 5e, 9 and 11, did the deglaciation reach the level of the warmest period of the present interglacial (Holocene, about 6 ka ago, cf. Section 5.2), and these events lasted only some 10 ka, i.e. the same span of time that has elapsed since the termination of the last glaciation.

5.4.2.4 The astronomical hypothesis

The obvious regularity of the deep-sea sediment records has been successfully correlated with secular changes in the Earth's orbit parameters to a degree that leaves only little doubt about some causal, yet not completely clarified link between them.

The attraction forces from the other planets cause secular changes of the Earth's orbit. The idea of the repeated glaciations being a result of these changes was first put forward by Croll (1875) and later formulated as a scientific hypothesis by Milankovitch (1930, 1941). Most recently, the orbit parameter and the consequent insolation changes have been calculated exactly by Vernekar (1972) and Berger (1976, 1978). The latter author included the higher-order terms necessary when going more than 500 ka back in time.

The three relevant orbit parameters are as follows (cf. Berger 1980b):

(1) The eccentricity, e, of the Earth's elliptic orbit varies with a mean period of 96 ka. Since the annual mean Earth-Sun distance is proportional to $(1 - e^2)^{1/4}$, the total annual insolation reaching the atmosphere is, to a first approximation, proportional to $(1 - e^2)^{-1/2}$, which can vary between 1 and 1.0027, i.e. by 0.27%. In the last 500 ka, however, it has only varied 0.1%, in the last 150 ka only 0.04%, which makes it difficult to accept eccentricity changes as a direct cause of the glaciations, in spite of the nearly perfect fit of the e-curve (Fig. 5.12(a)) to the long-term δ_c oscillations (Fig. 5.12(c)), when account is taken of a + 20 ka uncertainty on the 700 ka dating of the magnetic reversal.

(2) The obliquity, ε, i.e. the inclination of the equator upon the ecliptic (or the angle between the Earth's axis and the normal to the ecliptic) oscillates with a mean periodicity of 40.6 ka. At high latitudes, a high ε-value results in a high altitude of the sun at summer solstice, low at winter solstice. This situation favours summer melting on high latitude glaciers and thereby glacier retreat, all other factors being unchanged. And vice versa, a low ε-value favours glacier advance.
(3) the precession of the Earth's acis, i.e. its describing a cone relative to the fixed stars, takes 21 ka as an average. During that period the longitude, ω, of the perihelion (Earth closest to the Sun) relative to the present veral equinox, increases by 360°. For $\omega = 90°$, Northern Hemisphere summer solstice occurs when the Earth is in aphelion, corresponding to minimum angular velocity according to Kepler's second law, which means that the difference, ΔT, between the Northern Hemisphere summer and winter "half years" is maximum for a given value of e. At any time,

$$\Delta T = T_s - T_w = 1.273 \, T \, e \sin \omega,$$

T being the duration of the tropical year. At present $e \simeq 0.017$, $\sin \omega = 1$, and therefore $\Delta T = 8$ days, but in the period 130 - 90 ka B.P., when e was $\simeq 0.04$, ΔT oscillated between + 18.6 days; and the theoretical extremes are + 33.9 days. At present, the Southern Hemisphere is favoured because the perihelion occurs in early January; at that time of the year the extraterrestrial solar radiation is 7% higher than at the aphelion (early July; see Section 5.2 and below).

In Fig. 5.13 (from Berger, 1978), spanning the time from 200 ka B.P. to 50 ka A.P. (after present), the upper part shows the variations of the eccentricity e (dashed curve), the obliquity ε (full curve), and the deviation of the eccentricity-precession term $e \sin \omega$ (dashed-dotted curve). The lower part shows the deviations of the solar radiation intensity from the A.D. 1950 values for the Northern Hemisphere summer half-year at 80°N (full curve), 65°N (dotted curve), and 10°N (dashed-dotted curve). Obviously, the high latitude insolation curve follows the 41 ka quasi-periodicity of the obliquity, whereas the low latitude curve follows the 21 ka quasi-periodicity of the $e \sin \omega$ term in antiphase.

The 65°N summer insolation is particularly interesting, because the 60 to 75°N latitude belt contains the most important areas for deposition of huge amounts of continental ice. In Fig. 5.12 the curve (d) is shown for comparison with the δ_c profile along the Equatorial Pacific V28-238 core (c). Cross-spectral analysis of the last 400 ka of the curves in a frequency band corresponding to periods between 29 and 48 ka gives a correlation coefficient of 0.63, significant at the 98% confidence level, and a time lag close to zero. The lack of correlation in the lower part of the core may be due to a varying deposition rate prior to 400 ka B.P.

Further evidence for the validity of the astronomical hypothesis was given by Hays et al. (1976) in the form of spectral analysis of sedimentary records along two deep-sea cores (RC11-120 and E49-18) from the subantarctic Indian Ocean.

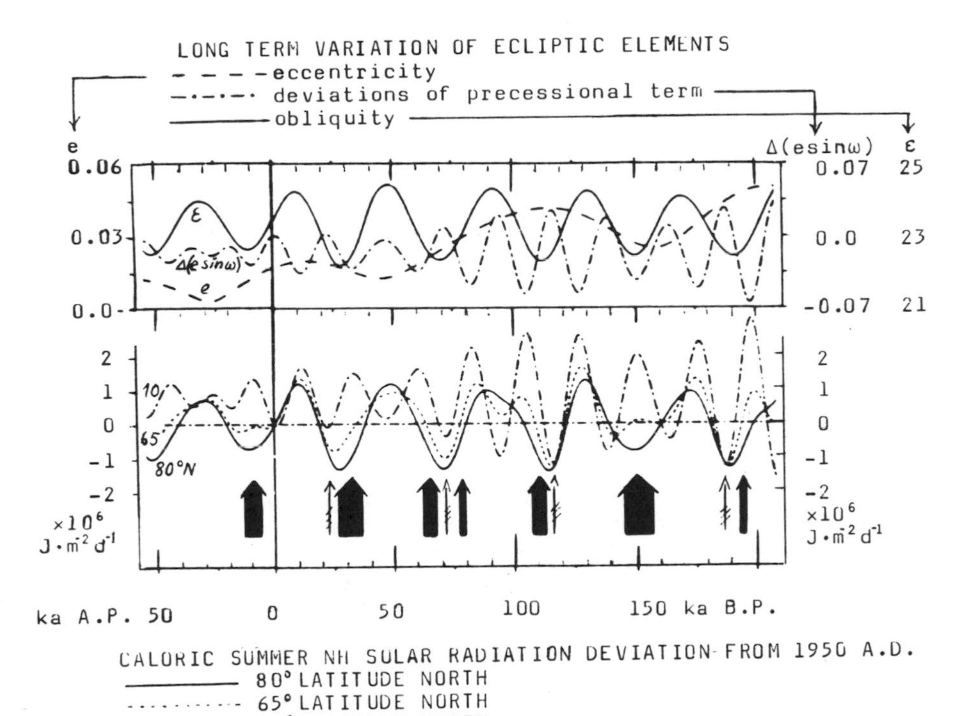

Fig. 5.13. Upper section: Earth-orbit element variations from 200 ka B.P. to 50 ka A.P. Lower section: Excess insolation fluctuations at three northern latitudes. The arrows indicate favourable conditions for glacier advance. From Berger (1978).

Taken together, they reach from the present back to late stage 13, i.e. back to 480 ka B.P. (Fig. 5.12(b)). The spectra show a dominant 106 ka peak, which may be associated with the 96 ka eccentricity peak; a minor, yet significant peak at 42 ka, corresponding to the 40.6 ka obliquity oscillation; and a split peak at 23.1 and 18.8 ka, corresponding to the 21 ka precession period.

As mentioned by Hays et al. (1976), one has to assume a non-linear response of the atmosphere-ocean-cryosphere-continent system (by internal feed-back mechanisms?), if the dominating 106 ka climatic cycle is to be ascribed to the less than 0.1% solar radiation changes due to eccentricity changes in the last 500 ka. but although an additional 101 ka contribution can be expected in case of non-linear interaction between 23.1 ka and 18.8 ka cycles (Wigley 1976), the long-term input to output signal ratio is still very low.

Most energy-balance modellers doubt whether the Milankovitch radiation changes are large enough to have produced the large fluctuations in the dimensions of the continental ice sheets; cf. Budyko (1969); and Oerlemans and v.d. Dool (1978). On the other hand, model calculations based on glacier mechanics considerations (Weertman 1976) indicate that they are large enough, if one makes the not impossible assumption that the accumulation and ablation rates, at least on the southern half of the American and Scandinavian ice sheets, were considerably higher than in Greenland and Antarctica today, and in fact double as high as the precipitation in north-east Canada today. Under such circumstances, it would take some 15-30 ka to build up the full sized continental ice sheets, and they could be extinguished during a 3 times shorter time interval (Weertman 1964), which is compatible with the rates of change in excess ice suggested by the δ_c curves in Fig. 5.12. Weertman (1976) points out that the compability of the Milankovitch hypothesis to glacier mechanics does not prove that the hypothesis is correct. Nor is it possible to use the glacier model to predict the future degree of glaciation, because the model is extremely sensitive to small variations in the accumulation to ablation rate ratio.

Mason (1976) reaches the same main conclusion from a meteorological point of view. He points out that the Milankovitch effect induces fluctuations in the annual incident solar radiation, e.g. from -2% at 25 ka to +1% at 10 ka B.P. at 65°N, which are much larger than any observed secular change in the "solar constant", and that the *seasonal* components fluctuate even more e.g. +4% in the summer half year at all northern latitudes 10 ka ago. The summer surface temperature at that time was estimated by Milankovitch to exceed present values at 65°N by some 4°C, whereas the annual temperature was less than a degree higher than today. Atmospheric model calculations suggest that due to the 2% radiation deficiency at 25 ka B.P. the annual temperatures should have been 1.4°C (Shaw and Donn 1968), or maybe 5°C (Manabe and Wetherald 1975) lower than today, the critical summer temperatures considerably lower.

Further, Mason (1976) calculates that the total radiation deficiency north of 45°N in times of ice sheet build-up is sufficient to compensate for the energy liberated in the atmosphere during the formation of snow, and that the radiation excess in times of ice sheet shrinkage is enough to supply the required heat for melting. However, he adds: "Because of the losses and redistribution experienced by the solar radiation after entering the atmosphere and the effects of all the feedback processes, the formation and melting of the ice can hardly be explained solely in terms of long-term fluctuations in the radiation incident on top of the atmosphere. This will be a task for realistic numerical models of the atmosphere-ocean-cryosphere system incorporating seasonal effects, variable cloud cover and albedo, a fully interactive radiation scheme and interactions between the atmosphere and the oceans which will involve computation of transport of heat and salt by ocean currents".

Full acceptance of the Milankovitch hypothesis as a well-founded theory implies, among other things, a clarification of *how* the small 100 ka imput signals may be transformed into fulminating output signals in the form of alternating glacial-interglacial stages. Such clarification requires however a much deeper understanding of the interactions between the various parts of the climatic system than we have today.

If, after all, the Milankovitch effect in terms of insolation changes proves to be too small to be held responsible for the great glaciation fluctuations, we would have to look for another causal link to explain the correlation between the astronomical forcing functions and the glaciation output. In the light of Section 5.4.1.2, indicating volcanic activity as a (the ?) cause for the Little Ice Age a few hundred years ago, one might speculate whether the planetary attraction forces interacting with those of the Sun and the Moon could, e.g., create tidal force fluctuations high enough to affect significantly the movements of the continental plates and thereby the volcanic activity.

5.4.2.5 Future trends

As long as the atmosphere and ocean circulation patterns and their responses to internal and external forcing functions are poorly understood, any attempt to predict the future climatic development will be tinged with speculation, even if the external forcing function were well known, as is the case from a Milankovitch point of view.

The best one can do for the moment is to check:

(a) if the long proxy data time series are, in themselves, predictable by standard statistical methods; and,

(b) if the future Milankovitch radiation distribution has enough precedents in the past to allow, by simple analogy considerations, an extension into the future of, e.g., the δ_c series.

(a) The maximum entropy prediction filter technique (Ulrych and Bishop 1975) has been applied on the last 400 ka of the two δ_c series shown in Fig. 5.12. Checks on the predictability of the series were performed by, first, smoothing the series by a 15 ka low pass digital filter; second, cutting off the last 125 ka of the data; third, running a 80-100 ka prediction filter through the remaining 275 ka of the series and beyond the 125 ka B.P. limit, thus "predicting" (dotted curves in Fig. 5.12(b) and (c)) the already known δ_c data after 125 ka B.P. Obviously, the correlation between the dotted and full drawn curves is not impressive, although the rapid decrease in δ_c around 120 ka B.P. as well as the minimum at approximately 70 ka are revealed in both cases. If, instead, all data prior to 340 ka B.P. are cut off, a similar procedure leads to the "predictions" backward in time shown by the dotted curves to the right in Fig. 5.12(b) and (c), of which the latter is absolutely poor, whereas most of stages 10 and 11 is fairly well extrapolated in Fig. 5.12(b), including the sudden drop at the stage 9/10 boundary and the shape of the δ_c maximum in stage 11.

Although maximum entropy predictions reveal the most probable development from any given point of time to judge from previous events, the δ_c series and, in particular, the V28-238 series, do not vary regularly enough to allow a δ_c prediction into the future with high confidence. It is therefore mainly of academic interest that application of the technique over the entire 400 ka interval leads to the dotted curves to the outer left in Fig. 5.12(b) and (c), both suggesting a relatively slow decrease in δ_c in the next 25 ka. On the other hand, it is remarkable that a very similar prediction has been obtained by Berger et al. (1980) using an autoregressive insolation model, which reproduces 87% of the δ_c variation in the last 500 ka.

(b) It appears from Fig. 5.13 that the summer insolation at 65°N will keep close to the present value in the next 20 ka, but it will increase at low latitudes and decrease at 80°N. The consequent radiative warming at low latitudes and cooling at high latitudes would seem to be in favour of an advance of polar glaciers. From 6 to 13 ka A.P. the difference between the summer insolation at 10°N and 65°N will be more than 10^6 J m^{-2} day^{-1} (equivalent to about 12 W m^{-2}) higher than today.

Let us go backward in time and select from Berger's insolation curves (part of which are shown in Fig. 5.13) the periods of time characterized by similar or more favourable conditions for glacier advance, i.e. 65°N summer insolation close to or lower than today, and at least 10^6 J m^{-2} day^{-1} difference between the 10°N and 65°N summer insolation curves. These periods are indicated by arrows in Fig. 5.13 back to 200 ka B.P., and in Fig. 5.12 back to 500 ka. Obviously, in the last 400 ka the above-mentioned situation occurred 10 times, and in all cases they were associated with significant δ_c minima. Only one δ_c minimum, 306-294 ka B.P., fell in a period of different radiation distribution (from 307 to 303 ka B.P. the summer radiation was lower than today at all norhern latitudes, cf. the crossed arrow).

We may conclude that 6-13 ka from now the Northern Hemisphere summer insolation will be distributed in a way that suggests a decreasing δ_C in the future, i.e. a glacier advance according to the widely accepted interpretation. In passing, we notice that this agrees with maximum entropy prediction. Another recent prediction (Kukla and Berger 1979) based on the future development of an empirically designed astronomic climate index estimates a deep δ_C minimum to occur already between 3 and 6 ka B.P. But neither the exact time of the onset, nor the ultimate extent of the future glaciation can be assessed from the available data - and, of course, possible anthropogenic effects in the future have been neglected.

If, in spite of these limitations, we go a step further and try to assess the future *climate* from the predicted trends in δ_C , we inevitably end up in speculation. It can only be established that, if the suggested build-up of several million km^3 of continental ice within the next some 5-10 ka should become a reality, it implies a climatic deterioration in North America, Europe and northern Asia that would severely aggravate the living conditions in many of the presently most densely populated areas.

It is not yet clarified, if a climatic transition can happen as abruptly as suggested by some proxy data (cf. for example Müller 1974, Woillard 1979), in spite of the more than 1 ka time constant of the oceans. But it cannot be excluded that under very peculiar conditions caused by, for example, intense volvanism (cf. Section 5.4.2.1), a thin permanent snow layer might be established over an area of the order of 1-2 million km^2 in sub-Arctic (Flohn 1974).

This would immediately change the general circulation pattern in the atmosphere and, thereby, the climate on a large scale. The other extreme would be a slow, maybe accelerating cooling in continuation of the general decrease in temperature averaging some 0.05°C per century that has characterized the climate since the Holocene optimum some 5 ka ago. No single generation would actually feel such long-term cooling, because it would be masked by much higher medium-term variations like, e.g., the "medieval warmth" and the Little Ice Age (cf. Fig. 5.10), and these variations themselves would be masked by short-term changes like the cool A.D. 1900-1930 and the warm A.D. 1930-1960 period.

The latter mode of gradual transition would of course involve the best perspective for the human society to adjust to the changing environments. Even in case of an abrupt deterioration, it might still be feasible to take effective measures to combat a widespread, growing ice sheet (Flohn 1978). But from a Milankovitch point of view it would be a fight against a symptom rather than the disease: the climatic conditions for a new glaciation would still remain for thousands of years.

5.4.3 *Anthropogenic impacts*

The increasing impact of human activities upon the atmosphere and the hydrosphere has caused serious concern, cf. Chapter 4. Many investigators consider the greenhouse effect of CO_2, H_2O, NO_x and other gases as the most imminent climatic threat to society as we know it today, in particular as regards the next few centuries. Model calculations of the ultimate consequence of the greenhouse effect are still immature, because important secondary effects, such as cloud cover changes due to atmospheric warming, cannot be adequately accounted for. This problem is linked to that of the predictability of climate, which perhaps is unsolvable in principle. However, the presently most advanced model calculations (cf. Section 4.2) predict that a 100% increase in CO_2 may raise the entire lower atmosphere temperature by 3°C on the average, and by 10°C at high latitudes. To use the considerable uncertainties involved as an argument for neglecting such results looks like high stake gambling, because the time constant of the main CO_2 sink (the oceans) is so high that it may take several centuries to re-establish the "natural" atmospheric CO_2-level, if possible at all (Oeschger et al. 1975). Nor should we console ourselves with the hope that the greenhouse effect may counterbalance the future cooling predicted by the Milankovitch hypothesis, if only for the reason that we are probably dealing with quite different time scales: a few centuries for the greenhouse effect, in contrast to several millennia for the Milankovitch effect.

When trying to assess the glaciological consequences of a possible 5-10°C warming of the polar regions, it is evident that the amount of sea-ice would decrease considerably, and perhaps leave an ice-free Arctic Ocean, which would, in itself, change the climatic boundary conditions (cf. Section 5.5). However, the reaction of the great ice sheets in Antarctica and Greenland is difficult to predict, due to insufficient knowledge about the dynamics and the boundary conditions of the ice sheets The existing three-dimensional ice sheet models (Budd et al. 1971, Jenssen 1977) are too simplified, e.g. they cannot even incorporate a general ice flow law, because the strain to stress relationship is extremely complicated. Hence, we are faced with more or less the same situation as in the case of climatic predictions, that any consideration is tinged with some speculation. Of course, it is possible to look into the past for geological or glaciological evidence of the consequences of precedents to the conditions expected in the future, but we may have to go so far back in time (early Tertiary?) to find 10°C higher temperatures in polar areas that the evidence for the reaction of possible former ice sheets is flimsy, at best.

TABLE 5.4. Volume of glacier ice

	Volume $\times 10^6$ km^3	Sea-level rise in case of melting m
East Antarctica	23	60
West Antarctica	7	5
Greenland	2.7	7
Other glaciers	0.24	0.6

In the following paragraphs, some thoughts on ice sheet stability will be briefly presented as a step towards estimating the impact of a climatic warming on the existing continental ice. First, however, it may be useful to consider the present magnitudes of the continental ice fields in terms of volume and equivalent sea level rise in case of melting (Table 5.4). The relatively small 5 m rise (Mercer 1968) corresponding to a total melting of the West Antarctic ice sheet is due to the fact that most of this ice sheet is grounded far below the present sea level, thus partly displacing ocean water. Obviously, the total amount of ice in "other glaciers" is an order of magnitude too small to account for the 5-10 m higher sea level during stage 5e some 125 ka ago, as estimated in Section 5.3. Consequently, the bulk of the equivalent 2-4 x 10^6 km^3 of ice must be ascribed to shrinkage of one or more of the three great ice sheets. Mercer (1968) takes the 5 m equivalent sea level change for West Antarctica as an indication of total disappearance of this marine ice sheet during stage 5e, the two other ice sheets being land based and therefore more stable. However, it appears from Table 5.4 that total disappearance of "other glaciers" in addition to a 20% shrinkage of the West Antarctic and Greenland ice sheets would be enough to account for a 5 m sea level rise, and it remains to be proven by adequate ice dynamic modelling that a 20% shrinkage would result in complete extinction of West Antarctica.

The East Antarctic ice sheet has existed for many millions of years (Frakes 1978), and it is generally considered extremely resistant to climatic changes. On the other hand, the West Antarctic and the Greenland ice sheets may be vulnerable to a drastic climatic warming and might shrink in three different ways:

(a) by increasing rate and area of ablation;

(b) by changing the ice flow parameters, i.e. making the ice less "viscous"; and;

(c) by changing the boundary conditions at the terminus and/or the ice-rock interface.

(a) There are some indications that the West Antarctic ice sheet was considerable larger during the last glaciation (Denton et al. 1971), and that it is still shrinking today (Whillans 1977). The present balance is poorly known, but, if negative, it cannot be due to surface ablation, because melting gives only an insignificant contribution ($\sim$ 0.5%) to the drainage of Antarctica. The average midsummer air temperature is well below the melting point along the edge, e.g. -5°C at sea level along the fronts of the Ross and Filchner-Ronne ice shelves (Tolstikow 1966). Consequently, it is questionable if even a 5-10°C warming would be able to amplify the surface melting and run-off by a factor of 30 needed to make it comparable with the present amount of meltwater running off from the Greenland ice sheet. Thus, it seems unlikely that West Antarctica could ever melt away directly.

In Greenland, on the other hand, surface melting in the marginal zones of ablation plays an important role in the mass budget. The run-off of meltwater amounts to more than 50% of the total balance, or some 270 km^3/yr. Nevertheless, the ice sheet is resistant to long lasting periods of warmer conditions than today, as shown by its survival throughout the Holocene climatic optimum. Furthermore there is clear evidence in the lowest part of the Camp Century δ_i record (unpublished) that at least North Greenland was covered by ice throughout stage 5e around 125 ka B.P. the problem is whether the ice sheet would be able to adjust to a 5-10°C climatic warming by retreating to a new steady state size.

Figure 5.14a shows how a valley glacier reacts to warming. Point 1 marks the position of the firn line separating the accumulation and ablation areas on the (full drawn) surface. At the onset of a warmer period, the firn line will be lifted to 2, thus diminishing the accumulation area and enlarging the ablation area. The glacier will therefore shrink, and the terminus will retreat uphill, until a new steady state is reached (dashed curve) with the new firn line 3 separating accumulation and ablation areas at the same ratio as prior to the warming, if all other factors, such as precipitation, cloudiness etc., remain unchanged.

The resistance of the glacier to warming thus depends on its capacity to retreat uphill to a higher mean surface elevation. A steady state ice sheet resting on a horizontal bedrock does not have this possibility. If, in Fig. 5.14b, the firn line is lifted from 1 to 2, shrinkage cannot re-establish the equilibrium between accumulation and the ablation, and the ice sheet will disappear. This is what happened to the North American and the Scandinavian ice sheets at the end of the glaciation.

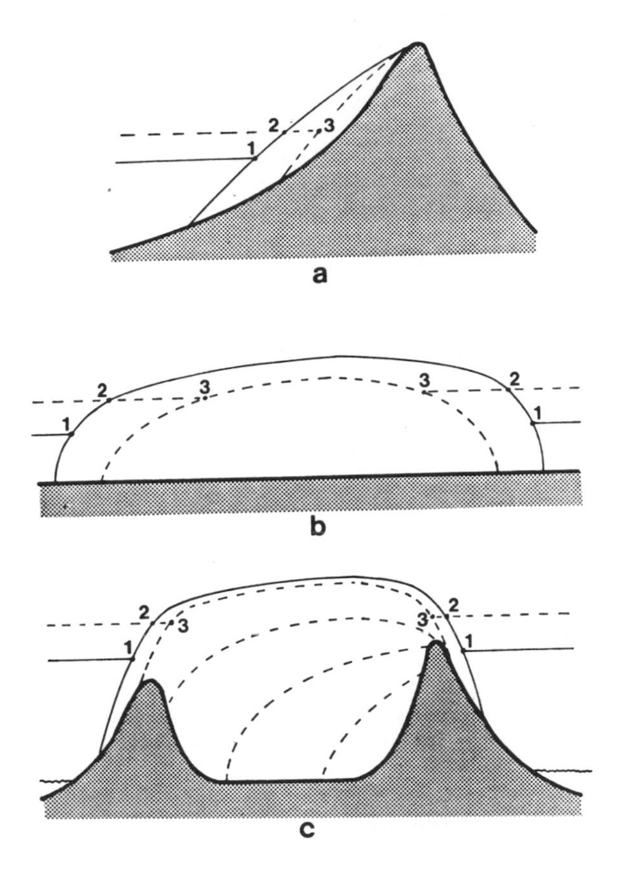

Fig. 5.14. (a) Valley glacier with firn line perpendicular to the paper at point 1 separating the accumulation zone above from the ablation zone below it. In case of higher mean summer temperature, point 1 will be lifted to 2 causing imbalance until the glacier has retreated by shrinkage to a new steady state (dashed) situation at higher mean elevation and with the firn line at point 3. (b) Steady state ice sheet resting on a horizontal bedrock. Lifting of the firn line and the consequent decrease of the accumulation to ablation area ratio cannot be compensated by shrinkage. The ice sheet will disintegrate. (c) East-West cross-section of the ice sheet in Mid-Greenland. Lifting of the firn line can be compensated by retreat of the edges uphill, until the western edge passes the subglacial mountain ridge.

The reason why the Greenland ice cap survived is that it was able to withdraw most of the edge from the low coastland to mountainous areas. In the simplified Fig. 5.14c, the full drawn curve shows the present surface in an East-West cross section of the ice cap in Mid-Greenland, resting on a bedrock that reaches elevations above 1000 m in the eastern part and above 750 m in the western part (with considerable local variations), whereas the central part of the bedrock is close to sea level. The above considerations suggest that the resistance of the ice cap against warming is limited to the point, when the western edge passes the top of the western subglacial ridge. Further shrinkage would apparently accelerate the disintegration until extinction, assuming an unchanged rate fo accumulation.

On the critical western slope, the elevation of the firn line averages 1600 m south of 71°N, and 1000 m at higher latitudes (Weidick 1976). The ablation area is less than 15% of the total, cf. hatched area in Fig. 5.15. But, if a doubling of the CO_2 content in the atmosphere should really cause an increase in summer temperature of 5°C in the South and up to 8°C in the North, corresponding to some 650 and 1000 m lifting of the firn line with unchanged precipitation, the ablation area would grow by a factor of 3, cf. the dotted area in Fig. 5.15, and moreover, the ablation in the low altitude areas would intensify. Hence, we might expect a total ablation of at least 3 times 270 km^3/yr, or 1.6 times the total accumulation today. This would not be an immediate threat to the existence of the Greenland ice sheet, because a negative mass balance of 500-600 km^3 per year corresponds to an annual shrinkage of only 0.02%, which in turn would cause a sea level rise of only 1.4 mm per year, and if the warming is associated with higher precipitation, these figures become even smaller.

Nevertheless, one cannot completely disregard the possibility that the shrinkage process will accelerate considerably and becomes irreversible, when the western ice margin retreat to lower altitudes behind the presently sub-glacial mountain threshold.

Admittedly, the above considerations imply some speculation, but they do indicate that the stability of the Greenland ice sheet should not be taken as granted, if the greenhouse effect turns out to be as serious as predicted. Although complete extinction of the ice sheet and the consequent 7 m rise in sea level would take several thousand years, according to Weertman's (1976) calculations, a "point of no return" might be reached soon after a drastic warming. This will have to be evaluated in a more rational way by (i) further studying the relationship between ablation and summer temperature; (ii) measuring the marginal surface and bedrock topographies, and, in particular, the ice thickness above the subglacial mountain ridge; and (iii) calculating by ice dynamic modelling the response of the ice sheet to any combined changes in the temperature/precipitation pattern, including the time constants involved in a possible disintegration.

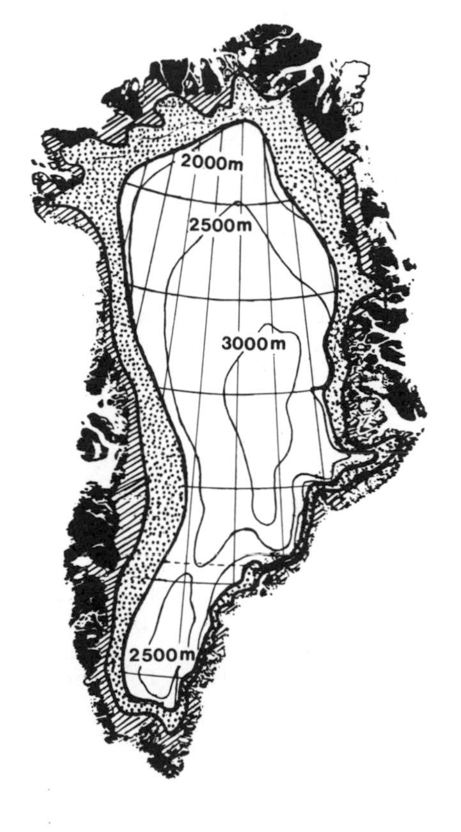

Fig. 5.15 The present ablation zones (hatched) along the margin of the Greenland ice sheet, and their expansion (dotted) in the case of a 5 - 8°C increase in summer temperature without other changes (e.g. in cloudiness and precipitation).

It is concluded that increased melting as a result of the greenhouse effect might constitute a threat to the Greenland ice sheet in the long run, but hardly to the Antarctic ice sheets.

(b) Faster creep and therefore higher discharge from the ice sheets will be the result of any warming of the ice (of the order of 10% per 1°C warming). However, depending on the accumulation rate, a temperature wave penetrates only very slowly downward from the surface. In fact, the bottom temperature at Camp Century (-13°C) is still today, 10 ka after the termination of glaciation, 10°C colder than that corresponding to equilibrium under the new ice-flow/surface-temperature/accumulation conditions established ten thousand years ago. And in the Byrd area in central West Antarctic only 20% of the glacial to interglacial temperature shift has reached a depth of 2000 m (Whillans 1977). Although this is estimated to have speeded up the ice flow by some 20%, a new warming of the same order, caused by the greenhouse effect, will need several thousand years to penetrate deep enough to further speed up the flow. We can therefore neglect the warming of the ice sheets in this context.

(c) At present, most of the East Antarctic and Greenland ice sheets are frozen to their bedrocks. If the bottom temperature rose to the pressure melting point, these ice sheets would rest on a thin film of water, which would enhance their discharge considerably. But, as mentioned above, this is an extremely remote possibility, because a shift to higher temperatures in the atmosphere would need many thousands or tens of thousand of years to reach the bottom layers.

The boundary conditions of the West Antarctic ice sheet are quite different. Most of the bedrock is at the pressure melting point, and the ice sheet is fringed by ice shelves floating on the ocean. They probably buttress the grounded ice sheet (Robin and Adie 1964), mainly where the ice shelves locally run aground on seabed shoals to form ice rises (Thomas et al. 1979) that function as separate ice sheets (Fig. 5.16). As mentioned above, melting from the surface of the ice shelves is negligible today and will hardly become important in a foreseeable future. However, at the subsurface of the ice shelves, i.e. at the ice-sea water interface, considerable amounts of sea water freeze on to the inner parts of the shelves, thus warming up the cold ice to the pressure melting point, and considerable melting takes place from the outer part of the subsurface. If the sea water warms up, the former process will slow down, and the latter will accelerate, all contributing to a retreat of the ice shelves. However, the buttressing effect will substantially diminish only when the ice fronts have passed the ice rises, as argued by Thomas et al. (1979), who also points out the negative feed-back implied in a faster outflow of inland ice that would tend to balance the disintegration of the shelves.

A full analysis of the future behavior of the West Antarctic ice sheet, including its response to a climatic warming, and the possibility of major surges call for more field data, in particular regarding the boundary conditions in very fast ice streams; better estimates of the future precipitation and temperature distributions; and more advanced ice flow models than those presently available. Until then, caution should be applied when evaluating the various attempts to sentence the West-Antarctic ice sheet to a speedy destruction by surges due to internal instability mechanisms (Hughes 1973, Thomas 1976, Weertman 1976,

Denton and Hughes 1981) or to climatic warming (Mercer 1978, Thomas et al. 1979).

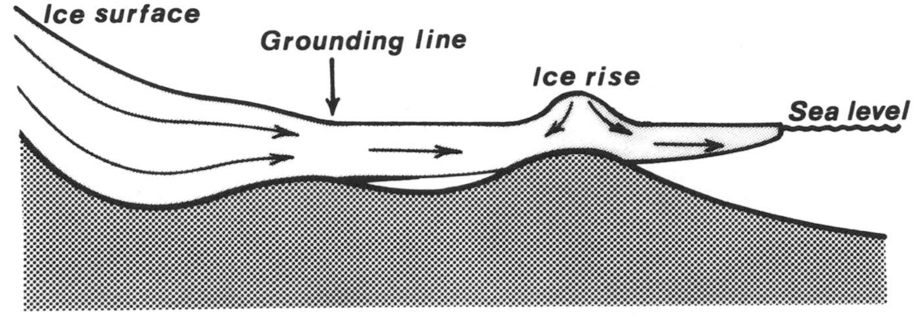

Fig. 5.16. Ice sheet (to the left) fringed by a floating ice shelf that is locally grounded on a seabed shoal and forms an ice rise that functions as a separate ice sheet.

5.4.4. *Summary and conclusions*

The future degree of glaciation will depend on natural and anthropogenic impacts on the atmosphere and the consequent climatic changes. The result of natural impacts in the past is known in the form of great glacier fluctuations, of which short-lasting advances seem to have been caused mainly by cooling due to volcanic activity, which is unpredictable at present. The long-term glacier fluctuations seem to be connected to predictable changes in the Earth's orbit parameters, and although the causal link between them is not yet completely clarified, the orbit parameters are developing so as to favour an increasing degree of glaciation in the future. The build-up of new huge continental ice sheets as known from the past would take many thousands of years, but large areas might become uninhabitable long before then, depending on the character of the implied climatic deterioration. This may develop smoothly through the next few thousand years, although the risk of a more abrupt shift cannot be dismissed.

However, man's impact may result in a climatic warming through the next few centuries, which may overcompensate the foreseeable natural cooling within the same period and perhaps for several centuries beyond it. If intense and sustained, the warming might be a threat, for different reasons, to the continued existence of the Greenland and the West Antarctic ice sheets.

It should be emphasized that the apparently opposite sign of the trends of the natural and artificially imposed climatic changes do not necessarily mean that one of them will eliminate the other one for ever, e.g. man's impact could be played out in some centuries, then leaving the climatic system dominated by natural impacts favouring glacier advance and perhaps a full glaciation as an ultimate consequence many millenia from now.

Finally, it should be stressed that the above analysis should not be considered as a prediction, but rather as a presentation of a possible future development that is serious and likely enough to demand intensified studies of climatic changes and their impacts on human society.

5.5 Ice-free Arctic and glaciated Antarctic

5.5.1 *Causes and time-scale of a possible disappearance of the Arctic drift-ice.*

The most fascinating (and also possibly the most controversial) problem of the future evolution of our climate is the possibility of a complete disappearance of the drifting ice of the Arctic Ocean. After demonstrating the sensitivity of the system, Budyko (1962) discussed the possibility of an artificial removal of the Arctic sea ice. Using a simplified model (1969, 1972) based on the heat budget of the Earth-atmosphere system, he suggested that a small increase of the solar constant or an increase of the atmospheric CO_2 content could lead rather rapidly to an ice-free Arctic Ocean, with an increase of surface temperatures of about 6-8°C in summer, 20-30°C in winter.

Since that time, arguments for and against such a drastic change have been ventilated. No complete model of the whole climatic system (including the physical and dynamical interactions and the feedback processes between atmosphere, drifting ice and ocean) has as yet been formulated. Recent model experiments by Hibler (1979) and Parkinson-Washington (1979) show that a model simulation of the non-linear interactions is possible; this is one of the most difficult but also most challenging tasks ahead. One of the first successful attemps to simulate the atmospheric circulation under the boundary conditions of an ice-free Arctic ocean has been made by Fletcher et al. (1973) based on the Mintz-Arakawa model (Fig. 5.17).

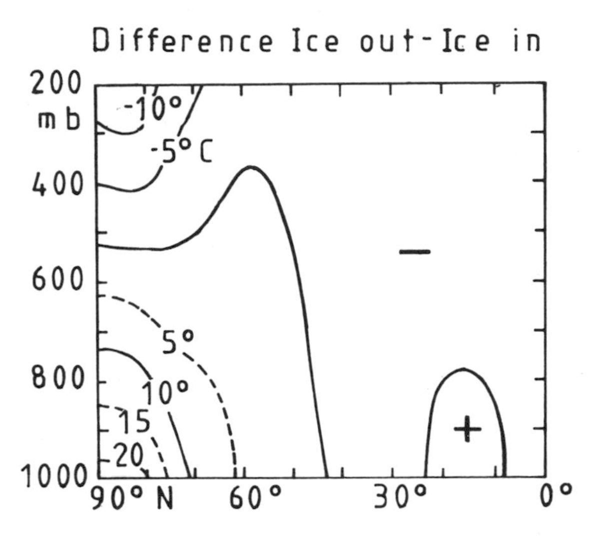

Fig. 5.17. Temperature change after introduction of an ice-free Arctic Ocean into an atmospheric circulation model (Mintz-Arakawa or RAND model, J. Fletcher et al.1973). Only changes north of latitude 60° are statistically significant; note strong destabilization in polar latitudes.

In this most recent version (1977), Budyko has again incorporated the well-known positive feedback mechanism between snow-cover, its albedo and temperature: infrared radiation cools the surface of a snow-cover and consequently the air above, the flat cold air dome expands marginally and leads to snowfall instead of rain, the snow-cover expands and more cold air is produced. This feedback mechanism works also - but somewhat more weakly - in the opposite direction: melting of the snow-cover reduces the albedo and thus the production of cold air. With this model, Budyko defines the average global air temperature difference between an ice-free regime and the present situation to be about 4°C. This is higher than the temperature increase in all other warm phases of the last two million years (Table 5.1). Fairly reliable temperature estimates from the vegetation pattern during relevant past epochs before that time (see Section 5.5.2) also suggest, at temperate latitudes of the Northern Hemisphere in the Late Tertiary, a temperature 4-6°C higher than the present value.

Under present climatic conditions, the *melting* process of the sea ice (Maykut-Untersteiner) lasts only 50-60 days (plus some 15-20 days for snow-melt); during that period, about 15-29 percent of the (average) mass of an ice-floe is melted. During the remaining 10 months, about the same amount is accreted by *freezing* from below. If one tentatively assumes, for a period of some years, a 10 percent increase of summer melting and 10 percent decrease of cold season freezing, the destruction of an ice-floe with a thickness of 2-3 m may take only a few years. If the melting season starts 1-2 weeks earlier, possibly as a consequence of advective warming, this should almost certainly lead to a substantial imbalance in the very sensitive mass budget. Here a negative feedback must be taken into account: the accretion rate during winter is negatively correlated with thickness, i.e. thin ice accretes faster than thick ice (Thorndike). In contrast to this a positive correlation exists during melting.

As demonstrated by the large fluctuation of the areal extent of the ice during the last millenium (Lamb 1977), at least some marginal (Atlantic) sections of the Arctic drift-ice can disappear rather rapidly. However, for the central core this argument is invalid: its maintenance since the Brunhes-Matuyama palaeomagnetic boundary 0.7 million years ago is verified by several independent lines of evidence; for the last 2.3 million years it is rather likely. On the other hand: if the global warming should exceed significantly the value during the warmest interglacial (see Section 5.3), this argument is no longer true. With a representative warming of 4°C or more, as during the Late Tertiary, an eventual disappearance of the central core of the drift-ice must be regarded as a necessary consequence. Assuming such a global warming of 4°C - derived from paleoclimatic evidence as well as from simplified models - as representative for an ice-free Arctic Ocean, we can estimate the equivalent level of virtual or real CO_2 content (Section 5.1). This level represents a threshold value above which the risk of such an evolution rises substantially. This risk level is given by an increase of the virtual CO_2-level (Table 5.1) by a factor near 3 (800-1150 ppm), of the real CO_2 level (taking into account the amplification of the greenhouse effect by other infrared-absorbing gases) by a factor between 2 and 2.5 (600-850 ppm). In this case, one should take into account (Budyko 1977), that with rising CO_2 content the climatic threshold should be located somewhat higher than with decreasing CO_2 content (hysteresis effect).

The atmospheric *heat budget* above a totally ice-free Arctic Ocean would be quite different from the present situation. Disregarding some earlier attemps, calculations presented by Vowinckel and Orvig shall be reproduced here as representing the most complete approach hitherto available. The heat budget equation for a column above the Arctic cap (75-90° Lat.) between the Earth's surface (Sfc) and the top of the atmosphere (TA) can be written as follows (R = net radiation flux, F = northward heat transport across 75°N by ocean (s) and atmosphere (a)):

$$R_{Sfc}\downarrow - R_{TA}\uparrow + F_a + F_s = 0.$$

Under equilibrium conditions the radiational heat loss of the atmospheric column ($R_{Sfc} - R_{TA}$) must be compensated by horizontal transports of heat both by air and sea ($F_a + F_s$). In the atmosphere, this heat transport consists of both sensible and latent heat (of water vapour). Table 5.5 gives estimates for these terms, assuming, in the case of an ice-free Arctic Ocean, either a cloudless atmosphere or a closed stratus cover. Both assumptions are unrealistic, but the role of clouds of different altitude, thickness and extension for the radiation fluxes is not yet sufficiently understood.

TABLE 5.5. Annual Radiation and Heat Budget of the Arctic cap (75-90°N) under different assumptions (Vowinckel and Orvig 1970).

			$R_{Sfc}{\downarrow} - R_{TA}{\uparrow}$	$F_a + F_s$
Present conditions	+ 3.3	+ 100.8	- 97.5	+ 97.5 W/m²
Ice-free, no clouds	29.4	59.1	- 29.7	+ 29.7 W/m²
Ice-free, closed status	42.7	92.9	- 50.2	+ 50.2 W/m²
Ice-free, seasonally varying clouds (1)	62.1	58.3	+ 3.8	- 3.8 W/m²

(1) see text

A more realistic assumption would be a nearly cloudless sky during summer (May-August), when the heated continents favour a thermal circulation with sinking motion above the cooler sea, while during the rest of the year a nearly closed cloud cover should prevail (last line of Table 5.5).
The last column $F_a + F_s$ is taken as a residual which would be difficult to split.

According to the most recent evaluation (Aagard and Croachman 1975), the actual oceanic heat flux F_s (10.6 W/m²) is greater than suggested by earlier investigations. The most interesting result of Table 5.5 (last line) is that the annual net radiation budget of an ice-free Arctic would be positive. In this case, heat would be available for export from the Arctic. This surplus of heat is easily understood from the consideration that during the *cold* season heat and water vapour would be exported in the upper branch of a thermally driven circulation from an ice-free ocean towards the sub-polar continents, which would certainly be colder. For an open Arctic Ocean a conservative estimate of R_{Sfc} is 45 W/m². Allowing 40% for evaporation during summer (including the small downward flux of sensible heat), the remaining

60 per cent (27 W/m^2 or $\sim$ 20 $kcal/cm^2$ per year) would be available for heat storage in the sea. This would yield in one single summer (of 4 months) a storage of 4 gcal/g at a 50 m mixed surface layer, which is equivalent to a warming of 4°C. Even if we assume much smaller values for the heat storage in the sea, a significant warming from year to year would result. During winter, infrared emission increases with temperature: if the surface temperature rises from -32°C to 0°C, the blackbody radiation increases by about 60%. But due to a much higher water vapour content, the atmospheric counter-radiation will also rise substantially: thus the net infrared radiation of the ocean surface is not expected to increase much. These estimates lead to the conclusion that after a limited number of years a new *equilibrium water temperature* would be reached which would be substantially higher than the melting temperature of salt water. Budyko's estimate of + 8°C appears to be not unrealistic. This value would also be consistent with the temperature estimated from Late Tertiary vegetation on the coasts of the Arctic Ocean (as the best availalble evidence).

It seems rather unlikely that to opposite hemispheric modes (permanently ice-covered versus permanently ice-free Arctic Ocean) could immediately follow each other. Since the adjacent continents - especially those north of the Polar Circle - will be in each case snow-covered during winter and develop a low-level layer of intense cold air, this air will, in a thermally driven circulation, move towards the ocean and cool the coastal waters. Thus the formation of a broad band of coastal ice during the winter months is unavoidable; the situation may resemble that now observed along the eastern coasts of Labrador or Siberia. It appears also not unlikely that during winter a new *seasonal ice-cover* may form (and melt during late spring), as happens to-day, e.g. in the Hudson Bay, in the Baltic or in the Sea of Okhotsk.

Using a recent interactive air-sea-ice model with limited spatial resolution, Manabe and Stouffer (1980) have shown that with an increase of CO_2 by a factor 4 the Arctic sea-ice disappears between July and October and reforms during early winter, but with only about half of the thickness in the standard case (Fig. 5.18). In this model the seasonal variations of average zonal temperatures are rather high; in the Arctic the warming would vary between about 1°C in July to 18°C in late autumn. In the southern hemiphere the warming would be reduced, due to the low intensity of the snow-albedo-temperature feedback above a very cold continental ice-sheet.

In a quite different model experiment, Parkinson and Kellogg (1979) have demonstrated that with a 5°C increase of surface air temperature the sea-ice similarly disappears during summer and reforms in autumn. In both cases, the situation in the Arctic might be comparable to the present subantarctic drift-ice, 85% of which is only seasonal (in the present Arctic less than 30%) with an average thickness of 1-1.5 m. Since neither model data nor paleoclimatic analogues are available, no details of a

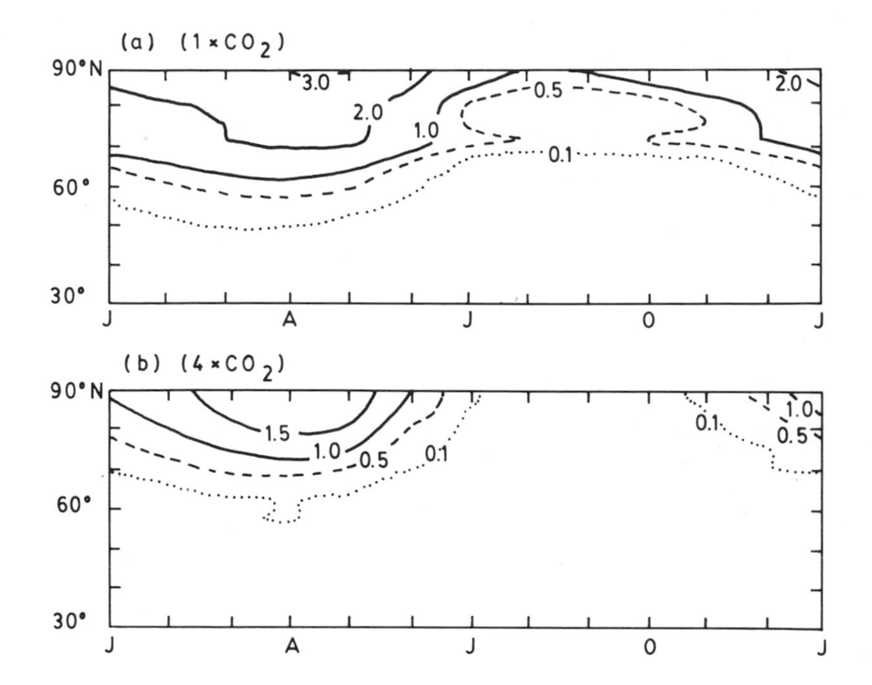

Fig. 5.18 Seasonal trend of the Arctic sea-ice thickness (in m), zonally averaged, after a climate model experiment by Manabe and Stouffer (1979). Above: standard case experiment; below experiment with a four-fold increase of CO_2 in the atmosphere. The dotted boundary delimits areas with an ice thickness above 10 cm.

climate with an Arctic Ocean which is only seasonally open can be given as yet. Based on the foregoing considerations, a rather rapid transition from a thin seasonal ice-cover to an all-year open Arctic Ocean (with some coastal ice during winter) may be envisaged (see also Section 4.2.7.3).

5.5.2 *Coexistence of an open Arctic Ocean and a glaciated Antarctic Continent during the Late Tertiary*

One of the main arguments against the possibility of an ice-free Arctic Ocean has been the existence of the highly glaciated Antarctic continent and its easily recognizable stability. Indeed it is not easy to imagine an asymmetric planet with one pole ice-free, the other covered under an ice-sheet more than 2 km thick. Nevertheless a similar asymmetry existed for several 10^7 years during the Permo-Carboniferous glacial period about 250 million years ago. At this time, one giant supercontinent ("Gondwana") was situated in the Southern Hemisphere and included its pole, while in the Northern Hemisphere the present coal deposits (in northwest and central Europe, in the Ukraine, in China, and in the Appalachians) were formed in tropical forest swamps similar to those now in southern Florida. It is not generally realised that the present boundary conditions also are asymmetric: one pole is situated on a large ice-covered continent, isolated from the continents by the circum-antarctic ocean, the other in a deep, nearly land-locked ocean with only a thin skin of floating ice. To-day this also causes an asymmetric circulation of atmosphere and ocean (Flohn 1978), most pronounced in the vicinity of the equator, where the "meteorological" equator is displaced to about latitude 6°N.

Looking into possible future climatic evolutions, it is of paramount interest that immediately before the onset of the Pleistocene exactly such a highly asymmetric, unipolar glaciation did in fact exist. As revealed by the results of the Deep Sea Drilling Program (Fig. 5.19), the Antarctic ice had developed to its present volume long *before* the Arctic drifting sea-ice formed. This lasted about 10 million years (i.e. five times as long as the whole Pleistocene with its about 20 glacials and interglacials). Kennett (1977) and Frakes (1979) have evaluated the available evidence, with only minor differences. During the last part of this time, the Pliocene (about 2-5 Ma (= 10^6 years ago), our ancestors, the early hominids, lived in a savanna environment in equatorial east Africa which apparently became gradually drier (Coppens 1978), and began to learn how to use pebbles as tools and weapons.

At the beginning of the Tertiary (65 Ma ago) Antarctica was situated near the South Pole, but connected with Australia as part of the former Gondwana continent. The meridional shape of this continent caused warm oceanic currents preventing any glaciation. After the begin of the Eocene (55 Ma ago), Australia separated from Antarctica and drifted northward with an average speed near 5 cm/a, while Antarctica remained nearly fixed in its present position around the South Pole. After a long period of increasing winter snow (possibly with local glaciations in West-Antarctica) and steadily decreasing water temperatures, the first significant drop of temperature occurred at 38 Ma ago, near the Eocene/Oligocene boundary, especially in the bottom temperature (5°C within less than 10^5 years, see Fig. 5.19). A similar sudden temperature decrease has recently been discovered in sediments of the same age around the North Sea Basin (Fig. 5.20) (Buchardt 1978). At this time substantial sea-ice was formed in the seas around Antarctica, coinciding with the formation of extended, but shallow glaciation in parts of the Antarctic continent (Kennett 1977). As under present conditions, the cold and dense bottom water spread outward into the other oceans leading to a world-wide crisis in the deep-sea fauna. A major *Antarctic ice cap* was formed during the middle Miocene

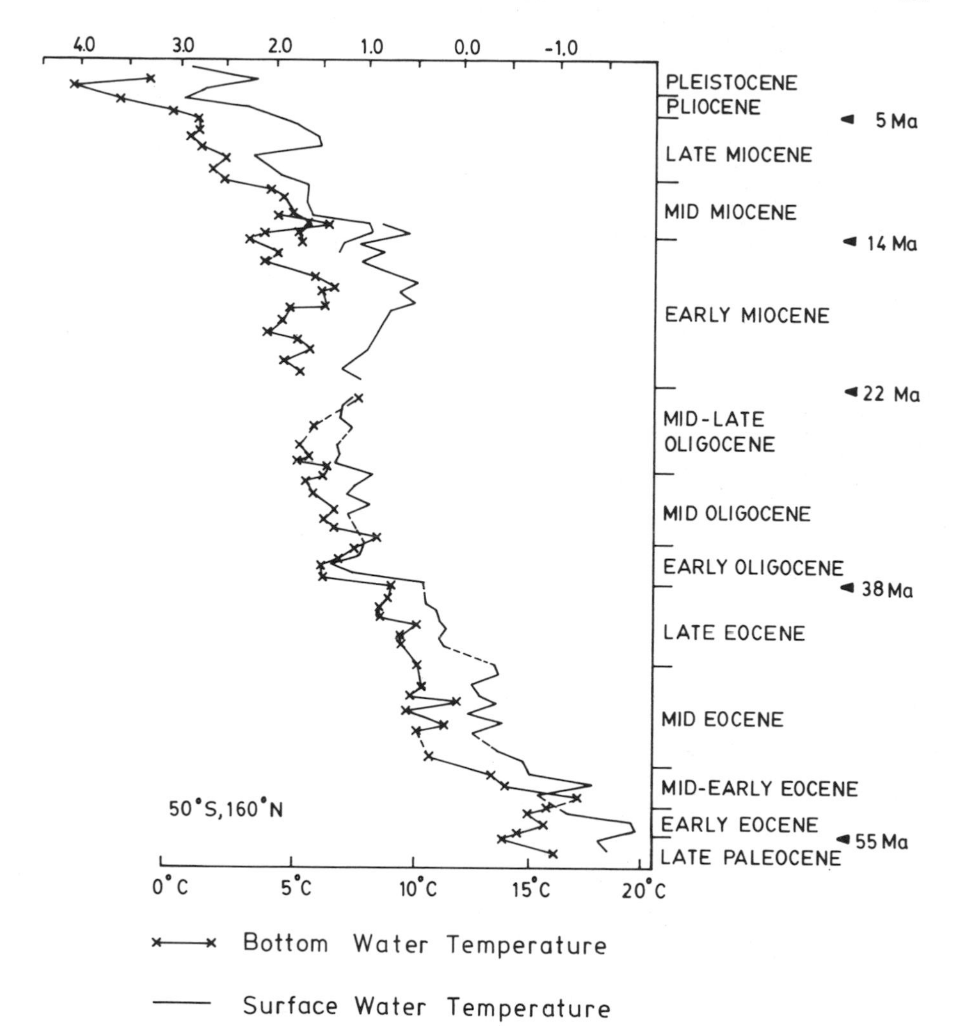

Fig. 5.19. Evolution of the surface and bottom temperatures - as derived from planctonic and benthic micro-fossils (see Section 6.3) - in the Subantarctic (about 50°S, 160°E) since late Paleocene ca. 58 million years ago (redrawn after Kennett 1977). Note the sudden drops of both records about 38 Ma and 13 Ma ago. Around 15 Ma the records of two adjacent cores overlap.

(14-12 Ma ago) as a permanent feature, which probably was at first still "warm" i.e. with temperatures near the melting point. This evolution occurred about simultaneously with a major increase of volcanic activity as represented in many ocean drillings (Kennett et al. 1977); a causal relationship has been supposed. During that period the present Rosa Ice Shelf did not

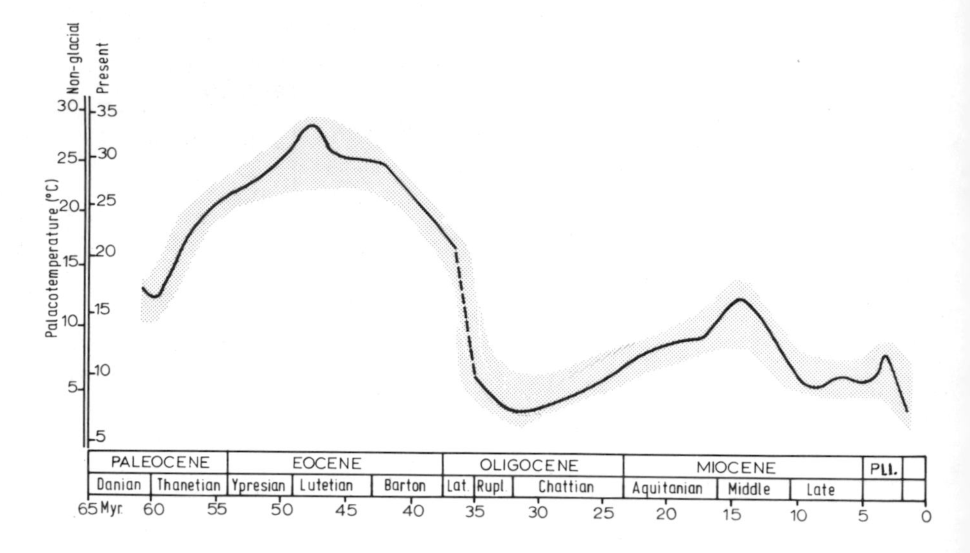

Fig. 5.20. Isotopic temperature estimates derived from molluscan shells in former shallow-sea deposits of the North Sea basin (Buchardt 1978). Shaded area represents limits of uncertainty; Myr = million years. "Present" temperature scale (at left) valid after accumulation of Antarctic ice during middle Miocene.

yet exist, and fossil pollen and algae indicate an open trans-antarctic sea (Brady and Martin 1979). After that evolution the first local mountain glaciers in southern Alaska appeared, as well as a first cold-water fauna in northernmost Japan, and a marked drop of surface temperature in the high-latitude Pacific.

The *highest glacial maximum* of Antarctica - now apparently as a "cold" and slow-moving ice-dome - was reached during the Late Miocene or Messinian (6.5-5 Ma ago), when the ice-volume was 50% greater than now. At this time, the height of the ice-dome must have been several hundred metres higher than now, and the Ross shelf-ice reached several hundred kilometres farther north. This peak was accompanied by a remarkable global cooling, by a 300 km northward spread of cold Antarctic surface waters and a high carbonate sedimentation rate at the equatorial Pacific, indicating strong upwelling of nutrient-rich cool water (Saito et al. 1975). Here 7-10 cyclic temperature changes were observed, with maxima as cold as during the peak glacials of the Pleistocene; this apparently indicates the intensity of equatorial upwelling. One of the most important consequences was a repeated glacio-eustatic drop of the sea-level of 50 m below the actual level, as produced by the storage of water at the continental Antarctic ice-dome. During these drops the

Gibraltar Strait - or its predecessor - fell dry and isolated the *Mediterranean* Sea, which evaporated completely 8-10 times to a depth of 3700 m and filled again, leaving a 300-500 m thick, laminated salt layer (Hsü et al. 1973, 1977). This lamination of the equatorial Pacific sediments indicates a cyclic behavior with a time-scale near 100000 years, probably caused, as during the Pleistocene, by astronomical changes of the Earth's orbit (Berger 1978, see also Section 5.4.2.2).

During this time and even after, the *Arctic* enjoyed a cool temperate climate with boreal forests extending up to the northernmost tips of land. During Late Miocene and Pliocene the whole shelf of Asia and Alaska fell dry, and the continents extended 200-600 km farther northward, e.g. in Siberia upto 81°N, in Canada upto 83°N (Hopkins 1967). The vegetation distribution in Siberia at a period shortly preceding the first formation of ice-sheets at the European continent (Late Pliocene, about 2.5Ma ago) has been mapped on the basis of rich data published by Russian authors (Frenzel 1968). No evidence of tundra nor wide-spread permafrost could be found. Summer temperatures have been estimated as higher by 3°C in western Europe, but by 4-5°C in eastern Europe; there, and in Siberia, annual and winter temperatures must have been much higher than now. Since at this time the relative and absolute heights of many mountains were significantly lower than now, as is clearly shown by the existence of smooth plateaus at the highest levels, oceanic rainfall could penetrate farther inland; precipitation has been estimated as 300-400 mm higher than now. In a few mountain areas (southern California, NW-Iceland) local glaciers formed simultaneously with these boreal forests - as today, e.g. in southern New Zealand - without significant large-scale climatic effects. A similar glaciation of parts of Greenland may be assumed, but due to strong glacial erosion no evidence has yet been found. Along the coast of Alaska boreal forests extended more than 800 km northward of the present limit of some of its trees; the fossil insect fauna at lat. 66°N resembled that now living at Lat. 48-50°N (Hopkins et al. 1971).

Late in this era, the ross ice-shelf in Antarçtica extended beyond its present level, and glaciation started about 3.6 Ma ago (Mercer 1976) in southern South America. Simultaneously with a northward expansion of the cold Antarctic surface water, ice-rafted pebbles indicate that Antarctic table icebergs also spread northward, farther than during the Pleistocene glaciation.

Of special interest is the climate of middle and lower latitudes at this time. Based on careful investigations, Lotze (1963) compiled maps of the position of the evaporite belt during past climates, i.e. of the *arid zone* of sebkhas, where soluble salts were sedimented in dry pans. These maps (Fig. 5.21) indicate a southward shift of its northern boundary from an average latitude near 47°N in the early and middle Tertiary to 42°N

—— Quaternary (0-2 Ma)
- - - Late Tertiary (5-15 Ma)
.... Early Tertiary (30-50Ma)

Fig. 5.21. Displacement of northern boundary of the Northern Hemisphere arid zone (evaporites = gypsum, salts etc.) combined after Lotze (1963). Crossing of boundaries in Central Asia and western North America mainly caused by uplift of mountains.

in the Miocene/Pliocene and to 38°N in the Pleistocene. The repeated desiccation of the Mediterranean during the Messinian only aggravated the arid conditions during a relative minor part of the whole duration of this quite asymmetric type of climate. A similar southward shift of the vegetation belt of North America has been documented by Dorf (1960) (Fig. 5.22).

During that time *southern Europe* was arid, and even south-central Europe was partly arid, with steppe or desert vegetation e.g. near Vienna (Schwarzbach 1974, Hsü 1977). While before events, tropical marine micro-fossils occurred in the Atlantic up to latitude 58°N, this boundary retreated during the desiccation of the Mediterranean at the eastern side of the Atlantic to about latitude 33°N. No such tropical species could enter the Mediterrean after its reopening at the early Pliocene (5 Ma ago), while in the Gulf Stream region they still reached latitude 50°N. As now, this indicated a marked longitudinal contrast, caused by the wind-driven surface currents of the oceans.

Many regional temperatures and precipitation estimates are quoted in the literature (e.g. Mägdefrau 1968, Schwarzbach 1974, Frakes 1979). However, since during this time most mountains existed only in a rudimentary form, such numerical data cannot

Fig. 5.22. Displacement of vegetation boundaries in North America (Dorf 1960), from middle and late Miocene to present. Left: southern boundary of subarctic forest; right: northern boundary of subtropical forests.

be taken as representative of near-future conditions. This is particularly obvious for examples in the case of the continental basins of Nevada or in the Mojave desert, both now arid in the rain-shadow of high mountains, which enjoyed, during the late Miocene, a rather moist maritime climate near sea-level. Summer temperatures were then 4 C lower, winter temperatures 8-10 C higher than now. During the same time *Central Europe* enjoyed a subtropical or (later) a warm-temperate climate (Mägdefrau 1968) with forest, which has been since transformed into lignite In the *Tropics*, the extension of savanna climates with seasonal rains was much greater than now, while the equatorial rainforest with all-year rain was apparently reduced. During the Miocene the vegetation patterns of the African continent (Maley 1980) revealed a marked asymmetry: while the southern Sahara was covered with a tropical humid or at least semi-humid vegetation, southern Africa and the Zaire basin were dry and sometimes full desert. A similar aridity has also been observed in northern Australia, especially after the Messinian peak of the Antarctic glaciation (Kemp 1978).

This evidence, still quite incomplete, seems to indicate that during the Late Tertiary (between 13 and 3 Ma ago) the hemispheric asymmetry of the general circulation was much more pronounced than now (Flohn 1978). With a few exceptions - notably the late Pliocene closing of the central American land bridge - the shape and position of the continents were then similar to present conditions. A large-scale glaciation of the northern continents did not start until 3.2 Ma ago (Shackleton 1977), the formation of the Arctic drift-ice only after that date, when the melting water had produced a shallow stable low-saline upper layer of the Arctic Ocean, about 2.3 Ma ago. during this long period the *simultaneous* existence of a continental *ice-dome* at the Antarctic and an *ice-free Arctic Ocean* is well established, with high stability in spite of climatic fluctuations of shorter time-scales. This situation must have caused marked circulation asymmetries of both atmosphere and ocean, which should be investigated with more sophisticated climate models.

5.5.3 *Implications of a hemispheric circulation asymmetry*

This factual situation, which lasted not less than the last 10 Ma before the Pleistocene, looks completely contrary to the view of the world familiar to the climatologist. A careful and critical survey of the vast geological literature on a global scale is needed. This task must be undertaken in cooperation between geology (especially paleobotany and paleooceanography) and climatology, with a full knowledge of the basic laws of planetary circulation and their consequences.

Atmospheric and deep-ocean circulation are controlled by the famous circulation theorem first formulated in 1897 by V. Bjerknes. It may be summarised as follows: the intensity of a thermal circulation between a heat source and a cold sink - here between the tropical belt and the pole - depends on the temperature difference between the source and sink, measured along a surface of equal density. On a rotating planet, this circulation is distorted, increasing with the rotation rate. In the surface mixing-layer of the ocean the circulation is mainly driven by the wind stress, in other words by the atmospheric circulation, while in the deep ocean density differences (caused by spatial variations of salinity and temperature) are the driving force.

A conservative estimate of the temperature of the atmospheric *column* above an ice-free Arctic Ocean leads to the result, that during summer it may have been only 2-3°C warmer, but during winter the stratification (now stable) would become extremely unstable with a surface temperature near $\pm$ 2°C (instead of -32°C now) and the disappearance of a marked temperature inversion. Most of the warming during winter was concentrated in the lowest layers where no comparative data from both poles can be given, due to the existence of a 3000 m high Antarctic ice-sheet. In the layer between 3 km and the polar tropopause (about 8 km) the average temperature increase above an ice-free Arctic can be estimated as at least + 2°C (summer), + 4°C (winter) and +3°C (annually).

In the atmosphere, the tropical circulation - which resembles a simple thermal circulation distorted by the fast rotation of the earth (near the equator with a speed of 465 m/s) into an elongated spiral - coexists and interacts with a complete different mode in higher latitudes. This mode consists of individual eddies - the cyclones and anticyclones of the weather map - travelling in permanent growth and decay from west to east, driven and controlled by the meandering upper westerlies. The average boundary between these two modes - the belt of subtropical high pressure cells - is determined by a baroclinic instability criterion (Smagorinsky 1963, Flohn 1964): the tropical circulation is largely stable, while the high latitude westerlies with their strong meridional temperature gradient are frequently unstable. This leads to a simple equation in which the latitude of the sub-tropical anticyclonic belt Φ_S, which controls the position of most climatic belts, depends on the ratio between the meridional temperature difference and the vertical temperature lapse rate, i.e. the vertical stability.

Korff and Flohn (1969) have verified this theoretical concept using long-year monthly averages of Φ_S and of the meridional temperature gradient at the 300/700 mb layer, in both hemispheres. This relation describes remarkably well (with correlation coefficients above 0.85) the seasonal variation and the actual hemspheric asymmetry of Φ_S (Fig. 5.23).

Since these seasonal and hemispheric variations are of the same magnitude as those anticipated in the case of an ice-free Arctic Ocean, this relation can be used for a first-order approximation of the expected displacement of the subtropical belt.

In this case, the vertical lapse rate within the troposphere is assumed to be constant. This assumption is not quite correct, since increasing CO_2 (and increasing evaporation of water vapour, see Manabe-Wetherald 1975) should lead to a slight decrease of stability. In the above-mentioned equation this (comparatively small) effect acts in a sense opposite to the decrease of the temperature difference and might thus slightly reduce the meridional displacement of Φ_S.

A simple extrapolation based on the temperature distribution in the layer 3-8 km above an ice-free Arctic, leaving temperatures above both equator and South Pole unchanged, would

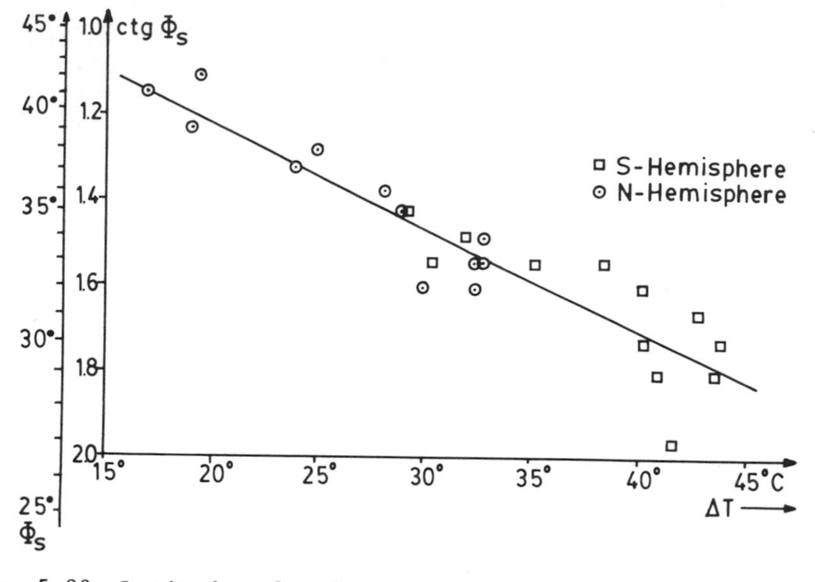

Fig. 5.23. Latitude of subtropical anticyclonic belt (Φ_S) in every month versus actual temperature difference Equator-Pole (300/700 mb layer). Circle (squares) = Northern (Southern) Hemisphere (Korff and Flohn 1969); actual data.

yield only a small northward displacement of the subtropical anticyclonic belt. However, the neglect of the Arctic layer 0-3 km in which the bulk of the warming (more than + 2°C) would occur, obviously leads to unrealistic results. We may estimate more realistic values for the average temperature increase of the whole troposphere (0-8 km) above the Arctic as follows: winter + 10°C, summer + 3°C, year + 7°C. This would result in an *annual shift* of the northern subtropical high-pressure belt from latitude 37° at present to latitude 41-43°, while the location of the southern subtropical belt should remain unchanged (31°S). During summer, this shift would reach probably not more than 100-200 km, but during winter the subtropical belt may displace some 800 km to the north (Flohn 1981). This should drastically reduce the extension of the subtropical belt of winter rains, which now are responsible for the water supply of California, the Mediterranean, Near and Middle East upto Russian Turkestan and the Punjab. The belt 45-50°N would be more frequently effected by summer droughts.

Such a displacement of the northern STA (subtropical anticyclone) without movement of the southern STA would also displace the average position of the "meteorological equator" to latitude 9-10°N, instead of 6°N at present. In this case, the seasonal displacement of the equatorial rain belt would be limited mainly to the belt between equator and latitude 20°N. A consequence of this would be a marked decrease of rainfall south of the equator, eventually leading to natural desertification in many countries of the belt 0-20°S. This would probably be aggravated by an increasing frequency and intensity of equatorial ocean upwelling,causing a reduction of maritime evaporation in this belt, especially during northern summer/southern winter, when the meteorological equator reaches its northernmost position. In this season, a nearly permanent occurrence of equatorial upwelling can be expected at least in the eastern and central parts of the Pacific and the Atlantic Oceans, while at the equatorial Indian Ocean no evidence for substantial upwelling is available at present. Since the position of the equatorial upwelling zone in the oceans is controlled by the reversal of Coriolis force at the mathematical equator, a meridional climatic shift of this belt is impossible.

This increasing frequency of equatorial upwelling would be caused by a greater predominance of the Southern Hemisphere circulation, in contrast to the weaker Northern Hemisphere circulation. Since the belt of upwelling lies just south of the equator, the weakening of the northern circulation is not expected to have had any effect south of the equator. Theoretical arguments would predict a decrease, or even the disappearance of equatorial upwelling (as e.g. during the Holocene moist period of the Eem - see Section 5.1 and 5.2) together with increasing evaporation - *only* in the case of the general weakening of the tropical easterlies on *both* sides of the equator. In the equatorial Pacific the regular occurrence of nutrient-rich equatorial upwelling during the Late Tertiary is shown to have occurred by a high rate of biogenic sedimentation.

In the present drought-ridden Sudan-Sahel belt, between about latitude 10°N and 18°N, a gradual increase of precipitation in the case of an ice-free Arctic *could* be possible. However, this conclusion is based only on one argument - the northward shift of planetary climatic belts. Actually the remarkable small summertime advance of the rainfall belt *in Africa* is most probably due to the subsidence-forcing role of the tropical easterly jet (Flohn 1964a): this effect should persist also during an ice-free Arctic Ocean. An other essential prerequisite would be - see Section 5.1 - the conservation of the reappearing vegetation which might be triggered by expanding summer rains, i.e. its protection against overgrazing and other man-induced desertification processes.

In the *monsoon area* of Southern Asia, the planetary circulation patterns are disturbed and largely controlled by the existence of the vast Tibetan highlands north of the Himalayas. Here, any rational foreshadowing of the effects of disappearing Arctic sea-ice seems to be impossible. During and since the Late Pliocene and Pleistocene, this giant mountain block has been lifted several kilometres. Visible evidence for uplifting is the Tibetan "Plateau" with its weak rolling topography, raised from near sea-level to an altitude of 4-5 km now. As an elevated heat-source, it controls the tropical easterly jet (Flohn 1968) and the powerful monsoon circulation between West Africa and the Philippines. In this extended area, the influence of an ice-free Arctic cannot be reliably estimated without a realistic interactive model with the present topography; this influence may be, during summer, comparatively weak. Climatological evidence from a period when the Himalayas did not exist (cf. also the model of (Hahn and Manabe 1975)) should certainly not be taken as representative for what might be expected under present boundary conditions. During winter a northward displacement of the subtropical jet (within the westerlies) of 5-6°C latitude might lead to a similar displacement of accompanying disturbances, perhaps with increasing precipitation in such remote area as Chinese Turkestan, but decreasing rainfall south of the Himalayas.

One of the first consequences of a complete disappearance of the Arctic drift-ice would be the formation during the cold season of a (reversed) frontal zone along the *northern coast of the continents*, between the cold air above the snow-covered land and the warm air above the open sea. Such a development should drastically enhance the rate of snowfall along the northern coast of the continents and Arctic islands. During summer, the low-level stratus should disappear, and the thermal circulation between the heated continents and the still relatively cool water should produce a predominance of anticyclonic situations above the Arctic Ocean.

Finally an attempt shall **be** made to estimate the climate in the area of the *European Community*. As in the preceding paragraphs, this will be done mainly in the form of a scenario based on available insight into the mechanisms of the general atmospheric circulation. Several results from advanced general circulation models (Manabe-Stouffer 1980, Manabe-Wetherald 1980, Wetherald-Manabe 1980, Manabe-Wetherald-Stouffer 1981, and others) should be taken into account, even if the model assumptions used in the Arctic cannot be considered as sufficiently realistic.

But since a really comprehensive model for this situation, with a seasonally (or permanently) ice-free Arctic Ocean, taking into account all feedback mechanisms between air, sea and ice, is not yet available and cannot be expected soon, no more than an educated guess can be given. This must be subject to revision in the light of the results of future realistic model experiments.

The coastal countries of the *Mediterranean* south of about latitude 40°N should become more or less arid: a reduction of winter rains together with increasing temperature and evaporation during summer could lead to near-desert conditions. However, any desiccation of the Mediterranean Sea (as during the Messinian 5-6 MA ago) can safely be excluded because of the depth of the present Gibraltar Strait. The northern coasts of the Mediterranean together with the Alps and south-central Europe (up to latitudes 48-50°N) should obtain a warm-temperate climate with some reduction of summer rains, i.e. with frequent warm-season droughts. Western Europe, the British Isles and southern Scandinavia should become partly semi-humid, most probably more than in the warm summers of this century (1921, 1976). While the vegetation period would be increased by 1-2 months, droughts should be more frequent than now. In more continental areas several hundred kilometres from the coast, the occurrence of cold winters with frequent snow-cover cannot be excluded. Above the North Sea and the Baltic as well as in the mid-latitude Atlantic the frequency of heavy gales and storm surges should diminish.

Based on the present pattern of climatic zones, which is not simply latitude-oriented but distorted due to the effect of continents on ocean currents and on temperature distribution, and on an extrapolation from Fig. 5.2.3, an attempt is made (Flohn 1980) to outline a schematic global model with an idealized continent showing the expected shift (Fig. 5.24). While the general pattern is displaced, north of about latitude 10°S, towards the north, the reduction of Zone 4 (subtropical winter-rains) and 6 and the disappearance of Zone 7 in the Northern Hemisphere are most remarkable; this would be accompanied by an expansion of Zones 3 and 5.

5.5.4 *Summary and Conclusions*

In contrast to the possibility of a new glaciation (Section 5.4.4) which could be initiated only, for example, by a series of volcanic eruptions surpassing the intensity and frequency of the activity observed during the last 500 years (probably the last 10000 years), the possibility of a man-induced destruction of the thin Arctic sea-ice cover is in man's own hands. The present discussion leaves little doubt but that with a massive increase of the CO_2 content of the air beyond a given threshold the probability will greatly increase of the development of an open Arctic Ocean and a consequent displacement of the large-scale climatic belts and hence of an alteration in the economic future in a period of fluctuating energy costs and our incomplete knowledge of the carbon cycle (especially in the terrestrial biosphere and the oceans, see Chapter 4) it is not possible at present to give a reasonable prediction of the time-scale of the climatic changes. On the other hand the relation between average temperatures and CO_2 content seems better established than the future evolution with time of the CO_2 content itself.

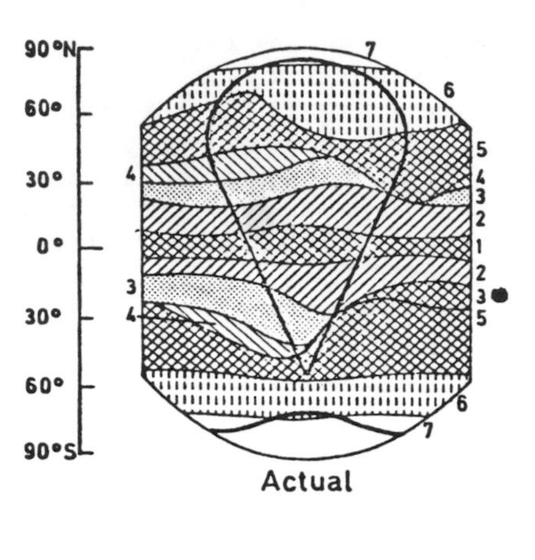

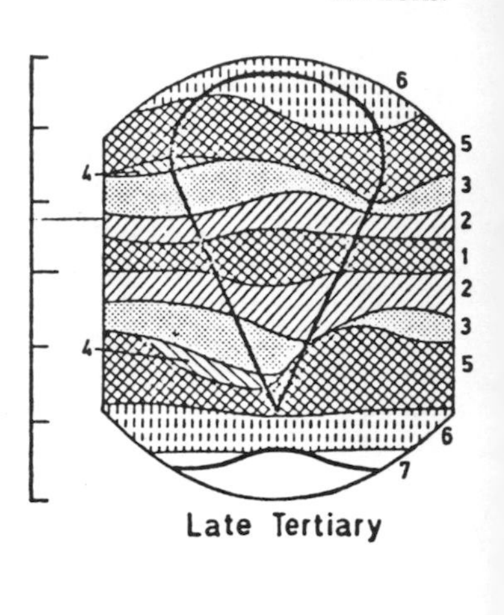

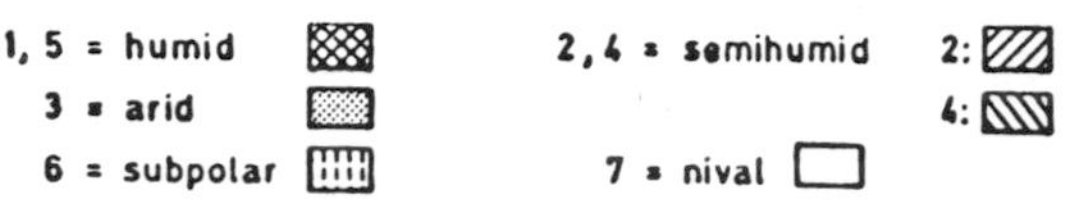

Fig. 5.24. Estimated patterns of climatic zones on a globe with only one idealized continent under actual (both poles glaciated) and Late Tertiary boundary conditions (only Antarctic glaciated). 1 (5) = tropical (temperate) zone with all-year rains, 2 (4) = rainy season during hemispheric summer (winter), 3 = mostly arid, 6 = subpolar with permafrost, 7 = permanent snow and ice.

With increasing CO_2 content and global warming, the probability of an ice-free Arctic Ocean increases not linearly, but - as a consequence of the non-linear positive feedback between snow and ice, albedo and temperature - distinctly faster, probably about exponentially. At first, the extension of sea-ice into the marginal seas (especially in the North Atlantic between Greenland and Franz-Josefs-Land) will be reduced, similar to what happened during the Viking colonization of SW-Greenland, the Holocene warm period and the Eem interglacial (Sections 5.1 - 5.3).

This will already cause a northward shift of the climatic belts most probably exceeding the transient shifts observed during the most extreme seasons of the short instrumental period of the last 100-300 years. Most likely it may also reduce the seasonally varying extension of the subantarctic sea-ice - but here our knowledge is quite limited, since the role of the "katabatic" winds blowing down from the enormous deep-freezer of the Antarctic ice-dome causing the formation of seasonal ice may be largely independent of a global warming.

The most fascinating possible evolution would involve

melting of the central core of the Arctic sea-ice in the area between Greenland, Alaska and Siberia, at first seasonally - as in the models of Manabe-Stouffer and Parkinson-Kellogg (Section 5.5.3) - and eventually during the whole year. The role of storage of solar radiation through warming of the upper shallow and low-saline layer has been mentioned (Section 5.5.1); probably this effect will lead rather soon to a reduction of the ice to the coastal shelf during winter. The time-scale of such an evolution has been estimated by Budyko (1977, Section 7.1) with a simple semi-empirical model, to be some decades (Fig. 5.25). Manabe and Stouffer obtained their results of the seaso-

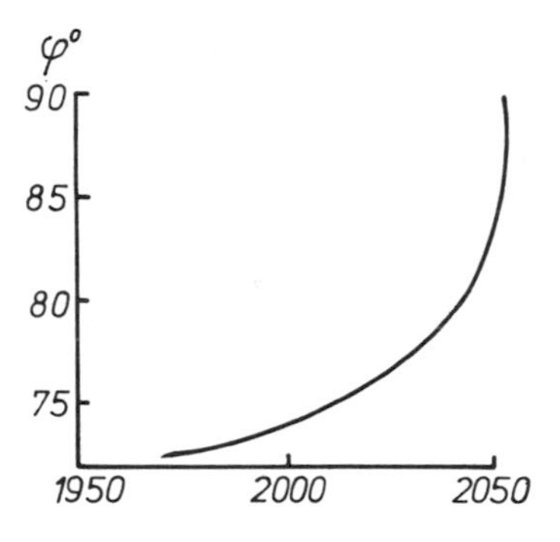

Fig. 5.25. Variations of the southern boundary Arctic sea-ice after Budyko (1977, fig. 34; see also p. 117 f). In his semi-empirical model an annual energy production rate of 6% is assumed (which should be above any realizable upper limit), together with an unspecified increase of CO_2, which is, in his consideration, only of secondary importance. Due to these debatable assumptions and to the deficiencies of each simplified model, the time-scale gives only a rough approximation; more realistic assumptions would tend to delay such an evolution.

nal melt of the sea-ice from an isothermal dry and motionless atmosphere (with 1200 ppm of CO_2 and an isothermal ocean) after 10 years of model time; they obtained it by an integration over the three following years. What would happen, if this model - which assumes the normal seasonal insolation cycle and a static shallow ocean which allows for correct seasonal heat storage - were run for additional 100 model years and increasing heat storage?

Such considerations, together with the observed results of the mass balance of multi-year ice-floes, lead to the result, that the polar air-ice-sea system is highly sensitive and could drastically change its status during a short time of not more than a few decades, once this evolution is initiated. It seems to be quite probable that a stage with seasonal opening and freezing would be only short-living and transitory. An open Arctic Ocean has not existed during the last 700000 years or so; however, it did exist, during a very long time, up to about 2.5 million years ago when boreal forests extended to its coasts (Hopkins 1967, Frenzel 1968) and Antarctica was heavily glaciated.

In a quite recent monograph (Paleoklimaty Pozdnego Kainozoya, Gidrometeoizdat 1983), Zubakov and Borsenkova have reviewed the available paleoclimatic and geological evidence for the late Cenozoic - i.e. the last 5-6 million years -, with a wealth of hitherto unavailable data especially from the U.S.S.R.

Several atmospheric effects opposing such a global warming (e.g. increasing cloudiness and low-level aerosol particles) have been seriously discussed in recent years. The present conclusion is that these are not negligable, but apparently unable to alter, in substance, the above-mentioned conclusions. This has been confirmed by an independent group of scientists, requested by the Science Advisor to the U.S. President, to look critically for unrecognized errors, misinterpretations and hidden effects. Certainly it cannot be excluded that a hitherto quite unsuspected factor may change the situation. However, in view of the extended discussion, during many years and among hundreds of responsible and critical scientists, such an unknown effect does not seem to be very likely.

Scientists are understandably slow to adopt conclusions which involve serious repercussions reaching far beyond their personal fields of knowledge and responsibility. In the present case, the findings on probable climatic trends do not depend solely on incomplete models which might well fail to take full account of a particular physical factor or of the complex feedback mechanisms involved. Furthermore, there is a relevant precedent in the relatively recent geological past, under boundary conditions (land-sea distribution, mountains) which did not differ radically from the present, at least towards the end of the period, about 3 Ma ago.

We may or may not like the conclusions - but we cannot just dismiss them out of hand.

Chapter 6

IMPACT OF CLIMATIC FLUCTUATIONS ON EUROPEAN AGRICULTURE

6.1 Introduction

The folklore of every European country abounds in traditional expressions of the importance of climate and weather to farming activities. An early and apt expression of the relationship is the observation of Theophrastus, a pupil of Aristotle, that it is the year which bears fruit and not the field, an empirical conclusion which has been confirmed by modern findings that the variations in crop yield from one season to the next can greatly exceed those between one soil type and another.

The relationship between weather and agriculture is dauntingly complex; it differs for pasture and tillage, for meat, milk and wool production, for cereal and root crops. Weather effects not only vary from crop to crop but even from variety to variety. Some periods in a plant's life are more vulnerable than others. Seeds, for example, will not germinate below a certain level of soil temperature or without a required quota of soil moisture; tender seedlings will perish in transient drought or by reason of a short sharp frost.

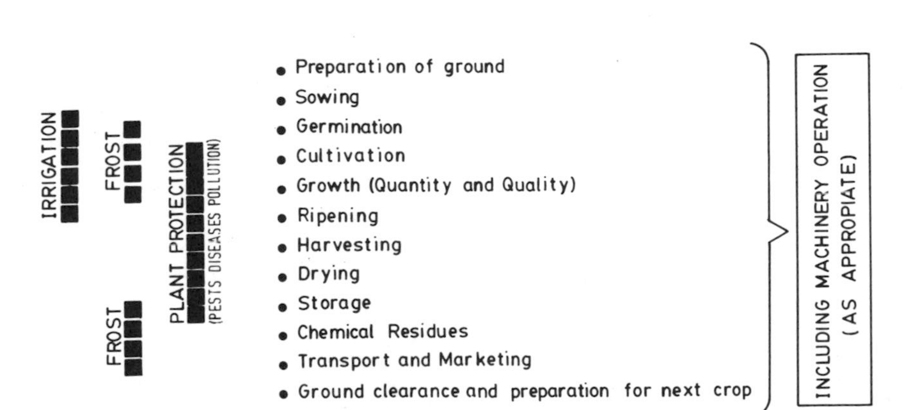

Fig. 6.1. Weather-sensitive aspects of spring-sown crops.

Figure 6.1 illustrates the successive stages in the life-cycle of a spring-sown crop which are variously dependent upon weather conditions. It is by no means exhaustive, for the current crop can be affected by the previous season's weather in respect, for example, of the quality of seed, of the temperature level and water reserves in the soil at sowing time, and the carry-over of disease and pests. A further complicating factor is the resilience of many growing things which enables them to make a remarkable recovery in favourable conditions at het end of an unpromising season.

Despite these complexities, detailed mathematical crop weather models haven been developed and are being continuously refined which simulate plant performance of specific crops in response to environmental factors, in terms of growth rate, development towards maturity and harvest yield. The input to these empirical-statistical models can be daily means or even hourly values of a range of meteorological parameters and not just monthly or seasonal averages. They have found promising application in forecasting probable yields during the current season and hence in the periodic estimation of regional food crop production.

Complex crop-weather models are, however, largely irrelevant to the analysis of harvest years before the present century because of the inadequancy of the available meteorological input. They are little more helpful in mapping the possible effects of expected future climatic trends, when the best we can hope for are indications as to whether the different seasons will tend, in the main, to become warmer or colder, drier or wetter, with perhaps some clues as to changes in the broad features of the atmospheric circulation.

Fortunately the primary weather needs of agriculture are for warmth and moisture, and the main changes in food production can be related to how well these requirements are met in the growing season. In drier areas the assessment of the adequacy of soil moisture requires a computation of the balance between incoming precipitation and losses due to evaporation and transpiration, but in most of Europe, where precipitation in the primary variable, changes in rainfall are normally sufficient as an indication of soil moisture variations.

We shall, therefore, focus in the main on changes in <u>rainfall</u> and <u>temperature</u> patterns and on their implications for <u>food production in Europe</u>. These implications will vary from region to region according to whether the main climatic stresses on agriculture arise from deficit or excess of heat or water. The effects of temperature changes will appear most clearly in northern regions (Iceland, much of northern Norway, Sweden, Finland and the USSR). On the other hand, in the drier lands of the Mediterranean regions and elsewhere in southeast Europe, water demand generally exceeds supply and this is the main factor determining agricultural production. In the rest of the continent

(Ireland, U.K., non-Mediterranean France, Benelux, Denmark, Germany and Central Europe generally), changes in the temperature and rainfall regime are both of significance. To this intermediate zone of complex and changeable weather and of high agricultural importance we shall necessarily devote our major attention (See Section 6.3).

A series of up-to-date review papers on world-wide aspects of the food-climate issue have been presented at an international workshop at Berlin, December 1980 (W. Bach et al. 1981).

6.2 Northern Europe

6.2.1 Temperature fluctuations and Nordic farming

At the best of times, farming is precarious in the cold northern fringe of Europe, say roughly north of 60°N, where winters are severe and extended, and where the most critical meteorological element for agriculture is temperature. A fairly close relationship has been shown by Bergthorsson (1967) to exist between temperature fluctuations and the frequency of famines as recorded in the Icelandic annals. In a area where all crops are grown close to their lower limits of thermal tolerance, quite a small temperature fall may involve the crossing of a critical growth threshold so that, as the records show, it may become uneconomical or even impossible to continue the cultivation of certain crops. Soon after the settlement of Iceland about 900 A.D. the cultivation of cereals, in particular of barley, was introduced over much of the country. The crop had its ups and downs in later years but it continued to be grown with some success for nearly 700 years. However grain growing virtually died out before 1600 A.D. and occasional later efforts to revive it have had only brief and scant success (Eythorssen and Sigtryggson 1971). The distinctly warmer period in the first half of the present century led to the reintroduction of barley, but the crop again lapsed in the colder, interlude which followed, and which also made potato cultivation more hazardous because of the increased incidence of frosts (Fridrikssen 1973).

Similarly in Scandinavia COsvald 1952) the significant rise in temperatures in the earlier part of this century meant easier conditions for agriculture, horticulture and forestry. No small part of the rise in rye and wheat yields during that period of milder, more maritime climate has been attributed to extra warmth in the vegetative period. During the 1930s a decade of hot summers in the Nordic countries led to correspondingly superb harvests.

As has often happened before, the complacency evoked by a series of favourable years was shattered by the severity of the winters from 1940 to 1942, among the coldest ever recorded in

Scandinavia, and by the general cooling trend of the following epoch which had unpleasant consequences for agriculture (Vargo 1973). In Lapland arable farming had gone into a clear decline by the 1960s, with falling yields and uncertain profits. The cultivation of wheat was abandoned in the early 1950s and that of rye at the end of the decade. Hay, barley and potatoes continue as precarious crops, subject to loss in the colder years.

Severe winters in northern climes can cause great damage by the "winter killing" of grass and cereals, especially in seasons of limited snow cover. The frequency of years with severe winter damage of cultivated pasture in Iceland in the 1960s trebled in comparison with the preceding decades, with up to 70% of grassland suffering damage in the north of the country in the harsher years. This diminution of the hay crop is seasons during which livestock must be fed indoors for longer periods undermines the economics of farming in areas where it is already borderline.

6.2.2 *The thermal growing season*

A second major aspect of the temperature problem in the northern countries is the restricted length of the period in which plants can remain in active growth. To examine this question more closely requires a discussion of the meteorological constraints on the length of the crop growing season which may be limited at either end by lack of adequate water, normally in semi-arid areas, or, as here, by low temperatures.

In the most favourable southern coastal lands of Iceland the average length of the frost-free period is about 150 days; this drops to less than 90 days at a point 35 km inland and at an elevation of some 100 m. It would be impracticable to base a general analysis of the thermal growing season on the period between the earliest and latest killing frosts, because of the paucity of relevant statistics and the great variation in frost hardiness of different crops. Similar difficulties preclude the use of soil temperatures, although readings at a depth corresponding to the seeding level of a particular crop might be most appropriate to determine the commencement of germination. The use of standard meteorological observations of air temperature is not only the most convenient approach but one that has found successful application in many countries.

A more difficult problem is the selection of a single representative base temperature for the growing season; clearly the beginning and end of the active annual growth of a number of crops and crop varieties cannot be too rigidly linked to one value of the mean air temperature. A criterion of 6°C was proposed by Angot nearly a hundred years ago, and has been widely tested in the interim. Experience has shown it to provide a remarkably valid index of the impact of the annual course of temperature first rises to 6°C and ends when it falls below that value later in the year.

Applying this definition to the annual mean temperature curve for Reyjavik (Fig. 6.2) we deduce, from its intersection with the solid 6°C line, that the average growing season in the period 1931-60 was about 150 days. We may also deduce the approximate effect of fall of 1°C throughout the annual course of temperature by noting the intersections of the curve with the present 7°C line (dashed). The result is a contraction of the already short growing season by nearly two weeks.

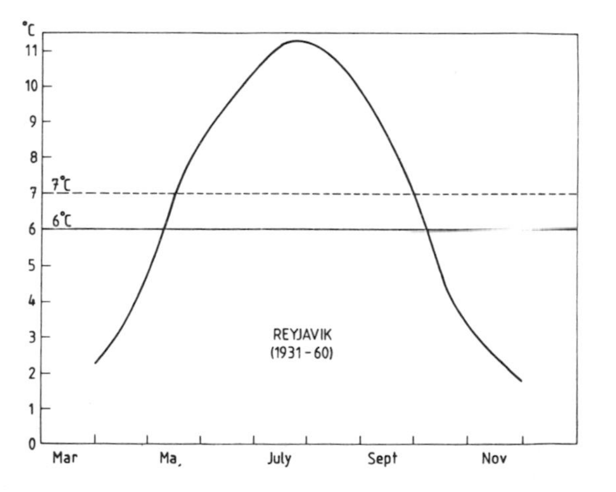

Fig. 6.2. Mean annual course of air temperature and growing season in Iceland.

Apart from the significant shortening of the growing season which results from falling temperatures, the reduced area bounded by the annual curve and the horizonal temperature line (Fig. 6.2) implies a corresponding reduction in the accumulation of daily mean temperature above the threshold value and hence in the energy available to the growing plant. In Fig. 6.2 the reduction in accumulated temperature units of "degree-days" corresponding to a fall of 1°C is approximately one-quarter. The closeness of the agreement of changes in the rather crude index of degree-days with statistics of hay losses in Iceland must be largely fortuitous; nevertheless, experience shows that a drop in the dry matter hay yield of the order of 1 tonne/ha or about 25% occurs for each 1°C deficit in the mean temperature. It follows that there are great fluctuations from year to year in the production of grass which is Iceland's most important crop. An average national hay yield which averaged 4.5 tonnes/ha in the warmer epoch of the present century dropped to about half in the subsequent cooler years (Fridriksson 1973).

The level of temperature which makes crop growing hazardous in Nordic countries also reduces the range and impact of diseases and pests. The most serious menace consists of attacks on overwintering crops by parasitic fungi (see Sections 2.3.3.1 and 6.3.4) which become active under snow cover, such as caused severe damage to winter wheat in the long harsh 1965/6 season.

6.3 Central Europe

6.3.1 The general climatic scene

Between the northern fringe of Europe where the main climatic problem of the farmer is low temperature and the Mediterranean region where water is often in short supply lies the most productive area of European agriculture, stretching roughly from 45°N to 60°N. It is blessed in the mean with a mildness of climate which constitutes, in the words of C.C. Wallen (1970). "one of the more spectacular anomalies in the world climatic pattern". Average rainfall is adequate not only on the oceanic fringe but also well into the continental mainland, since the west to east gradient of rainfall is quite modest. The adequacy of the mean does not imply that European agriculture cannot suffer, sometimes acutely, from climatic problems involving both temperature and rainfall. At one extreme drought and damagingly high temperatures occasionally occur; however the more common problems of the farmers are posed by cold and wet conditions.

6.3.2 Temperature

The problems which the northern countries experience in growing any cereal crops whatever recur in this warmer zone when attempts are made to introduce crops which call for more warmth than is comfortably available in the growing saseon. Raising maize for grain in England is an example of cultivation close to the lower thermal tolerance of the crop. In such boderline cases even a slight warming of the climate could lead to a remarkable potential extension of the crop.

On the basis of accumulated temperatures during the months May-October above a base temperature of 10°C (in place of the 6°C base appropriate to grass and other crops adapted to lesser temperatures), Hough has mapped the fairly restricted lowland areas in south-eastern England in which current varieties of maize could be grown successfully and harvested at a moisture content of 40% in at least nine years out of ten (Anon 1971). His results show that the area suitable for raising maize for grain in England would be dramatically extended if the mean temperature in each growing month were increased by as little as 0.5°C. Naturally a similar cooling would lead to a distinct diminution of the chances of growing corn successfully so far north.

A crude rule of thumb as to the effect of temperature change on the poleward boundary of growth of particular crops may be deduced from a study of the mean Summer isotherms. These run roughly from west to east in much of central Europe, with a north to south temperature gradient of 2°C per 300/500 km. The order of southward regression of the thermal boundary of a crop is thus some 100 km for eah 0.5°C of climatic cooling. A similar northward extension would be made possible, other things being equal, by a warming of 0.5°C.

In contrast to a crop like maize, one might expect that slight climatic variations would have only minor repercussions on the growth of crops growing well inside the extreme limits of their environmental tolerance, and hence not exposed to any serious heat or moisture stress. Pasture, for example, is so ideally suited to the climate of Ireland that it is difficult to see how it might be affected by any but the most sensational change in Irish weather. Yet the slight temperature changes during the past century have had appreciable repercussions on the length of the grass growing season, and consequently of the period during which it was necessary to feed stock indoors on purchased or specially grown fodder.

Around the turn of the century the growing season in Ireland, as defined by the 6°C criterion, averaged 8 months. Later, in the epoch of early springs and mild autumns, it increased to almost 9 months. This resulted in an economically significant reduction of 25% in the period of winter stock feeding. In the most recent years, the average length of the growing period has partially regressed to an average of slightly under 8½ months. Similar changes have been noted in other parts of western Europe

How a small temperature variation can appreciably change the growing season will be clear from Fig. 6.3. This shows, in respect of Dublin and Vienna, two locations with the same mean annual temperature, the rather different course of monthly mean temperature. As before, the effect of a fall of 1°C on the length of the growing season can be roughly estimated by comparing the intercept of the annual temperature curve on the solid 6°C isotherm and the dotted 7°C isotherms. Because the annual range of temperature in Ireland is much smaller than in Austria the slope of the curve where it intersects the isotherms is considerably less steep and the effect on the changed length of the chord is greater. The contraction at either end of the growing season for a given fall of temperature is much more marked in a maritime climate than in an continental one.

This sensitivity of the growing season in western Europe to temperature change has repercussions also for hill farming, for the preceding argument applies whether the fall in temperature arises from a climatic fluctuation or because of increased elevation. Indeed the rate of contraction of the growing season above mean sea level is not only more rapid in the maritime countries of western Europe than in the more continental parts, but it is, by reason of the exceptional temperateness of the local climate, more rapid than anywhere else in the world. The average shortening of the upland growing season in Britain is about two weeks for every 100 m of elevation; In Switzerland, as in the New England states of the U.S.A., it is about one week. Cooling with height further involves a reduction in the "accumulated temperature" during the shortened growing season, which means an additional reduction of potential productivity. In both respects, the implications of climate change are particularly important for hill farming in the maritime countries of western Europe.

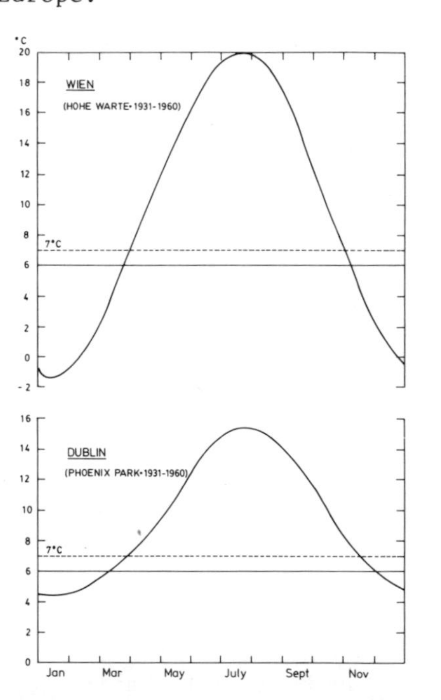

Fig. 6.3. Mean annual course of air temperature and growing seasons in Austria and Ireland.

In discussing the effects of temperature change on the growing season, in the context of Fig. 6.2 and 6.3, we assumed a uniform alteration of temperature troughout the year, i.e. a displacement of the annual curve as a whole without change of shape. However, it is clear from the figures that the temperature regimes in winter and summer have little direct effect on the dates of the beginning or end of growth, which are determined purely by the spring and autumn weather. A modest padding of the "shoulders" of the temperature curve without other change would be sufficient to lengthen the growing season appreciably.

For this reason, records of the course of seasonal temperature (Fig. 2.9) are more significant for agriculture than those of the trend of annual temperature. The average lengthening of the growing season which was widely experienced earlier this century was due in about equal measure to earlier and warmer springs and to mild long-lasting autumns. The recent recession was caused by an average trend towards colder springs; on the whole, autumn has tended so far to maintain a remarkably high level of temperature. Any trend towards markedly colder autumns would have serious repercussions for agriculture.

6.3.3 Deficit or surplus of water

There are many parts of the world in which lack of moisture can be a major cause of crop failure and the death of stock, and where drought is a recurrent natural disaster and often the precursor of famine. Over a great part of Europe, however, and particularly in the maritime northwest, rainfall is usually adequate for agriculture and its fluctuations from one growing season to another are rarely excessive. Indeed seasons which are drier than normal are traditionally regarded as one of the lesser climatic tribulations of the European farmer. "It is", wrote Boate (1755) over 200 years ago, "a common saying in Ireland that the very driest summers there never hurt the land; for although the corn and grass upon the high and dry ground may get harm, nevertheless the country in general gets more good than hurt by it, and when any dearths fall out to be in Ireland, they are not caused through immoderate heat and drought, as in most other countries, but through too much wet and excessive rain".

The Atlantic fringe of Europe is, of course, blessed with a particularly abundant and dependable rainfall, but folk wisdom as expressed in proverbs reflects the same experience in other parts of western Europe. L.P. Smith quotes the saying that "drought never ruined the English farmer", while Le Roy Ladurie summarizes a host of rural adages from the continental mainland in the words: "a year of rain is a year of pain, while a very dry year is one of good cheer". Nevertheless the impact of the severe drought of 1975-6 is sufficiently fresh in farming memory to show that western Europe is far from immune from the effects of occasional damaging drought.

The relationship of moisture deficiency to plant growth is extremely complex and involves many physical and biological factors. Agricultural drought arises when the supply of soil water available to a crop over a critical time period is inadequate: it thus involves both the plant species and the soil type. Rainfall is the most important single factor determining drought, but not merely the rainfall within the growing season, since the carry-over of soil water reserves from previous months can make an important contribution to the moisture balance.

The timing of drought in relation to the different stages of plant growth is of major importance; for this reason, the distribution of rainfall during the growing season is at least as important as its total quantity. Early and severe moisture stress will inhibit germination and can kill off plants at a very vulnerable stage. Prolonged drought during periods when growth is normally most vigorous may greatly reduce yield. Even short-lived but sharp water stress at a stage when cereal kernels are soft and moist can shrivel the grains and reduce the harvest.

On the other hand, relatively mild and short-lived moisture stress may cause no permanent damage; its effect is often compensated by subsequent recovery under favourable weather conditions. Indeed it has been found that short periods of water stress early in the life of a crop can even lead to increased yields, probably due to a greater uptake of nutrients.

Farm animals weakened by drought are more vulnerable to certain diseases and pests. Crops too which have been subjected to water stress are liable to suffer more damage from aphids. In both cases, some compensation is provided by the reduced economic losses due to the numerous moisture-loving diseases and pests.

The heavy impact of drought in 1976 arose from the fact that in many parts of western Europe it was particularly long sustained, with few relieving rainy spells. Its duration and intensity varied from one area to another, but over much of the region it represented the culmination of a markedly dry spell shich had already dominated the year 1975, with its especially dry and sunny summer, and in places extended back to the autumn of 1974. There was in consequence a cumulative effect, both in the depletion of soil moisture and in demands upons stored water in reservoirs and aquifers.

The first half of 1976 was particularly dry in many part of western Europe. Figure 6.4 shows the total rainfall during the eight-month from December 1975 to July 1976, expressed as a percentage of the 1931-60 average. The detail in the diagram would show some change depending on the sequence of months selected for analysis (thus, for example, rainfall in the Netherlands in the period February to August 1976 was only 41% of normal), but the broad rainfall picture is representative. The occurence of heat-wave conditions in high summer aggravated the deficit of soil moisture.

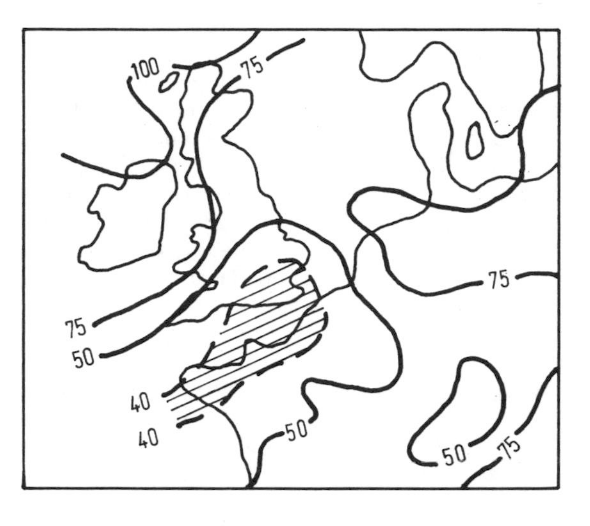

Fig. 6.4. The drought of 1976: Rainfall December 1975 - July 1976 in per cent of normal. (WMO Bulletin, July 1977)

There was a sharp reduction in the harvest in all the affected area. Compared with the 1974 season, described as one of normal weather, the yield per hectare of the French potato crop showed a drop of 10% in 1975 and 36% in 1976. In Belgium the yield of potatoes, which had been excellent in 1974, fell to almost half in 1976; this was the result of the most serious deficit in spring and summer rainfall for over 50 years. In Denmark, where the summer of 1976 was the driest since records began over one hundred years ago, the shortfall in the potato crop between 1974 and 1976 was 39%; the corresponding figure in the Netherlands was 30%, in the Federal Republic of Germany 24%, in the United Kingdom 32% and in Ireland, on the outer fringe of the affected areas 10%. Wheat yields showed similar but generally lesser falls: U.K., 23%; France, 18%; elsewhere, between 10 and 15%. Yields of food crops were generally good to the east and south of the drought area, so the shortfall in Europe as a geographical whole was not severe.

There were other problems for agriculture apart from poor crops and milk yields in this remarkable dry spell, during which many long-standing records of low precipitation were broken. In common with industry and the private user, the farmer was subject to the severe restrictions on the use of water which were found necessary in many countries. The lack of rainfall had numerous side-effects. Thus it was found in 1976 that the rate of application of herbicides had to be increased in order to achieve effective weed control. Furthermore, chemical residues proved to be more persistent not only in harvested crops but also in the top layers of soil, so that succeeding crops suffered from phytotoxicity. As is usual in prolonged dry spells, the fire hazard increased sharply and there were numerous and damaging outbreaks of forest, bush and grassland fires.

The experience of 1976 strongly suggests that present-day European agriculture, although, in some respects, better protected against weather hazards compared with earlier times, may be becoming progressively more vulnerable to seasons of prolonged dry weather. It is true that facilities for irrigation are now more widely available, but the heart of the problem is that it is precisely at the times when irrigation is most urgently needed for the farmer's crops that water is in shortest supply. The ever growing comsumption of water by industry and by the general public means that on occasions of shortage agriculture must compete for supplies with many other interests. Furthermore the need of modern intensive agriculture itself for water has increased very considerably. In addition to irrigation, water is required for drinking by livestock, for cooling milk and for air conditioning, for washing of livestock, vegetables, premises and machinery, for chemical sprays and for frost protection.

At the same time, the power supplies so essential to intensive agriculture are endangered in drought in cases where dependence is mainly on hydroelectricity. Finally, the problem of disposal of agricultural wastes and of avoiding pollution of rivers beyond acceptable levels, already a headache even in normal seasons, becomes more acute as the water flow declines in prolonged dry weather.

Seasons of more rainfall than the farmer would wish for have, in the past, represented a more frequently recurring phenomenon than damaging drought in northwest Europe. The problems which they bring for agriculture are therefore more familiar, and the harmful effects of excessive rain are easier to summarize. At worst it can deny the farmer access to his land because of flooding or because of the untrafficability of lanes or tracks. Wet days and waterlogged soil can hinder or prevent field work, and are equally a problem whether in preparing the ground, in constructing seedbeds, in sowing and tending crops, in spreading fertilizers, in applying weedkillers, fungicides or insecticides. Above all, there is the problem of saving crops, particularly of hay or cereals, in "sopping" harvest times.

There is a British saying that a moist season which is good for grass is good for little else. Most agricultural crops have their growth retarded and their yields seriously affected if they suffer waterlogging for any length of time. In many countries, the menace of direct flooding has been considerably reduced by drainage or controlled run-off, but in some districts land drainage in upland catchments leads to increased risk of flood damage in the lower reaches of the river. Cereal crops can be damaged by lodging in seasons of heavy precipitation. Wheat does poorly when ripening is delayed by late rains. Certain fungal plant diseases such as the downy mildews and certain stock maladies are particularly virulent in wet seasons.

As in the case of temperature fluctuations, the impact of excessive or deficient rainfall is inadequately represented by annual or growing season totals; the distribution of the rainfall throughout the year is of vital importance.

6.3.4 Other significant weather factors

While rainfall and temperature are the primary weather variables which affect agricultural activities there are other parameters which are also significant. Sunshine is needed in the ripening process and for improving the quality of crops such as fruit, vines, sugar beet, honey, etc. Frequent and intense hailstorms cause direct damage, while conditions of high humidity favour many fungal attacks.

Snow and high winds are two weather elements which deserve more than passing mention since their impact on west European agriculture can be considerable.

6.3.4.1 Snow

Experience during the long harsh winter of 1978-9 in western Europe illustrated once again the extent to which snow can bring disruption, hardship and extra costs to regions in which farmers are unaccustomed to coping with it or ill-equipped to do so. Snow falling even in modest quantities can give rise to traffic chaos in areas not organized to deal with it; the resultant disruption of traffic can, for example, prevent the arrival at farms of fuel oil, animal foodstuffs and other supplies, and the despatch to market of stock or milk. The handling of farm animals, especially sheep, is seriously affected. Unaccustomed snow makes life particularly difficult for upland farmers. The drifting of snow in windy weather can aggravate matters, especially where farm buildings were not designed to allow for it. Narrow country roadways, closely contained by earthen banks or bush fences, are liable to become completely blocked by deep snow drifts.

A less obvious problem of snowy winters is the damage which can be caused to winter cereals and grasses by low-temperature parsitic fungi which thrive under the snow corver. In Finland, pink snowmould (*Fusarium nivale*) can reduce winter crops of rye and wheat by some 25%; in the Netherlands it has been found to cause severe lodging in the same cereals after snowy winters.

On the other hand, snow cover can play a benign role by serving as a blanket to prevent excessive cold from killing overwintering crops, meadows and autumn-sown grain and oil crops. In its absence, the damage caused by low temperatures can be severe and widespread. On one night in January 1942 a frost of about -28°C accompanied by strong winds killed off the unprotected wheat crops in large parts of southwestern Sweden. After harsh winters of little snow which permit the frost to penetrate deeply into the soil, the retardation of vegetative growth in the following summer can be marked, especially in peaty soil.

On the January chart of mean sea level air temperatures, a large part of western Europe is encompassed between the isotherms of -2°C and +4°C, i.e. from a winter temperature level which implies virtually constant snowcover to one at which precipitation is much more likely to fall as rain rather than sleet or snow, and where snow rarely lies for long on the ground. Variations within this border zone in the frequency of snow and level of snowcover depend in part on relatively minor variations in the atmospheric circulation from one winter to the next; in general a colder winter, especially in the western sector, will mean more snow. It is of interest to do rough analysis of the probable consequence for snowcover of a small change in the winter temperature regime.

The mean winter isotherms which, over North America, run roughly in a west to east direction are deflected far to the north as they cross the Atlantic, and then plunge sharply in a roughly north to south direction on reaching western Europe. Thus the 0°C mean sea level isotherm for January which flanks the south coast of Iceland continues in a northeasterly direction well into the Arctic Circle, then veers abruptly down along the Atlantic coast of Norway, skirts the east coast of Denmark and continues southwards before it is deflected sharply eastwards as it reaches the central European mountains. According to Wallén, permanent snowcover in the European lowlands in the months January/February may be said to be located east of a line running along the Scandinavian west coast and continuing southwards more or less along the Oder-Neisse line, i.e. a little to the east of the 0°C isotherm. The January temperature gradient is oriented east to west and is approximately 2°C per 500 km. The broad effect of a 1°C fall in winter temperature would thus to be displace the zone of permanent snowcover in high winter to the west by something of the order of 250 km, and, of course, to increase the frequency of snow to the westward of the new limit and to lower the altitude of the snowline. Similarly, a winter warming of 1°C would tend to dissipate the snowcover over about the same distance to the east.

6.3.4.2 Winds

The maritime fringe of northwest Europe is the part of the densely populated world which is subject to the highest level of sustained wind speed. The hurricane zones of the subtropics experience stronger spot winds but average out well below the mean

January wind speed of 10 m/s, which prevails in the western half of Ireland and the northwest of Scotland. Throughout much of western Europe, the advantages of shelter for agriculture far outweight its disadvantages; for most crops it not only increases yield-by protecting plants and blossoms, lessening fruit drop, preventing lodging-but it also improves quality. In combination with a high level of rainfall, the wind levels make "driving rain" a particularly troublesome problem which affects livestock in the open and makes the thermal insulation of buildings more difficult than in most parts of the world. The problem of exposure on windswept uplands is severe.

Any change in the general circulation which reduced the frequency and intensity of high winds would be favourable to agriculture, for apart from the direct harm which they cause, it would reduce the impact of sea salt damage to plants around the coasts. It would further prevent the compounding of ill-effects to agriculture which occur when strong winds accompany hail or bitterly cold weather. It is true that fresh winds discourage the activity of the insect vectors which spread the virus diseases of potatoes and other crops, but the wind levels in the Scottish and Irish seed potato growing areas could drop by over half without endangering the quality of the product.

6.3.5 *Wheat and potato crops as affected by climate*

Any general analysis of weather effects on agriculture is necessarily coarse-grained since each farming activity has its own particular environmental requirements. The weather which favours one crop at a particular stage of its growth may hinder the development of another. The sequence of weather needs of particular food crops can be most usefully illustrated by taking a closer look at wheat and potatoes, two very dissimilar crops which are nevertheless basic food stuffs which have played a prominent role in European agriculture, and which take on even greater importance whenever, as in wartime, other food supplies become endangered.

6.3.5.1 *Wheat*

As in the case of other crops, high yields of wheat depend on a number of factors apart from suitable weather: good seed stock, suitable soil maintained in good tilth, the best level and balance of fertilizers, effective control of pests and diseases, the skill of the farmer, etc. That weather in particular is a major factor emerged clearly in the last century from a study of the long series of yield returns from the famous Rothamsted wheat plots, in which other conditions were kept uniform. In the most favourable weather sequence of 1862-3, a dry November-

December and a wettish January were followed by a markedly dry spring. There was adequate rain in the May-June period to support active growth, after which the weather up to harvest was warm, bright and dry. The total rainfall during the growing season was less than 600 mm, and the crop yield exceeded the average by more than 50%. In contrast, the disastrous wheat harvest of 1879 which scarcely touched half of the normal yield was the result of an uninterrupted sequence of wet and sunless months, with a total rainfall in the growing season of over 900 mm.

This British example is representative of the main problem of growing wheat in the central belt of Europe, i.e. the risk of excessive moisture (WMO 1975). This contrasts with the drier Mediterranean lands, for, although wheat is relatively modest in its water requirements, drought can hinder ploughing, delay germination and kill the ear. In the extreme northern countries, the lack of warmth and light is the primary obstacle, causing the dreaded "green years" when the wheat fails to ripen and rots in the ear because of cold and lack of sun.

More important than the total rainfall during the growing season in the main Europe wheat belt is the distribution of dry and wet spells during the life of the plant. Plentiful moisture in the early summer when the tillers are pushing into ear is beneficial to both autumn- and spring-sown wheat. On the other hand, abundant and persistent rains at either end of the growing season can be disastrous. Apart from the difficulty of preparing the ground for the autumn sowing of the crop, heavy rains in autumn and winter cause waterlogging of the soil, drown the seeds, retard root development and tillering and weaken the plant structure. In a well-drained soil, the roots will continue to grow throughout a reasonably dry winter and will provide valuable insurance against dry weather later. Excessive rain later in the growing season is worst of all for it can undo the good resulting from earlier favourable weather and, if it continues to harvest time, give no opportunity for a late recovery.

In general then, a dry but not droughty year and particularly a dry and sunny summer benefit both winter and spring cereals, particularly if they are backed by adequate reserves of soil moisture. A heat wave, on the other hand, can cause summer heat stress, lead to the scorching of the still immature grain and bring on an early and light harvest. Undue cold can also cause damage at different stages of growth; a particularly harsh winter can lead to soil heaving and also cause direct damage to varieties of lesser hardiness, while severe frost can do great harm at the sensitive heading and flowering stages. On the whole, a winter season of average coldness is best for wheat in central Europe since unseasonable mildness favours weeds and pests.

To a certain extent, present day grain harvesting methods can undo some of the worst consequences of bad weather. The combine harvester has the ability to pick up lodged crops, but more slowly and at greater cost, apart from the loss of yield from poorly-filled yields which results from early lodging; a crop which lodges early and severely can be reduced by a much as 1 tonne/ha due to less favourable light interception and reduced assimilation.

6.3.5.2 Potatoes

In contrast to wheat, one of the man's oldest staple foods, the potato is a relative newcomer to the European agricultural scene. The first consignment was brought to Spain from South America just over four hundred years ago. It was received in Europe with the reluctance which might be expected for a type of food quite strange to palates accustomed to grain and dairy products; nor did it endear itself to the scholarly in view of its close botanical relationship to certain dubious and even highly poisonous plants,such as the erotic henbane and the deadly belladonna or nightshade.

The comparitive rapidity with which the potato was adopted in Ireland, was due, in the first place, to the climatic differential which worked against grain crops and particularly wheat in the moist Atlantic fringe of Europe. The grain crop was often deficient because the growing season did not peak to a warm, dry, sunny harvest time. In the wet years, when the cereals faltered or completely failed, the sturdy potato, at that time free of major disease, continued to flourish and proved itself an invaluable stand-by for the masses. By the opening years of the eighteenth century, the potato, which at that period did not keep for more than six months, was well established as the winter food of the Irish poor and would shortly be widely adopted in some other parts of Europe.

It would be difficult to understand the varying impact on European agriculture of disastrous weather seasons in the eighteenth and ninteenth centuries without keeping in mind the different climatic reactions of the wheat and potato crops, and of their respective vulnerability to certain weather-influenced diseases, as discussed in the following section.

Meanwhile, it should be noted that the potato crop's ready tolerance of wet weather is not its sole climatic merit. It has no fussy requirement for well-prepared seed beds, and while vulnerable to early and late frosts, it is capable of growing and giving a reasonable yield in any except the very coldest or very driest climate, nor has it a series of growth stages with corresponding specialized weather requirements. Indeed. it is the most widely cultivated crop which comes nearest in toughness and vigour tc a weed.

6.3.6 *Climate and the incidence of agricultural diseases and pests*

Apart from the direct effects of weather on farming, a factor of major importance for agricultural production is the influence exerted by climatic fluctuations on the uneasy balance between crops and farm animals on the one hand and a malevolent variety of diseases and pests on the other. This balance is upset in a season when the weather sequence is unusually favourable to the multiplication of a parasite or an increase in its virulence, particularly if conditions at the same time tend to weaken the resistance of the host. In some seasons, the reason why certain parasites flourish may be less that the weather helped them to increase and multiply than that it discouraged the growth of competing organisms. In either case, the cumulative effect of a sequence of even mildly favourable years may be an explosive multiplication of a particular pest. Wind transport of insects and disease spores from a region of multiplication or successful overwintering to a region of vulnerable crops or animals can be an important factor in epidemiology.

Significant examples of weather influences on plant pathology are not hard to find. Recently the desert locust, which appeared some years ago to be virtually wiped out, has taken advantage of more frequent rains and relaxed vigilance to renew its ravages. Nearer home, the Colorado potato beetle has shown its ability to multiply explosively in favourable meteorological and biological conditions. On the other hand, its numbers decline in cool moist summers, as in France in 1946. Severe winters with limited snow cover permitting low soil temperatures, such as occurred for example in Europe in 1962-3, cause high mortality among all species of beetle and fly which over-winter as eggs or larvae in the ground.
Grasshopper plagues build up in a succession of hot dry seasons, mainly because the eggs and nymphs escape destruction by parasites and diseases which flourish during cool wet conditions.

Animals affected by severe weather, either directly or by reason of shortage of forage, become more vulnerable to pests and disease. Drought not only predisposes cattle to illness but also increases the risk of gastroenteritis as the responsible larvae are ingested with close-cropped grass. In wet seasons, the level of liver fluke disease in sheep and cattle rises sharply. Wind assists the local spread of food-and-mouth disease in rainy weather.

We shall examine in a little more detail the role of the environment in the development of some major diseases of the wheat potato crops which, in different ways, are highly sensitive to weather conditions and whose impact responds significantly to climatic fluctuations.

6.3.6.1 Black and yellow rusts of wheat

The problem of rust epidemics, long a serious menace to the wheat crop, and their relationship to weather conditions has recently been well reviewed by Zadoks (1966). His analysis of the serious epidemic of yellow or stripe rust (Puccinia striiformis) which attacked a large part of the European wheat crop in 1961 showed that it was favoured by the high level of disease in the previous year, and by the following succession of weather conditions which led to further multiplication and carry-over:
a) a wet summer in 1960, favouring the growth of a volunteer crop;
b) a long mild autumn with many dewy nights;
c) either a mild winter or continuous snow cover to protect the inoculum;
d) a short and mild dry spell at the end of winter;
e) an early warm and moist spring in 1961.

A quite different kind of weather mechanism is involved in the build up of epidemics of black or stem rust (Puccinia graminis), where the most important factor is the long-distance transport of disease spores from the warmer southern districts. Two main migration tracks in Europe are indicated (Fig. 6.5). The western path runs broadly norteasterly from Morocco through the Iberian Peninsula towards Scandinavia. The main eastern track runs from the lower Danube valley towards the Baltic, with possible branches to both east and west.

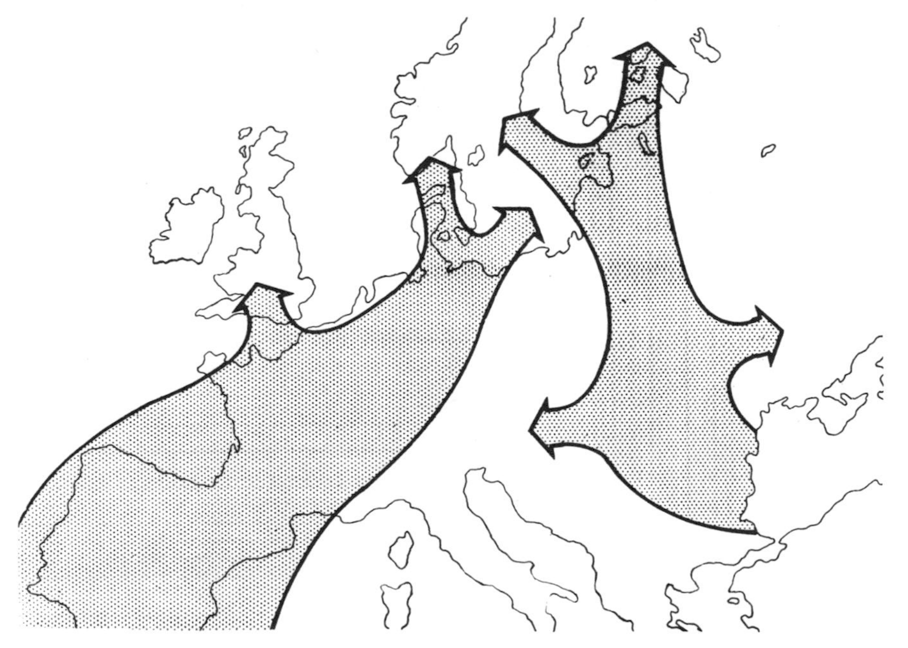

Fig. 6.5. Mean migration tracks of black rust of wheat (after Zadoks).

Clearly, the character of the European wind pattern at the critical crop stages and the main areas of deposition of the spores, as brought down by gravity or scoured by rainfall, are of vital importance in determining the level of black rust damage in the wehat crop.

6.3.6.2 Potato blight

Potato blight, caused by the fungus, Phytophthora infestans, has been a serious scourge of the European potato crop since 1845. The inoculum is carried over from one season to the next in infected tubers, and the disease builds up in a number of successive generations in periods of favourable weather. These consist of lengthy spells of saturated or nearly saturated air with modest temperatures, not falling below 10°C and preferably nearer 15°C, and the presence of a layer of free water on the plant surfaces to permit infection. Under European conditions these requirements are frequently met within the warm sectors of a series of open wave depressions consisting of maritime tropical air. When anticyclonic conditions prevail in continental Europe, such invasions of maritime tropical air are likely to affect only the Atlantic littoral, but in seasons when then depression tracks are further to the South (Fig. 6.6) their influence can penetrate into much of central Europe and set off a widespread attack of potato blight.

Shallow thundery depressions which are not uncommon in continental Europe in summer can also create the weather conditions which favour blight.

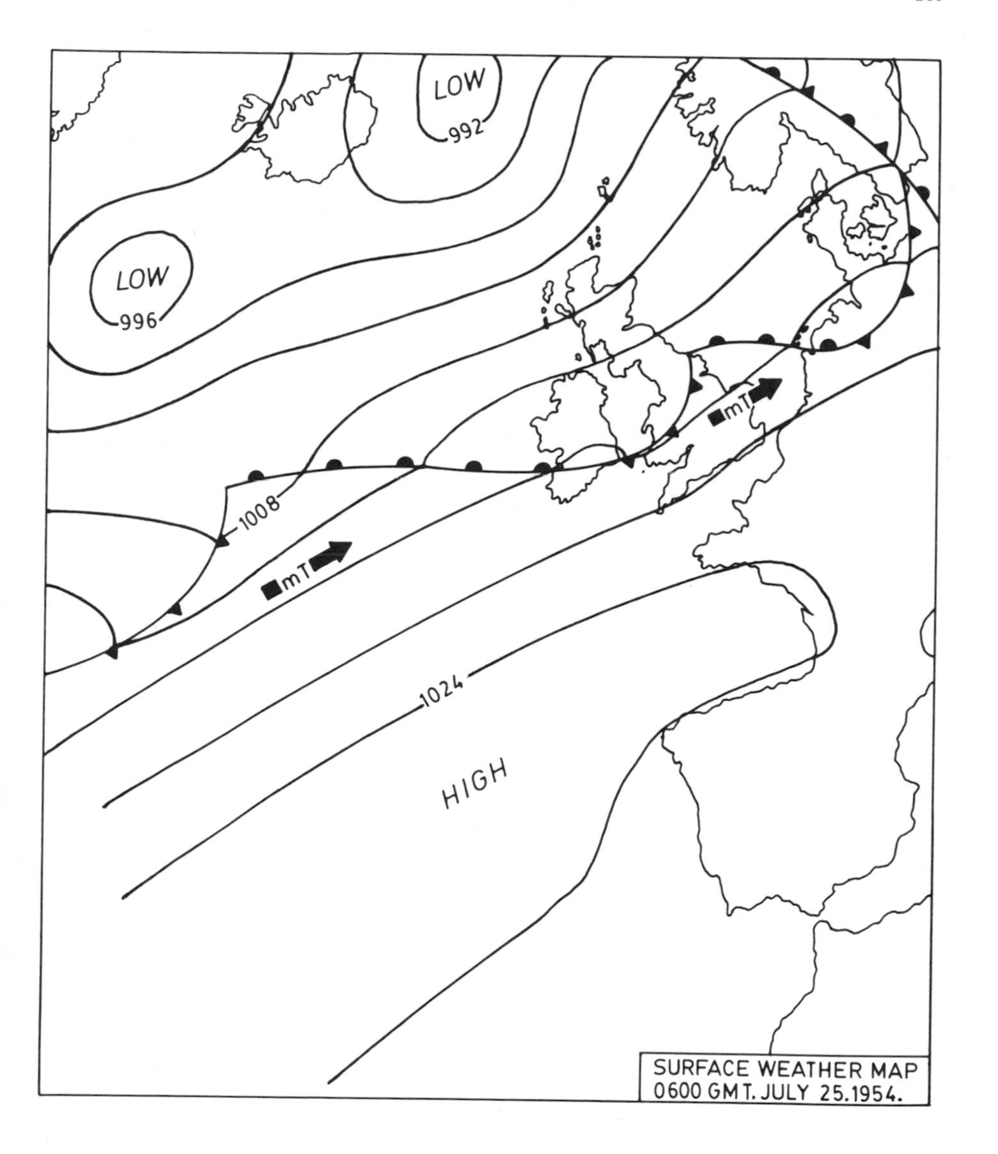

Fig. 6.6. Synoptic weather situation (mT = maritime tropical air) favourable to potato blight in NW Europe.

6.3.7 *European food production and the weather: the lessons of history*

Looking back over the European food harvests of the last two or three centuries and seeking to relate them to seasonal weather one comes face to face with certain difficulties. Not only do the meteorological data become more sketchy the further back one goes, but reliable agricultural statistics particularly for the earlier years are even thinner on the ground. One season's returns were not necessarily independent of another's; a poor harvest often tended to lead to deficiencies in the following year because of scarcity of seed to sow and the poor quality of the seed.

Even where there is an adequate supply of broad meteorological information, the data may mask certain short-term weather "episodes" which influenced a particular critical harvest. Thus the failure of the French wheat crop in the prerevolution year of 1788 was due in part to broad weather patterns such as the rains of October-November 1787 which hindered sowing, the mild winter which followed to encourage weeds and pests and finally a spring drought, but possibly in equal or greater degree to a couple of critical short-term weather incidents, such as the scorching of the grain by a blast of dry shrivelling heat at a particularly sensitive stage, and the damage caused by the famous hailstorms of July 1788. The social consequences of the wheat deficiency were aggravated by the severe winter of 1788/9 (Neumann 1977).

The pattern of a critical sequence of weather types, none of major significance in itself but cumulative in effect, recurs in many of the historical years of European food shortage. Further, one finds that a single bad harvest did not necessarily have severe consequences, unless it came after a lulling sequence of good years had built up false confidence. The worst impact, as one might expect, came from sequences or an unusually high frequency of years of poor farming weather. Indeed, there exists a general tendency for years or seasons with excessive weather anomalies to repeat and/or to occur in clusters (see Chapter 1).

Some of the best available series of agricultural statistics for earlier years refer to grain prices. The cost of food is not necessarily a true index of harvest fluctuations resulting from favourable or unfavourable growing seasons. War is only one of the factors which can distort grain prices. Nevertheless, the graph of nearly four centuries of German rye prices (Fig. 6.7) reflects to a surprisingly good degree not just local food conditions, but indeed the major ups and downs of the west European grain harvests, when compared with the parallel data assembled by Le Roy Ladurie (1972) and others (Hoskins 1968, Jones 1964, Slichter van Bath 1963). One must keep in mind that only after a particularly bad harvest did the price rise manifest itself immediately; normally a shortage became fully effective only in the following year. A sequence of poor harvests over a large part of Europe had a cumulative effect on grain prices, marked on the graph by a peak spanning several years.

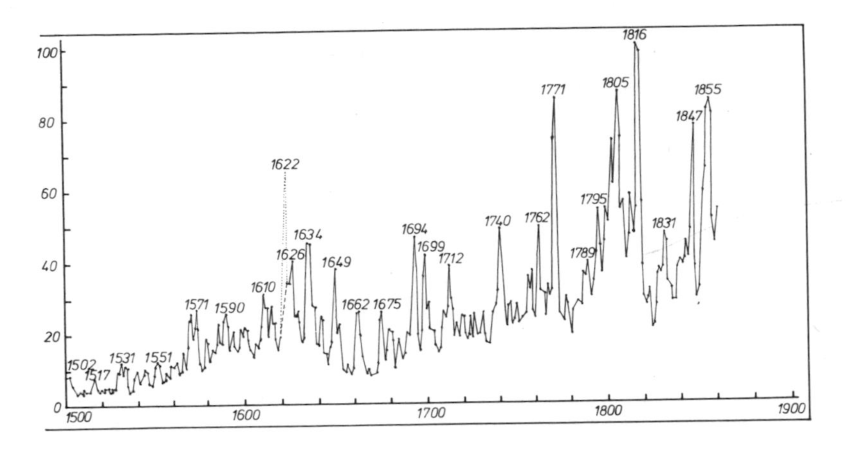

Fig. 6.7. German rye price index (average from 5 cities, preliminary values, after Flohn) reflects good and bad growing seasons (1622 peak: currency devaluation, 1634 war crisis; 1950 = 100).

The average price level in Fig. 6.7 showed no marked trend during most of the two centuries prior to about 1740, but subsequently tended to rise reaching a climax at the time of the Napeleonic wars. In both periods, there were sharp fluctuations from one year to another. The valleys reflect the good years of prosperity and comparative plenty. Even in the seventeenth century, which Le Roy Ladurie characterises as one of the grimmest periods in the historiography of European famine, the records indicate some lowlands of relative food adequacy which are confirmed in Fig. 6.7: the lull of the 1650's, the spell of good harvest between 1665 and 1672, and the generally bountiful years of the 1680s (cf. section 2.3.3).

Each of these periods of relief ended with a renewal of bad weather and food problems, the most disastrous being the infamous "black decade" which brought the seventeenth century to a close (cf. section 2.3.3.1). One of the worst famines western Europe had known since the early Middle Ages began, according to Le Roy Ladurie, with the harvest of 1693. France was converted into "a great desolate hospital without provisions". In Scotland, the Jacobites remembered this cold and barren decade as "King William's years", preferring to attribute the disastrous weather to God's anger at an unwelcome change of monarch; the consequent misery is said to have weakened Scottish resistance to the Act of Union. Catastrophic famines in Finland and the

Baltic countries have recently been described (Neumann and Lindgren 1979, 1981).

A succession of generally good harvests followed the turn of the century but the decade closed with a number of difficult seasons. Again the climate relented in the long balmy period of some twenty years centred about 1725. In Sweden, for example, Utterström (1961) reports that this spell of exceptionally mild weather favoured the growing of grain and the pasturing of cattle; these decades helped to provide employment and to improve public health and longevity, so that the era was marked by a sharp rise in the population of Sweden.

The complacency evoked by this sustained period of benign weather was abruptly shattered throughout western and northern Europe by the notoriously disastrous winter of 1739-40 (Manley 1958).

This had been preceded by a wet and stormy season which led to a poor late harvest. Towards the end of December 1739 an abnormally cold spell set in which continued unabated for two months, to make the winter "one of the most hellish in the memory of man". The major rivers froze over as did the Zuyder Zee and the kattegat. Manley's (1954, 1974) tables of mean air temperature in central England show that the January-February readings in 1740 were more than 6°C below the contemporary decadel average. Little rain fell in the first half of 1740 and drought in many areas added to the problems caused by continued cold, with temperatures well below normal in every month of 1740 except September. According to Manley's figures, the September mildness was a shortlived interlude; the October temperature plummeted by nearly 9°C to the lowest figure for that month in over 250 years' records. In fact the 1739-40 "giant among hard winters" was a prelude to some 18 months of cold harsh weather; the year 1740 as a whole had the lowest mean annual temperature in Manley's entire series.

The demographic crisis which affected much of western and northern Europe in the early 1740s was particularly acute in Ireland (Drake 1968). Potatoes had become the staple winter food of the Irish poor; in the deceptively mild epoch of the earlier 1700s it was common practice not to harvest the crop in the autumn but to leave it in the ground to be dug out as required during the winter. In the bitter cold of the 1739-40 winter the Irish potato crop was destroyed in the ground. The resultant mortality in what became known in folk memory as "the year of the slaughter" was comparable to that in the great famine of the following century. The lesson of this dreadful winter was quickly learned. From 1741 onward potatoes were harvested and safely pitted or stored before the winter set in. New longer-lasting varieties, such as the superb "Irish Apple", were developed which helped to accelerate the adoption of the potato elsewhere in Europe, where it made a substantial contribution to blunting the edge of famine in the following years.

For the years from 1740 to the end of the Napoleonic wars the price graph in Fig. 6.7 is partly distorted by the disruptions of war. It does, however, show the effects of good harvests in the well-documented warm decade centred about 1780 and the climacteric year of 1816 in a decade of poor summers in which bad weather, poor harvests and distress occurred over a wide part of the world and led to the highest peak in Fig. 6.7.

Human memory of abnormal weather tends to be shortlived; each disastrous season is recalled only until another more recent event replaces it as "the worst in the recollection of the oldest inhabitant". Nevertheless the year 1816 maintains its reputation as "the year without a summer", "poverty year" or "eighteen hundred and froze to death" (Landsberg and Albert 1974, Milham 1924). It has been the subject of countless articles and has recently been comprehensively reviewed by Post (1977). The chief weather abnormality of the year on both sides of the Atlantic was the occurence of a succession of very cold months, particularly in the summer half-year; July 1816 returned the lowest mean temperature in England of any July in Manley's long series of data. In Europe, the substained cold weather and persistent rain of this drenching summer hampered first the growth and then the ripening of the wheat crop, and finally caused it to sprout in the ear. In a pithy review of the character of successive harvests, the "Gardeners' Chronicle" (1846, p. 308) had this to say of 1816: "Extremely cold and wet throughout. One of the worst harvests ever known. Bad times". The situation was aggravated by the economic recession which followed the Napoleonic wars and by the fact that 1816 was the climax of a series of five growing seasons of inclement weather over much of the world. After this, mass emigration to the U.S.A. started especially in areas with small farms (at the lower limit of subsistence) in Western Germany and elsewhere, with further peaks after bad harvests in the following decades.

Although the potatoes grown in this harsh season were small and watery they made a substantial contribution towards lessening starvation in Ireland and in other countries where they were extensively grown.

However, the dual safeguard against bad seasons of potatoes and grain was being gradually eroded in Ireland where by 1845 the vast majority of a swollen population of over 8 millionm people depended for their food on the potato alone. How this vulnerable economy collapsed in ruins in 1845-7 is a vivid illustration of the dangers of monoculture and of complacency (Bourke 1964).

The summer of 1845 in Europe was a dreary one, cold and wet, with the path of atmospheric depressions displaced considerably to the south. There was a tendency to attribute the failure of the potato crop to the unseasonable weather, although experienced growers recalled that the crop had stood up well against even worse conditions in 1816. In fact, the potato in Europe was being attacked for the first time by a plant disease which had raged in the United States of America since 1843. It first came to notice late in June 1845 in West Flanders where trials were being carried out of potato varieties newly imported from North and South America. In conditions of wind and weather which favoured its spread, it appeared within three months in most of Europe (Fig. 6.8).

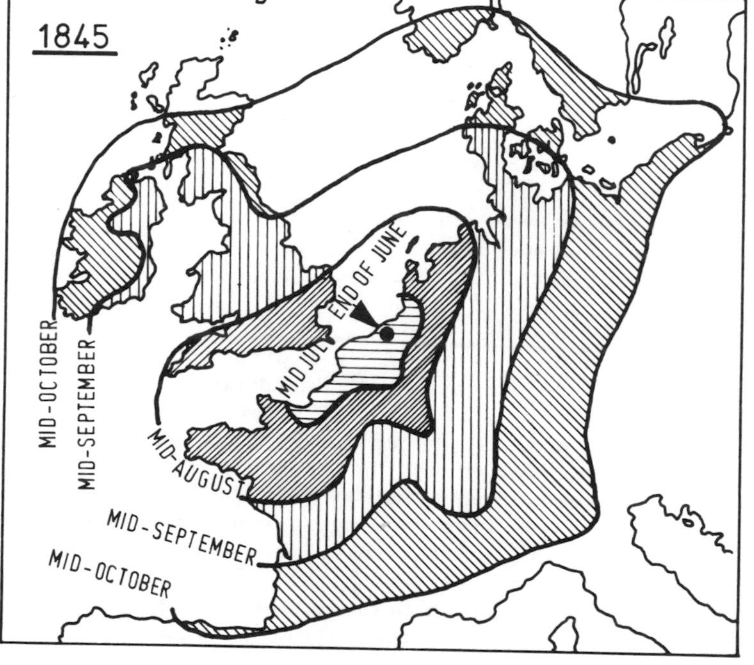

Fig. 6.8. The spread of potato blight in 1845.

The ravages of disease on the 1845 potato crop combined with a poor grain harvest caused great distress in Belgium and elsewhere in the affected parts of Europe. The effect on the lives of the poor in Switzerland is vividly described in Jeremias Gotthelf's well known novel, "Käthi die Grossmutter", published in 1847. The disease came later to Ireland in 1845, and dealt the crop only a glancing blow, so that the potato harvest fell below normal by no more than about one-quarter.

Thus far the weather had played only an ancillary role by favouring the development and spread of the new disease. Now however occurred a sequence of weather types which led step by step to inevitable disaster in Ireland and great distress in Highland Scotland. A mild winter allowed disease-infected tubers which had been discarded in the fields during the 1845 harvest to survive and to carry over infection on a large scale to the following year's crop. A sodden spring delayed the sowing of grain and potatoes in 1846, and a period of drought and record high temperatures in June of that year further retarded the growth of the crops. Over mainland Europe and much of England continuous anticyclonic conditions and scanty rainfall limited both grain production and potato disease in 1846, but from July onward depressions skirting the Atlantic coasts brought ideal conditions for the development of potato blight in Ireland, so that a crop which was highly susceptible to the new disease was wiped out at a stage when tuber formation had scarcely begun. Nor dit the weather relent even at that late stage. Fierce autumnal gales delayed relief ships bearing food from America; a winter of abnormal severity in 1846-7 increased the problems of a starving population and hindered the relief work of the Society of Friends and others. The main consequences of the Irish potato famine are well known (Woodham-Smith 1962): widespread starvation and fever, and an immediate fall in population due to death or flight of nearly two million souls, followed by a century of continuous haemorrhage by emigration, which makes the Irish demographic graph unique in the world (Fig. 6.9).

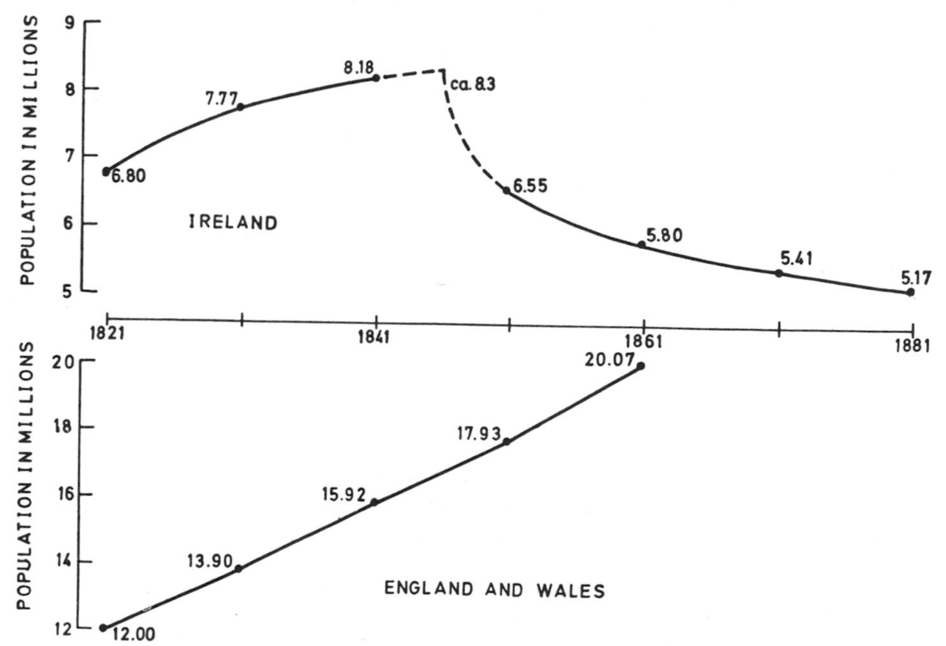

Fig. 6.9. Population changes in Ireland (above) and England + Wales (below) during the 19th century.

Just beyond the end of the graph in Fig. 6.7 lies the decade of the 1860s, one of the hottest and driest on record. There were prolonged droughts in much of central and southern Europe. Shortly afterwards came the "sopping harvests" of 1879 and 1880. In the most recent hundred years, there were still considerable fluctuations from one growing season to another, but due to improved traffic and trade conditions, a bad harvest, although still capable of causing local hardship and deprivation, no longer led inevitably to acute food shortage in Europe.

6.4 Mediterranean Europe

6.4.1 The climatic and agricultural scene

The Mediterranean region, which for the present purpose consists of Spain, Southern France, Italy and Greece, deserves special attention since both climate and agriculture differ materially from the rest of Europe. We shall focus attention on particular practical problems of major importance to regional agriculture.

The characteristic aspects of the Mediterranean climate which have the greatest impact on agricultural production are three-fold: a marked tendency to water shortage during the whole year (In addition to the regularly dry summer season), high values of solar radiation and the occurrence of violent or extreme weather phenomena. First in importance of these "episodic" destructive phenomena is hail which, because of its frequency and intensity over wide areas, represents a serious economic problem on a national and regional scale. Damage may also be caused by severe and localised rainstorms; a rainfall total of almost 950 mm in a 24 hour period has been recorded at an Italian observing station. Farmers suffer also from whirlwinds and from excessive heatwaves with occasional days on which the temperature climbs to 50°C. As elsewhere in Europe, spring frost can cause great damage to crops.

Wheat and maize are extensively grown in the Mediterranean area: and in regions of suitable climate and soil, such as the Po valley, rice is an important crop. The climatic needs of the wheat crop have been discussed earlier (Section 6.3.5); reference should be made, however, to the early Italian work on meteorological aspects of wheat growing (Azzi 1921) which laid the foundation for later developments.

Again while it is not proposed to add to the earlier discussion of pathogens and pests, it should be recalled that some of the earliest work on forecasting a plant disease with the help of weather data related to attacks of the mildew of the vine (Plasmopara viticola) in France and Italy.

Certain crops which are particularly adapted to the Mediterranean climate are discussed later, i.e. the vine and, in lesser detail, the citrus tree and the olive.

6.4.2 *Water deficit*

One of the major problems of farmers in the Mediterranean climate is the recurrent difficulty of providing for the water requirements of crops. In this climatic region, annual or monthly rainfall figures cannot safely be used as even a crude index of water availability in the soil. For an accurate analysis of soil moisture in the root zone of crops it is necessary to take into consideration losses by evapotranspiration and the various relevant characteristics of the soil.

A reasonable approximation to a balance sheet of soil moisture availability may be drawn up by comparing, for successive periods of five or ten days, the precipitation P with the evapotranspiration E, as calculated with the help of one of the established formulae. e.g. that of Fenman, which has been found to give reasonable results in the Mediterranean region. Assuming average soil characteristics and a representative crop, one may calculate the cumulative "water deficit", $D_c = \Sigma(E-P)$. Run-off is allowed for in this computation by ignoring any surplus above the saturation level of the soil. The calculation commences from a conventional date, generally the first of January and continues until the end of the rainy season.

The critical date is that on which the value of D_c becomes positive, i.e. the onset of the "arid season". The higher the positive values of D_c during the arid season, the more the soil progressively dries out and the greater the deviation between actual and potential evapotranspiration. The actual evapotranspiration then depends less on climatic factors and more on characteristics of soil and crop, until the final stage in which plant growth and crop production are seriously affected.

In the Mediterranean region, where rainfall is irregularly distributed throughout the year, with greatest amounts concentrated in the final three months and very little falling in summer, severe water deficits can occur during the growing season even though the annual total may appear adequate.

In the Mediterranean region, where rainfall is irregularly distributed throughout the year, with greatest amounts concentrated in the final three months and very little falling in summer, severe water deficits can occur during the growing season even though the annual total may appear adequate.

Figure 6.10 illustrates this phenomenon in a schematic way, and indicates the extent to which the soil moisture deficit varies between regions and from one season to another. The diagram shows the annual trend of the water deficit in the years 1968-72 in two representative farming areas: eastern Sicily, with a typically Mediterranean climate and the Po Valley with an almost continental climate and rainfall both during winter and summer (see also Section 3.2.2). Although the mean annual rainfall and the conventional indices of aridity (such as that of De Martonne) are about equal in the two areas, the course of the water deficit, as it affects farm crops, differs radically and indeed the types of agriculture followed are correspondingly different.

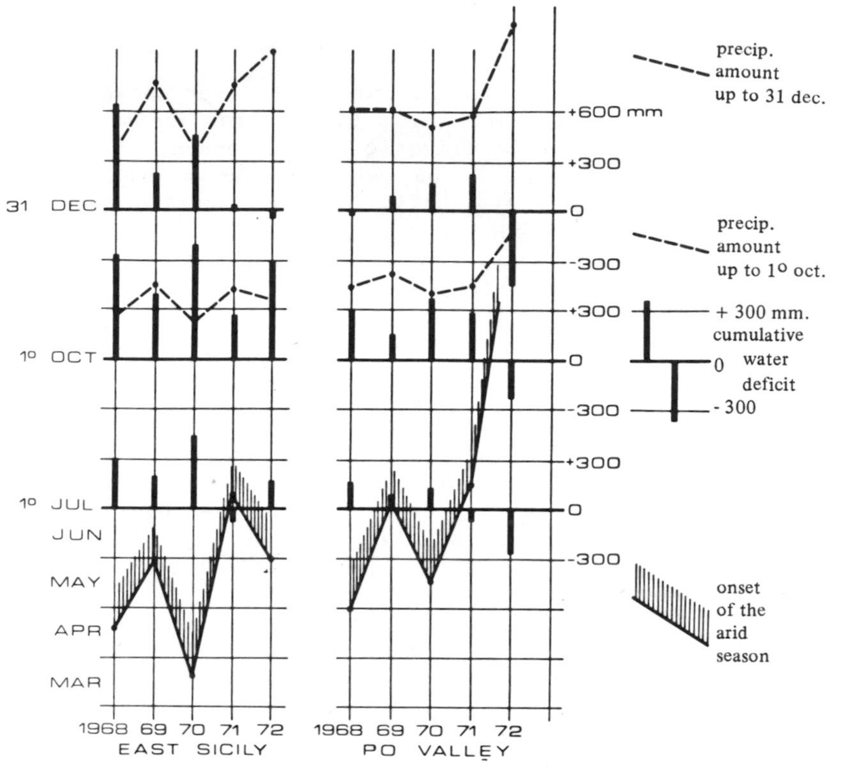

Fig. 6.10. Annual course op precipitation, of the water balance in the soil, and of the consequent variation of the date of onset of the arid season in East Sicily and in the Po Valley, 1968-72.

Figure 6.10 shows also how the extent to which the arid season overlaps the growing season in the Mediterranean region varies from one year to another. It is important to keep this fact in mind in trying to evaluate the impact of possible climatic variations on Mediterranean agriculture. Such impact depends less upon variations in the annual totals or averages of the different meteorological parameters, than on the distribution of these parameters in the course of the year, and this in turn is controlled by seasonal and irregular changes in the pattern of the general circulation of the atmosphere.

The relationship of dynamic meteorology to the annual course of the different meteorological parameters, and the precise impact of the latter on agricultural production in the Mediterranean countries, requires further research as a prelude to better understanding of climate variability and its economic consequences.

6.4.3 Solar radiation and temperature

More incoming energy from solar radiation reaches the earth's surface in the Mediterranean area, both seasonally and annually, than elsewhere in Europe. This is due only partly to the latitude factor, by reason of which the intensity of radiation reaching the outher atmosphere is at a mazimum in low latitudes: an important element is the generally lower cloudiness of the Mediterranean region, particularly in the summer months.

The level of radiation is an important factor in crop production, affecting as it does both quality and quantity. A simple approximate index of radiation is the curation of bright sunshine, which is recorded at many meteorological observing stations. Mean annual sunshine reaches about 3.000 hours in the southern coastal regions of Europe, corresponding to almost 150.000 cal/cm^2 of global radiation per year (about 6.3 gJ/m^2 per year or about 200 W/m^2).

The fluctuations from one year to another in the amount of solar energy reaching crop level have a significant effect on agricultural production. Considering the monthly values over a period of twenty years, the variation at a single observing station, between absolute maximum and minimum readings, fluctuates according to area, by 30-40% of the average in winter and by 15-20% in summer. On an annual basis the variation at an individual station rarely exceeds 10%.

Figures for radiation at the earth's surface in Italy reached unusually low levels in the year 1972. The deficit below the average level in the previous ten years was 7.3% over the country as a whole and reached 12.5% over the Po Valley; the corresponding decreases in net radiation (which includes the effective outgoing radiation of the Earth's surface) were 9.5% and 16% respectively. This marked lack of sunshine, which was accompanied by an abundant if not exceptional rainfall, had a depressing effect on agriculture, although not to the degree which might have been expected. After making allowances for crop yield trends, production per hectare was down by 3.5% for wheat, by 9.7% for olives, and by 2.4% for oranges. The yield of grapes actually increased by 7.2%, but the quality, as was to be expected, was very poor.

Air temperature and net radiation are closely interconnected, and it is often difficult to distinguish between the effects on plant growth of the two parameters. Furthermore the effect of temperature is far from simple or linear; from one stage of growth to another its effect may change radically. In general, higher temperatures accelerate the vegetative cycle and lead to an earlier date of maturity and harvest.

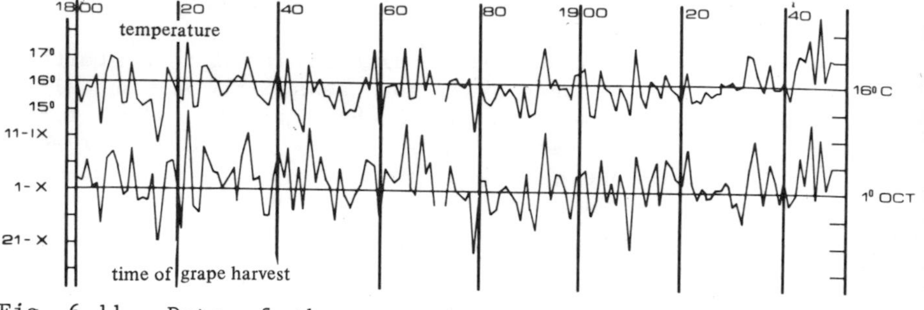

Fig. 6.11. Date of the grape harvest and mean temperature of the growing season, France, 1800-1950.

Figure 6.11 shows, for each year in the period 1800-1950, the mean air temperature during the growing season of the grapes and the date when the grape harvest in a region of France began (the earlier dates are higher on the vertical scale). The coefficient of coincidence, which measures the extent of variation of the two curves in the same sense, is as high as 75%. However, a corresponding parallelism does not apply to annual yields or the quality of vine, both of which are related in a complex way to a number of weather factors, as will be discussed further in the Section 6.4.4. Here, as elsewhere in agriculture, knowledge of the fluctuations of a single climatic factor, such as mean annual temperatures, is never a sufficient basis for deducing the consequences for agricultural production.

Records of the dates of the beginning of the annual wine harvest exist for various parts of Europe. These lists, which in some cases date back to the sixteenth century (Legrand 1977, Ladurie and Baulant 1980), can be used more readily as an aid to the study of past variability of seasonal temperatures, than as indicators of climatic effects on agricultural production.

6.4.4 *Weather and the vine*

To evaluate the implications of specific climatic changes for agriculture requires a knowledge of the response of farm crops and animals to different meteorological factors. Although considerable research has been undertaken on this problem, much remains to be done before objective quantitative yardsticks are available to relate climatic conditions to the growth and yield of different crop species and, in particular, fruit trees. The present state of knowledge may usefully be illustrated by discussing the climatic requirements of the vine in order to ensure a good production of wine as regards both quality and quantity.

The vegetative cycle of the vine extends over two years although there is a harvest every year. While the developments which lead to the current year's harvest are taking place, the foundation for the harvest of the coming year is already being laid in spring in a latent form. The rudimentary shoots usually appear in April, then they gradually enter a dormant period which begins in August and ends in November. During the winter months they prepare for the season of active growth which generally begins in April and ends with the ripening of the grapes and the harvest in September/October.

The climatic factors which affect to varying degrees the course of this lengthy gestation, are many and complex. As a first practical approach they may be listed under three main headings:

a) sun: e.g. solar radiation, sunshine, or average temperature (the latter being used with due caution);

b) water: e.g. water deficit, the ratio of actual to potential evapotranspiration, or monthly precipitation (again applied with caution);

c) heat: as expressed, for example, by an index such as the sum of degree-days, i.e. the excess of day-by-day temperature above specified threshold values.

Studies have been carried out in France to determine the more significant links between the above three groups of meteorological parameters and the quantity and quality of the grape harvest, the latter being measured by the alcohol content of the wine and inversely by its acidity.

The results for the various phases of the two-year growth cycle of the vine are presented schematically in Fig. 6.12. The sign of correlation, inverted in the case of acidity, is indicated, and more significant correlations are enclosed in circles.

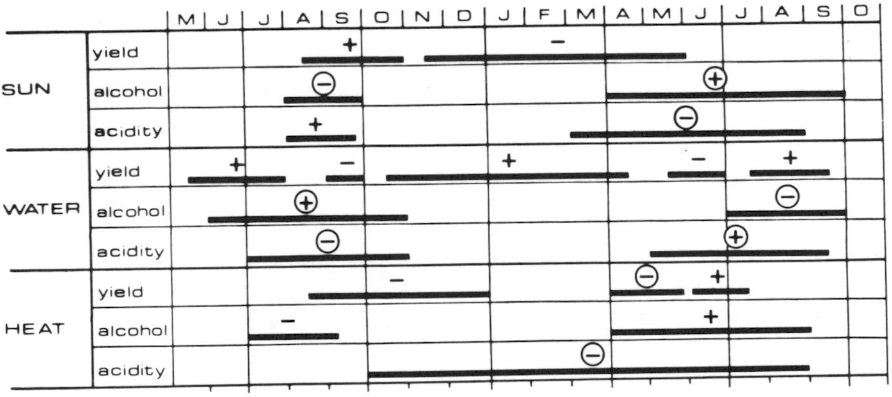

M J J A S O N D J F M A M J J A S O
SUN yield alcohol acidity
WATER yield alcohol acidity
HEAT yield alcohol acidity

Fig. 6.12. Tentative hypothesis of the correlation between certain meteorological factors and the quantity and quality of the grape harvest in France. The more significant correlations, positive or negative, are encircled.

It should be noted that changes in average values of the meteorological parameters rather than their absolute values have been taken into account, and that the averages differ for each area of production and each vine species, partly because the onset and duration of the successive growth stages for which the averages are calculated, differ according to plant type and zone.

A further complication is that the various meteorological factors are not only interrelated among themselves but also interact in their effect on plant life, largely because each affects the different stages of growth and so contributes to altering the lenghts of the periods during which the other factors are operative. Two main consequences result from this. Firstly, one cannot validly draw up a rigid calendar of the requirements of plant life for sun, water and heat; the available climatic information must be used in a flexible way, relating growth phases of specific crop varieties to different meteorological patterns. Furthermore, it is clearly not possible to specify the plant-climate relationship by means of a system of linear equations with independent variables. Promising results have been obtained in the USA and elsewhere, from computer studies of detailed mathematical models which try to match the complexity of the real-life situation. There is distinct hope that such models, by establishing significant indices of partial correlations, may suggest practical ways of improving our analysis of the agricultural consequences of altering the average value of one or more climatic parameters.

6.4.5 *Interannual climatic fluctuations and Mediterranean agriculture*

In evaluating the impact of climatic variation on agricultural production, especially in regions where certain parameters, such as water supplies in the Mediterranean zone, are frequently close to threshold values, it is clearly not enough to consider only the stability or long-term drift of average values of the different climatic factors. In fact, the degree of variation from one year to the next is of equal or even greater importance, since the efficiency of an agricultural system with respect to the climate in which it has developed, is closely related to characteristic limits of the fluctuation of climatic parameters around their long-term average values.

For example, a variation of 10 - 15% in the annual precipitation total may not in itself involve serious consequences for agriculture even in dry regions. On the other hand, the frequency with which extreme anomalies occur in the meteorological values of a single season is important. This is because of the fact that the compensatory mechanism of a plant enables it to cope, at the expense of some loss production, up to a certain limit; on the other hand, beyond a further limit (e.g. the permanent wilting point in the case of sustained water shortage) the plant dies and the harvest is completely lost.

A realistic evaluation of climatic impact is best based, not on a comparison with a hypothetical "average year", but by considering a sequence of seasons, covering say, ten to twenty years, and determining the frequency and severity of the unfavourable seasons and their effect on production. This approach, which is widely used to arrive at operational planning decisions, such as the suitability of a site for development as an orchard, or for growing particular crops, seem a reasonable method for estimating the agricultural effects of a climate change.

Figure 6.13 is designed to illustrate the foregoing argument; it is based on a number of highly simplified hypotheses, none of which is however such as to affect significantly the broad conclusions which can be drawn from it.

Let us consider a particular climatic element X (average temperature of the growth season, annual rainfall, etc.) and suppose that the crop under consideration is perfectly adapted to the current climate, so that the mean climatic value X_o is optimum for it, and that the yield curve (the curve (b) in Fig. 6.13) for the individual years is symmetrical and similar to the curve, assumed to be Gaussian, of the climatic element X (curve (a)). The two limits of yield mentioned above, that is a significant loss of yield and the total crop destruction, thus coincide with single and double values of the standard deviation, σ.

Taking as the unit of yield measurement the optimum harvest corresponding to X_o, then it is clear that the expected total return in a spell of 20 years is not 20 but about 20% below that value, since over such a period one is liable to experience 5 or 6 years of poor harvest and one of complete crop destruction.

Let us assume a change of climate from that which corresponds to curve (a) in Fig. 6.13 to one in which the new average of the element is indicated by X_I and the new standard deviation by σ_1. In Fig. 6.13 (c) illustrates a case in which the X curve is displaced without change of shape and hence $\sigma_1 = \sigma_o$;(e) shows an alteration on the standard deviation ($\sigma_1 > \sigma_o$) without affecting the average value, so that $X_I = X_o$. The corresponding yield curve over a period of 20 years are given in figures (d) and (f) respectively, in which variations in the frequency of good, bad and disastrous seasons are also shown.

One may deduce from Fig. 6.13 that the drop in production due to a displacement of the climate average, tends to be modest. In fact, the reduction in the total 20-years return arising from a displacement of less than half the standard deviation does not exceed 5%. To take a concrete example, let us consider annual rainfall for which the average standard deviation in the

Mediterranean countries is some 40%. The impact of a climatic variation which changes an optimum annual precipitation of 750 mm/year by $\pm$ 150 mm/year would be to reduce the combined harvest over 20 years by 5.1%. The consequence of a change in the annual variability could be greater: if the standard deviation in the above example were to increase by 20% from 300 to 360 mm/year, the yield over the 20-year period would decrease by 6.5%; on the other hand a reduction of variability by 20% would increase the long-term yield by about 8%.

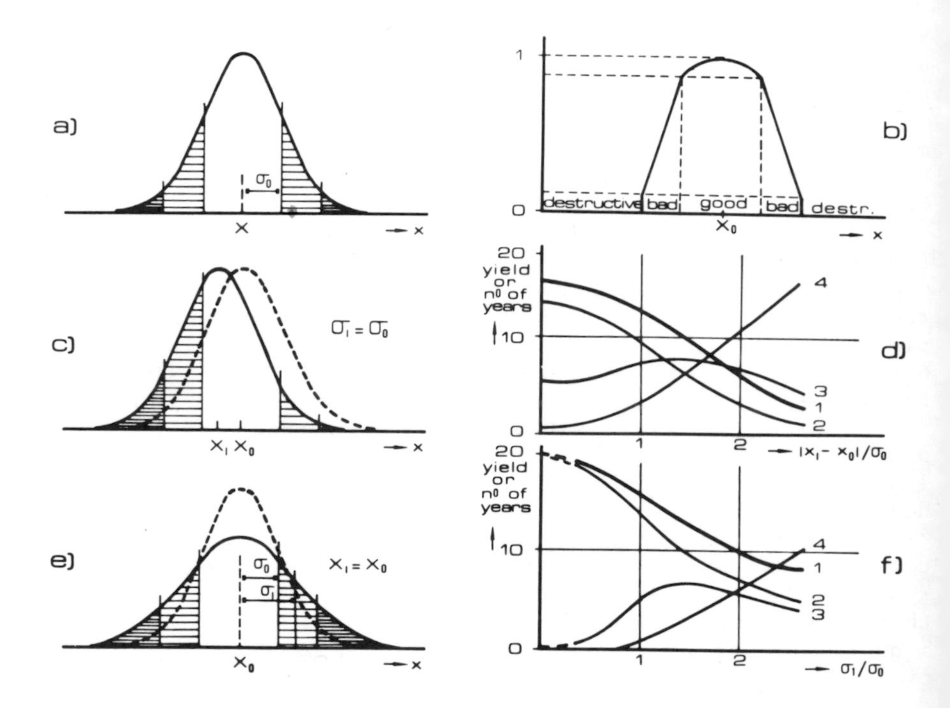

Fig. 6.13. Theoretical relationship between climatic fluctuations and the global crop yield over a 20-year period. Sub-figure (b) shows the distribution of years of good, poor and zero yields, corresponding to the distribution of a significant climatic element shown in curve (a), curve (e) illustrates a climatic change in which the standard deviation is altered but the average remains constant. Curve (c) shows a climatic change in which its average alters but the standard deviation does not; (d) and (f) show the distribution of crop yields corresponding to (c) and (e), respectively. Here, 1 = global yields in a period of 20 years, 2 = number of years of good yields, 3 = number of years of bad yields, 4 = number of years of zero yields.

Table 6.1 which forms a practical example of the theory of Fig. 6.13, shows the percentage variability of the average national yields over the period 1947-76 of three representative Mediterranean crops (vine, olives, oranges) in four regions (Spain, France, Italy, Greece). The actual variability of yields is critical for the maintenance of stable agricultural price structures and for the provision of storage facilities. In this regard, the more significant index is the average variability from one year to another, F (see Section 3.1), which in a truly Gaussian distribution is related to the standard deviation by the equation $F = 2\sigma/\sqrt{\pi} = 1.13\ \sigma$.

TABLE 6.1 Per cent variability of yields per unit area, corrected for historical trend. σ = standard deviation, F = interannual variability.

	Spain		France		Italy		Greece	
	σ	F	σ	F	σ	F	σ	F
Vine	16.6	15.9	12.1	13.4	9.3	11.3	11.1	10.0
Orange	15.3	15.5	-	-	8.5	8.1	11.4	12.5
Olive	24.8	29.3	-	-	24.9	35.1	36.1	42.8

In Table 6.1 the values for percentage variability refer to yield per unit area and not total production, and they have been rectified to eliminate trends due to social and environmental changes, and to the introduction if improved crop varieties and better methods of cultivation. Thus, what they mainly reflect is the response to meteorological conditions. As regards the olive crop, it should be noted that the average production in the odd years is radically superior to that of the even years: by 12% in Spain, 32% in Italy and 35% in Greece.

It must be emphasized that the variabilities as presented in Table 6.1 are averaged over large areas, and represent a considerable smoothing out of values from a diversity of local climates, soils, exposures, etc. Thus, for example, the indices for Italy are lower, not because local crops react less fully to meteorological changes, but because the geographical structure of the country is such that some independence exists between the seasonal weather in different parts and hence there is a tendency towards overall compensation in the national crop yields.

6.4.6 *Hail*

The most damaging episodic weather phenomenon for Mediterranean agriculture is hail.

It is also one of the more difficult weather types to record with accuracy because of its extreme variability in space and time, and because of its limited extent (normally a few square kilometres) and short duration (often only a few minutes). One of the fundamental problems in establishing a climatology of hail and in studying its variability, is the virtual impossibility of constructing a valid "hail index" which would take account of its main characteristics: its frequency, its extent and its intensity, as related to the kinetic energy of impact.

Clearly these are also the main elements which contribute to the "damage potential" of hail in respect to farm crops; one refers to damage potential and not to actual damage since the latter involves additional factors, such as the kind of crop and its stage of development at the time of the hail fall. Actual damage to the crop due to hail leads to economic loss, a loss which depends however upon many more factors such as the state of the market, price levels as dictated by supply and demand, etc.

The longest and most complete set of data relative to hail impact on Mediterranean agriculture consists of economic information supplied by governmental control agencies and insurance companies. All available Italian records of this kind have been critically analysed and compared with corresponding physical and meteorological data, so as to develop a hail index which has been found satisfactory in practice, provided it is applied to a large enough area and to a period of not less than one year.

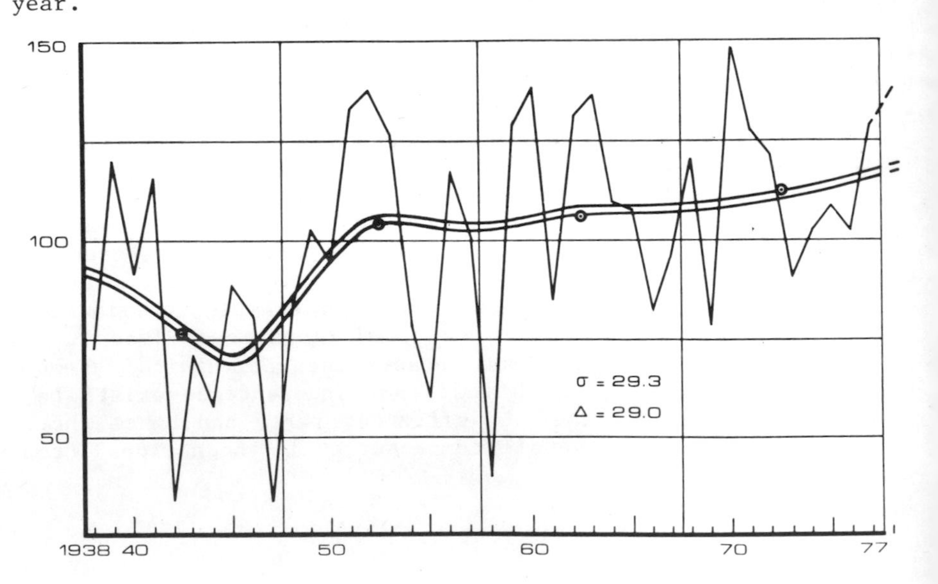

Fig. 6.14. Variation about an average value of 100 of the annual hail index for Italy, 1931-77. The double curve shows the broad trend of the index over the period.

over the forty-year period shown as 100 on the vertical scale. Certain interesting aspects of the variability of the hail index appear: (a) a long-term trend based on the mean values for each of the four decades and indicated by the double curve, shows how the basic pattern has varied; (b) a very high dispersion of the annual values, with a standard deviation of 29.3%, reflects an extremely variable effect on agricultural incomes; (c) a mean interannual variability of 29% is lower than might be expected in the light of the standard deviation and tends to confirm the existence of a basic trend (d) in the consecutive 20-year periods 1938-77 there are significant differences not only in the average values (90.4 and 109.6) but also in the variability as indicated by the standard deviations (31.7 and 27.5).

Incidentally the latter change is in the reverse sense of what might have been expected, in that a notable increase of hail in the later period corresponded with lower annual variability.

6.4.7 *Conclusions*

In Section 6.4, concerning the impact of climatic fluctuations on Mediterranean Europe agriculture, we have examined those climatic peculiarities which chiefly characterize such regions in their relationship to agriculture: the possible scarcity of water in the soil as regards the water requirements of farm crops; strong solar radiation with much sunshine and high temperatures; the frequency of meteorological phenomena harmful to crops, e.g. hail, which is one of the phenomena which causes the greatest losses in production.

There then follows a development of the discussion of some examples relative to such characteristics and to the impact that climatic fluctuations have on farm crop production.

Such a discussion may be useful in order to prove the necessity of a further examination of the various aspects of the complex relations that exist between crop-production, climate and climatic fluctuations, through an organic and co-ordinated research programme. Such researches should not only have a cognitive aim but should also be applicable and help in decision-making; they should serve above all to estimate the possible future agro-climatic scenario in Mediterranean regions.

6.5 Looking to the future

That the weather has played an important role in past food crises in Europe is abundantly clear. When, however, we seek to use history to provide guidance for future planning, a question must immediately arise as to the degree of relevancy, if indeed any, of examples taken from the past. Modern farming is better cushioned against the more extreme weather-induced fluctuations in agricultural production. Social and economic conditions are much improved compared with those of our forefathers who were ill-equipped to meet the effects of any form of inclemency.

The relevance of the agrometeorological past to current European food production is discussed in the following paragraphs in terms of three sub-questions:

a) Are years of such severe weather as that shown by past instrumental records liable to recur again?

b) If so, could extreme weather, in the altered circumstances of modern agriculture, cause crop losses by direct damage on anything like the scale of the past? Given current plant protection and quarantine procedures, could weather-favoured pathogens spread as in earlier years and cause comparable losses?

c) In the less vulnerable socio-economic circumstances of present day Europe, could the consequences of reduced food production due to weather conditions be as serious as in the past?

6.5.1 *The recurrence of years of extreme weather*

The three hundred years or so of instrumental weather records for limited areas which are available to us represent, in fact, only a tiny sample of the range of annual weather sequences which European agriculture may be called upon to face. As L.P. Smith (1965) has pointed out, even if we limit our analysis of a particular year just to the four seasons, and to the average temperature and rainfall in each, expressed in terms of warm/ normal/cold and wet/normal/dry, this crude classification alone provides for 9^4 or 6.561 possible annual combinations. It is therefore highly improbable that we have encompassed in our mere three centuries of records examples of the very worst extremes of climatic stress. The frequency with which weather observations surpass previous extreme values even at long-established observing stations indicates that years at least as difficult for farmers as the most trying seasons of the past must be expected to recur. This conclusion is valid irrespective of any superimposed climatic trend which may occur, for the random fluctuations from one year to the next are more wide-ranging than any likely trend, and are likely, especially in the coming decades, to involve more acute problems for human affairs and the food-producing economy. This is certainly not to minimise the importance for agriculture of the slower and steadier climatic changes which, in the long term, may radically alter farming practice, but the time-scale of such changes is expected to be such as to permit of gradual adaption, painful and expensive though it may be, to the altered environmental conditions. One cannot similarly adjust to rapid weather fluctuations.

The risk of increased variability of the climate is thus a cause of more immediate concern, not least because of the disruption it would cause to agricultural policy. A prime objective of crop husbandry in recent years has been to iron out random variations in food production from one year to the next, so that distribution, storage and marketing might become more orderly processes, with fewer gluts and shortages, and a more stable price structure. Excessive weather fluctuations, with more droughts, floods, extreme temperatures and other anomalies could make this objective unattainable.

Changes in the atmospheric circulation pattern, even if relatively shortlived, can produce radical displacement of the centres of meteorological action, and so vary the wind flow and with it the rainfall and temperature pattern, so as to create serious problems for a way of agriculture which has evolved in relationship to an accustomed climate. On a major and sustained scale, such a change could convert what are currently highly productive regions into problem areas.

6.5.2 *Modern agriculture and its vulnerability to weather*

Modern European agriculture is separated from the era of notoriously disastrous harvests by up to two centuries of agricultural reasearch and development. In earlier years, when drainage was primitive or non-existent, much of the land was sour and often waterlogged. Today, systematic drainage carries away unwanted water, permitting not only better cultivation and earlier sowing but also a wider use of modern machinery. On the other hand, irrigation is widely practised to counter dry spells. The judicious use of shelter-belts protects against wind damage in exposed areas. Cold-resistant crop varieties have been developed, and protection against frost damage is more widely practised. Stiff-strawed cereal varieties have been introduced which are less liable to lodging.

Efficient time-saving machinery enables the farmer to carry out necessary operations quickly and to take advantage of favourable if transient weather. Other factors contributing to increased food production have been care in the preservation of farmyard manure, and increased use of artificial fertilizers and of lime, where appropriate. Last, but far from least, spectacularly high-yielding varieties of many staple food crops have been developed which are specifically bred to extract the maximum advantage out of the climatic and other conditions of the area in which they are grown.

By reason of the remarkable changes brought about by modern agricultural technology, the farmer is no longer completely at the mercy of the seasonal weather. Moderate deviations from climatic expectastions may inconvenience him but will rarely cause serious disruption or loss. A good harvest is the normal expectation; it is not felt to call as urgently for a hearfelt thanksgiving service as in the case of our immediate ancestors.

There is, however, a serious risk of complacency arising from the achievements of modern agriculture. The steady rise in U.S. grain production in the climatically favourable period of some 15 years up to 1971 convinced many commentators that the weather factor had been virtually eliminated from cereal growing, thanks to plant breeding and selection, and to better management practices. The global food crisis which was triggered off by adverse weather in 1972 shattered this illusion in a most dramatic way and led to the convening of the World Food Conference of 1974. The effects of unfavourable seasons on farming had indeed been reduced but were still far from being eliminated.

An examination of U.K. national yields in recent years suggests a coefficient of variation of about + 8% for cereals and + 16% for sugar beet and potatoes. In this fairly compact area, the bad years show a drop of about one-sixth compared with the better grain harvests (which nowadays are increasingly the expected harvests) and the corresponding short-fall in root-crops can be one-third. In an area as large as western Europe, a modest estimate of the occasional fall below optimum might be 5 - 10% for cereals and 10-20% for potatoes and root crops.

One must not place undue reliance in the belief that the weather, however extreme in one part of the world, is likely to average out over the main food-producing areas and thus tend to nullify the total yield loss. The inter-dependability of weather around the globe has long been recognised. Recently the Economic Research Service of the U.S. Department of Agriculture (1975) carried out an analysis of crop yield trends and variations over the period 1953-73 in twenty-five regions comprising the world's major grain producing areas. When cereal yields fell because of adverse weather in one region, they were found to be likely to be lower in other regions too. The correlation was not high, but poor years such as 1972 seemed to be experienced in many of the world's grain sources at the same time. Simultaneous occurrence of adverse weather anomalies in distant areas of the globe with associated oceanic anomalies-known as "teleconnections"-may also lead to reduced fish catches as during 1972, when the Peruvian anchovy fishery broke down (Flohn 1978).

We may also note briefly certain potential dangers for the future in the very developments which have brought modern agriculture to its present peak. The more efficiently farming is organised to make the very best use of its customary weather assets the more sharply is production likely to fall if the climate deviates appreciably from the expected. Crop varieties which are finely tuned to a narrow range of temperature and of moisture availability can be bred to give remarkable yields in the conditions to which they are programmed, but in more extreme weather can fare worse than the varieties which they superseded. The new high-yielding strains require more fertilizer support and a greater water supply for maximum yield and are sensitive to deficiencies on either of these scores.

The exposure of the new agriculture to sudden sharp loss due to disease or pest is a cause of concern to those involved in crop protection (Feekes 1967, Ten Houten 1974). The recent revival in Africa and the Near East of the desert locust menace illustrates how a problem apparently overcome or at least brought under control may flare up suddenly, particularly after a few critical seasons. The risk of serious attacks by plant parasites is paradoxically increased by the very practices which are used today to promote food production: introduction of new varieties, more intensive cropping, the growing of more than one crop per year, the wider practice of irrigation, increased use of fertilizers. Several significant recent disease outbreaks show the phytosanitary dangers of monoculture, i.e. the growing of huge areas of the same crop and often of the same variety in place of the diversified patterns of traditional farming. The history of the Dutch wheat cultivar Heines VII, for example, illustrates how a severe disease epidemic can follow the abrupt break down of resistance in the face of a new race of a pathogen Careful trials of this most promising wheat variety over several years proved it to be resistant to all known strains of the yellowrust disease. Introduced commercially in 1952, it confirmed its promise in the field and rapidly became the most widely grown variety in the Netherlands. In 1955, by which time 81% of the total Dutch wheat area was sown with Heines VII, it was everywhere heavily attacked by yellow rust, and in 1956 more than two-thirds of the winter wheat area was destroyed. Shortly afterwards the new destructive rust race, dispersed on the wind, was found in much of Europe from southern Germany to Sweden, and as far west as Ireland.

The Heines VII recalls some aspects of the historic potato blight epidemic of 1845-6. An even closer analogue is provided by the tobacco blue mould outbreak of 1958-61. This fungal disease, not previously known east of the Atlantic, spread from a small source in England in the autumn of 1958 to appear in a few years in every tobacco growing region throughout Europe, in North Africa and deep into Asia. The rapidity with which the disease spread across extensive areas and the damage which it caused is a reminder of the continued menace of pathogens, granted suitable weather and favourable winds. As illustrated in an earlier section, climatic change may radically disturb the disease-crop relationship for better or worse, either by altering the ambient in which the pathogen works out its life cycle (see Fig. 6.6) or by changing the wind pattern on which it depends for long-distance transport (Fig. 6.5).

It has been claimed that farm mechanization has contributed to disease risk, as, for instance, in the spread of certain *Fusarium* diseases as a result of combine harvesters blowing infected husks over the fields. In addition, the combine leaves more residual straw on the field and sheds more grain, both possible factors in fomenting disease.

Certainly mechanization is the major factor which has made modern farming so energy-intensive and thus robbed it of much of the self-reliance of the past. The farmer has become increasingly dependent upon good supply lines and open communications for deliveries of oil, fodder and food, for servicing of tractors and other machinery, for moving of livestock, etc. The disruption of transport by snow in the 1978-9 winter caused severe problems especially on isolated upland holdings, and dispelled, at least temporarily, the false sense of security which had been built up by a preceding series of mild winters.

6.5.3 *The consequences of crop losses in the socio-economic circumstances of the Europe of today*

Our ancestors were involved in an agrarian economy, largely at a subsistence level, which operated on a very fine margin between adequacy and shortage. Today all this is changed over practically the whole of Europe. No longer does a bad harvest, except perhaps in wartime, involve shortages of food or even famine. A higher level of prosperity, with industry more important to European economy than agriculture, is only one of the factors which has brought this situation about. The development of massive centres of agricultural production in North and South America and in Australia-New Zealand, coupled with progress in sea and land transport, made imported food supplies available at low prices. The general diet became more diversified, with less dependance on cereals and potatoes. Meanwhile ouput per acre, in western Europe as in some other areas, increased spectacularly, tripling since the end of World War II.

The position from a global point of view is not so reassuring. Masked behind the temporary inconvenience of local surplusses of some agricultural products is the permanent problem of a continuous rise of world poplulation and the recurrent food crises which arise in some areas. Total world food production may, in the opinion of the U.S. Department of Agriculture, Economic Research Service (1975), be expected to keep a half step ahead of population in the next few decades. Half a step is scarcely a distance to inspire confidence, especially if the forward march can be erratic and interrupted by bad weather; and indeed the U.S. report forecasts that there will be times and places of critical shortage.

Quite apart from these haphazard reverses, there is some question as to whether the remarkable rate of increase in food production in recent years can maintain its momentum and continue to nose ahead of requirements. The stimulus of agricultural mechanization, for example, is unique in any one area and cannot be repeated. The gap between the productivity of the better farmers and the target results as obtained in agricultural experimental farms is constantly narrowing. Exponential growth cannot be maintained indefinitely.

To discuss in any detail the implication for Europe of food shortages arising elsewhere in the world is outside the scope of this work. Giraudoux claimed, in a striking phrase, that the prosperous are in a position to watch catastrophe from a terrace. Irrespective of whether such a stance were considered acceptable or even possible in the world of today, the consequences of any future drop in European food production would clearly be intensified by concurrent shortage elsewhere in the world.

The general food situation may be compared to that of a man walking along a sea shore of irregular depth. If he is already immersed up to his neck, even a quite shallow depression will take him out of his depth; this was the food situation in earlier times in Europe and which still obtains in some parts of the world where the population lives close to the food production level of the best harvest years. If, however, the water reaches no higher than the paddler's knees, the occasional depths may slightly inconvenience him and perhaps cause him to stumble, but will involve no serious danger-unless the rising tide should catch him unawares.

Even in the worst circumstances, with disease and pest adding to crop losses caused by unfavourable weather, it seems improbable that a peacetime Europe would be forced to go hungry in the foreseeable future. It is, however, almost inevitable that occasional situations of food shortage involving inconvenience and expense will recur from time to time. These will bear little resemblance to the European food crises of earlier centuries or to those that still occur in other parts of the world. This is not to say that they would be accepted with complete equanimity by populations which have a much higher level of expectation than their forefathers.

6.6 Summary

The impact of weather on farm production is complex; it varies for instance, from one crop to another and with the stage of growth of plants. The primary climatic needs of agriculture are for warmth and moisture.

The significance of temperature change is most marked in the cold northern fringe of Europe (Section 6.2); a particular problem in this area in years of late springs and cold autumns is the further shortening of an already critically brief thermal growing season. In the Mediterranean region (Section 6.4) water shortage is the main factor limiting agricultural production;

problems arise also from the high levels of solar radiation and from violent weather phenomena such as hail storms.

In the most productive zone of European agriculture, roughly between 45°N and 60°N, both temperature and rainfall can present serious problems, as can snow, frost and high winds (Section 6.3). Quite small temperature changes can significantly affect the length of the crop growing season in the maritime areas of western Europe; this has important repercussions for upland farming. Wet seasons have been the particular bane of farmers in the past; however growing competition for water supplies from other interests suggests that, in coming years, European agriculture may become progressively more vulnerable to seasons of prolonged dry weather, such as 1976.

A more detailed analysis is presented of the climatic requirements of the wheat and potato crops, and of the effect of weather on their exposure to major diseases and pests. In the Mediterranean zone, the reponse of the two-year vegetative cycle of the vine to climatic conditions is analysed.

A review of the European food harvests of the past two or three centuries reveals certain recurrent themes. In some years, such as 1788, the eve of the French revolution, harvest failure was due less to the kind of abnormal weather which shows up in annual or seasonal statistics than to a critical sequence of damaging weather episodes. The notoriously disastrous winter of 1739-40 is a classical example of how the complacency and relaxed vigilance evoked by a sustained period of benign weather can be abruptly shattered by a single bad season; several similar if less spectacular examples have occurred in the present century. The infamous year 1816 which brought disaster on both sides of the Atlantic not only illustrates the cummulative effect of a series of poor summers, but warns once again how the interdependability of world weather can lead to simultaneous harvest losses over a large part of the globe. The lessons of the potato famine of 1845-7 are still valid today; the new agriculture is potentially vulnerable to sudden sharp loss due to plant disease or pest.

European agriculture could almost certainly learn to adapt, although with inconvenience and expense, to gradual long-term climatic drift (Section 6.5). The risk of increased variability of climate between one season and the next presents more immediate problems particularly in an economy seeking to minimize random variations of annual food production, and so to rationalize distribution, storage, marketing and price structure. Seasons of even moderate fall of food production in Europe could raise problems in a world context, quite apart from the reactions of a home population which has a much higher level of expectation than their forefathers.

REFERENCES

CHAPTER 1

Bolin, B.: 1975, in The Physical Basis of Climate and Climate Modelling. ICSU-WMO, GARP Publication Series No. 16.

Hastenrath, S. and Heller L.: 1977, Quart. Roy. Meteorol. Soc. 103, 77-92; cf. Monthly Weather Rev. 106, (1978) 1290-1298.

Lamb, H.H.: 1972, Climate: Present, Past and Future, Vol. I, Methuen, London, XXXI + 613 pp.

Lamb, H.H.: 1979, Climate: Present, Past and Future, Vol. II, Methuen, London, XXX + 835 pp.

Manabe, S., Bryan, K. and Spelman, M.J.: 1979, Dynamics Atmos. Oceans 3, 393-426; Corr. l.c.4, (1980) 137.

Mitchell, Ir. J.M., Stockton, Ch. W. and Meko, D.M.: 1979, in B.M. McCormack and T.A. Seliga (eds.) Solar-Terrestrial Influences on Weather and Climate, D. Reidel, Dordrecht, pp. 125-143.

Shutts, G.J. and Green, J.S.A.: 1979, Nature 277, 355-358.

Smagorinsky, J.: 1974, in W. Hess (ed.) Weather and Climate Modification.

Washington, W.M., Semtner, A.J., Meehl, G.A., Knight, D.J. and Mayer, Th. A.: 1980, Phys. Oceanogr. 10, 1887-1908.

SELECTED MONOGRAPHS AND CONFERENCE PROCEEDINGS

Allison, I. (ed.): 1981, Sea Level, Ice and Climatic Change, Intern. Assoc. Hydrol. Sciences, IAHS Publ. No. 131, XV + 471 pp.

Bach, W. et al. (eds.): 1979, Man's Impact on Climate, Development in Atmospheric Science 10, Elsevier, Amsterdam XXIII + 327 pp.

Bach, W., Pankrath, J. and Williams J. (eds.): 1980, Interactions of Energy and Climate, D. Reidel, Dordrecht, XL + 569 pp.

Bach, W., Pankrath, J. and Schneider, S.H. (eds.): 1981, Food-Climate Interactions, D. Reidel, Dordrecht, XXXI + 504 pp.

Bach, W.: 1982, Gefahr für unser Klima?, Verlag C.F. Müller, Karlsruhe, XXI + 317 pp. (in German); English edn., D. Reidel, Dordrecht, (1984).

Baumgartner, A. and Reichel, E.: 1975, The World Water Balance, R. Oldenbourg, München, 179 pp.

Berger, A. (ed.): 1981, Climatic Variations and Variability: Facts and Theories, NATO Series C 72, D. Reidel, Dordrecht, XXVI + 795 pp.

Bolin, B. et al. (eds.): 1979, The Global Carbon Cycle, SCOPE 13, J. Wiley, 491 pp.

Bryson, R.A. and Murray, Th. A.: 1977, Climates of Hunger, University of Wisconsin Press, XV + 171 pp.

Budyko, M.I.: 1974, Climatic Change, Leningrad (in Russian); English trans. Am. Geophys. Union 1977, VII + 261 pp.

Budyko, M.I.: 1982, The Earth's Climate: Past and Future, Int. Geophys. Series, Vol. 29, Academic Press, New York, X + 307 pp.

Centre National d'Etudes Spatiales (ed.): 1978, Évolution des Atmosphères Planétaires et Climatologie de la Terre, Nice, XVIII + 574 pp.

Clark, W.C. (ed.): 1982, Carbon Dioxide Review; Oxford University Press, XIX + 469 pp.

Frakes, L.A.: 1979, Climates Throughout Geologic Time, Elsevier, Amsterdam, XII + 304 pp.

Frenzel, B.: 1967, Die Klimaschwankungen des Eiszeitalters, Vieweg, Braunschweig, XI + 296 pp.

GARP (Global Atmospheric Research Program): 1975, The Physical Basis of Climate and Climate Modelling, in B. Bolin (ed.), World Meteorological Organization, GARP Publication Series 16, XXIII + 265 pp.

Gribbin, J. (ed.): 1978, Climatic Change, Cambridge University Press, XI + 280 P.

Hare, F.K.: 1977, Climate and Desertification. In Desertification: Its Causes and Consequences, Pergamon Press, London, pp. 63-120.

Hess, W. (ed.): 1974, Weather and Climate Modification, Wiley, New York, 1025 pp.

Kellogg, W.W.: 1974, Effects of Human Activities on Global Climate, World Meteorological Organization Technical Note No. 156, WMO-NO. 486, XVIII + 47 pp.

Kellogg, W.W. and Schware R.: 1981, Climate Change and Society, Consequences of Increasing Carbon Dioxide, Westview Press, Boulder, Colorado.

Lamb, H.H.: 1972,1979, Climate: Present, Past and Future, Methuen, London, Vol. I (1972) XXXI + 613 pp, Vol. III (1977) XXX + 835 pp.

Lamb, H.H.: 1982, Climate, History and the Modern World, Methuen, London, XIX + 387 pp.

Le Roy Ladurie, E.: 1971, Times of Feast, Times of Famine, New York, 426 pp.

Lockwood, J.G.: 1979, Causes of Climate, Edward Arnold, London, 260 pp.

National Academy of Sciences: 1975, Understanding Climatic Change, A Program for Action, Washington, D.C., XV + 239 pp.

Oeschger, H., Messerli, B. and Svilar, M. (eds.): 1980, Das Klima. Analysen und Modelle, Geschichte und Zukunft, Springer-Verlag, Heidelberg, IX + 181 S.

Pittock, W. et al. (eds.): 1978, Climatic Changes and Variability - A Southern Perspective, Cambridge University Press, XXIII + 445 pp.

Roberts, W.O. and Lansford, H.: 1979, The Climate Mandate, W.H. Freeman, San Francisco, VIII + 197 pp.

Von Rudloff, H.: 1967, Die Schwankungen und Fendelungen des Klimas in Europa seit dem Beginn der regelmäßigen Instrumenten-Beobachtungen (1670), Vieweg, Braunschweig.

Schneider, S.H. and Mesirov, L.E.: 1976, The Genesis Strategy, Plenum Press, New York, XXI + 419 pp.

Schönwiese, C.D.: 1979, Klimaschwankungen. Verständliche Wissenschaft, Band 115, XII + 181 S.

Schwarzbach, M.: 1974, Das Klima der Vorzeit, Eine Einführung in die Paläoklimatologie,3. Auflage Stuttgart, VIII + 380 pp.

Stumm, W. (ed.): 1976, Global Chemical Cycles and Their Alteration by Man, Dahlem Konferenzen, Vol. 2, 385 pp.

Takahashi, K. and Yoshino, M.M. (eds.): 1978, Climatic Change and Food Production, University of Tokyo Press, XI + 433 pp.

Tickett, C.: 1978, Climatic Change and World Affairs, Harvard Studies in International Affairs No. 37, Pergamon Press, 78 pp.

Wigley, T.M.L., Ingram, M.J. and Farmer, G. (eds.): 1981, Climate and History. Studies in Past Climates and their Impact on Man, Cambridge University Press, XII + 530 pp.

Williams, J. (ed.): 1978, Carbon dioxide, Climate and Society, Pergamon Press, X + 332 pp.

Williams, M.A.J. and Faure, H. (eds.): 1980, The Sahara and the Nile, Balkema, Rotterdam, XVI + 607 pp.

World Meteorological Organization: 1979, Proceedings of the World Climate Conference, Geneva, WMO-No. 537, XII + 791 pp.

Van Zinderen Bakker, E.M. (ed.): 1978, Antarctic Glacial History and World Palaeoenvironments, Balkema, Rotterdam, VIII + 172 pp.

CHAPTER 2

Alexandre, P.: 1978, in Transgressies en Occupatiegeschiedenis in de Kustgebieden van Nederland en België (Interdisciplinair Colloquium), 5-7 September 1978, Rijksuniversiteit.

Alenius, P. and Makkonen, L.: 1981, Arch. Meteor. Geophys. Bioklim. B 29, 393-398.

Beveridge, W.H.: 1921, Economic Journal 31, 421-453.

Beveridge, W.H.: 1922, J. Roy. Statistical Soc. 85, 418-454.

Boece, H. (Hector Boethius): 1536, The History and Chronicles of Scotland. English translation (J. Bellenden), Edinburgh, 1821.

British Association for the Advancement of Science, Aberdeen Meeting: 1963, The North-East of Scotland, Aberdeen (Central Press), 256 pp.

Christensen, A.E.: 1938, Nordisk kultur 2: Befolkning i middelalderen, Copenhagen, Oslo, Stockholm, 1-57 pp.

Dansgaard, W., et al.: 1975, Nature 255, 24-28.

Van den Dool, H.M. Krijnen, H.J. and Schuurmans, C.J.E.: 1978, Climatic Change, 1. 319-330.

Douglas, K.S., Lamb, H.H. and Loader, C.: 1978, A Meteorological Study of July to October 1588: The Spanish Armade Storms. Research Publication No. 6 (CRU RP 6). Norwich (University of East Anglia, Climatic Research Unit).

Douglas, K.S. and Lamb, H.H.: 1979, Weather Observations and a Tentative meteorological Analysis of the Period May to July 1588. Research Publication No. 6a (CRU RP 6a). Norwich (University of East Anglia, Climatic Research Unit).

Flohn, H.: 1949, Vierteljahrsschrift naturforsch. Ges. Zürich 95, 28-41.

Flohn, H.: 1967, in Die Schwankungen und Pendelungen des Klimas in Europa seit dem Beginn der regelmässigen Instrumenten-Beobachtungen (1670), by H. von Rudloff, (Vieweg) Braunschweig 81-90 pp.

Flohn, H.: 1978, Promet, Number 2/3, Dt. Wetterdienst 1-20 pp.

Gottschalk, M.K.E.: 1971, Stormvloeden en Rivieroverstromingen in Nederland: Deel I - De Periode voor 1400.

Gottschalk, M.K.E.: 1975, Ibid. Deel II - De periode 1400-1600.

Gottschalk, M.K.E.: 1977, Ibid. Deel III - De Periode 1600-1700. Van Gorcum, Assen/Amsterdam.

Graham, H.G.: 1899, The Social Life of Scotland in the Eighteenth Century, Black, London (2 volumes).

Griswold, W.J.: 1978, Climatic Variation and the Social Revolutions of Seventeenth Century Anatolia. (Unpublished manuscript).

Henderson, E.: 1879, The Annals of Dunfermline: AD 1069-1878, John Tweed, Glasgow.

Hoel, A. and Werenskiøld, W.: 1962, Norsk Polarinstituttets Skrifter, Nr. 114. Universitetsforlaget, Oslo.

Hollstein, E.: 1965, Bonner Jahrbücher 165, 1-27.

Holmsen, A.: 1961, Norges Historie, (Universitetsforlaget), Oslo Bergen 480 pp.

Holmsen, A.: 1978, Hva kan vi vite om Agrar-katastrofen i Norge i middelalderen? (Universitetsforlaget), Oslo-Bergen-Tromsø 128 pp.

Huber, B. and Giertz-Siebenlist, V.: 1969, Sitz Ber. Osterr. Ak. Wiss. Wien (Abt. I) 178 (1-4), 37-42.

Hughes, M.K., Kelly, R.M., Pilcher, J.R. and La Marche, V.C. (eds.): 1982, Climate from Tree Rings, Cambridge Univ. Press, 223 pp.

Ingram, M.J., Underhill, D.J. and Wigley, T.M.L.: 1978, Nature 276, 329-334.

Kington, J.A.: 1975, Meteorological Magazine 104, 33-52.

Koch, L.: 1945, The East Greenland Ice, Meddelelser om Grønland, Band, 130, Nr. 3. Copenhagen.

Labrijn, A.: 1945, Het Klimaat van Nederland Gedurende de Laatste Twee en een Halve Eeuw, De Bilt (Koninklijk Nederlandsch Meteorologisch Instituut, KNMI No. 102).

Ladurie, E. Le Roy: 1967, Histoire du climat depuis l'an mil. Flammarion, Paris, 366 pp.

Ladurie, E. Le Roy: 1971, Times of Feast, Times of Famine, Doubleday, New York, 426 pp.

Lahr, E.: 1950, Un siècle d'observations météorologiques appliqués à l'étude du climat luxembourgeois, Bourg-Bourger, Luxembourg.

Lamb, H.H.: 1965, Palaeogeogr., Palaeoclim., Palaeoecol. 1, 13-37 pp.

Lamb, H.H.: 1972, British Isles Weather Types and a Register of the Daily Sequence of Circulation patterns 1861-1971, Geophysical Memoir 116, London.

Lamb, H.H.: 1977, Climate: Present, Past and Future. Volume 2: Climatic History and the Future, Methuen, London and Barnes and Noble, New York, 835 pp.

Lamb, H.H.: 1978, in Transgressies en Occupatiegeschiedenis in de Kustgebieden van Nederland en België - Interdisciplinair Colloquium 5-7 September 1978, Rijksuniversiteit Gent.

Lamb, H.H.: 1982, Climate, History and the Modern World, Methuen, London, 387 pp.

Lamb, H.H. and Johnson, A.I.: 1966, Secular Variations of the Atmospheric Circulation Since 1750, Geophysical Memoir 110, London.

Libby, L.M.: 1974, correlation of Historic Climate with Historic Prices and Wages, Santa Monica, California, R. and D. Associates, unpublished memorandum.

Lindbekk, K.: 1978, Lofoten og Vesterålens historie: 1500-1700, Dreyer, Stavanger.

Manley, G.: 1952, Climate and the British Scene, Collins - New Naturalist series, London, 314 pp.

Manley, G.: 1953, Quart. J. Roy. Meteor. Soc. 79, 242-261.

Manley, G.: 1957, Scottish Geographical Magazine 75 (1), 19-28.

Manley, G.: 1974, Quart. J. Roy. Meteor. Soc. 100, 389-405.

Matthews, J.A.: 1977, Boreas 6, 13-24.

Müller, K.: 1953, Geschichte des badischen Weinbaus, Schauenburg, Lahr in Baden, 283 pp.

Nicholas, F.J. and Glasspoole, J.: 1931, British Rainfall, 299-306 pp.

Parry, M.L.: 1978, Climatic Change, Agriculture and Settlement, Dawson, Folkestone, and Archon Books, Hamden, Conn. 214 pp.

Pfister, Chr.: 1975, Agrarkonjunktur und Witterungsverlauf im westlichen Schweizer Mittelland 1755-1797, Geographisches Institut der Universität Bern, Berne, 279 pp.

Pfister, Chr.: 1978, in proceedings of the Nordic Symposium on Climatic Changes and Related Problems, Copenhagen 24-28 April 1978, Dan. Meteor. Inst. - Climatological Papers No. 4, 1-6 pp.

Rotberg, R.J. and Rabb, Th. K., (eds.): 1981, Climate and History, Studies in Interdisciplinary History, Princeton Univ. Press, Xii + 280 pp.

Von Rudloff, H.: 1967, Die Schwankungen und Pendelungen des Klimas in Europa seit dem Beginn der regelmässigen Instrumenten Beobachtungen (1670), Vieweg, Braunschweig, 370 pp.

Salvesen, H., et al.: 1977, Norwegian Archaeol. Rev. 10 (1-2), 107-154.

Schweingruber, F.H.: 1975, Helvetia Archaeologica 21 (6), 2-15.

Schweingruber, F.H.: 1976, Röntgenuntersuchungen an jahrringen, Neue Züricher Zeitung Nr. 180, 34 pp.

Schweingruber, F.H., Bräker, O.U. and Schär, E.: 1979, Boreas 8, 427-452.

Sinclair, Sir John: 1791-9, Statistical Account of Scotland, Edinburgh.

Sinclair, Sir John: 1825, Analysis of the Statistical Account of Scotland, Edinburgh, (2 volumes).

Solvang, J.: 1942, Korndyrking nordom Malangen i eldre tid. Håløygminne, 6 (3), 249-255. Hålogaland Historielag, Harstad, Norway.

Wigley, T.M.L., et al.: 1979, Geographical Patterns of Climatic Change: 1000 BC to 1700 AD., Quaternary Res.

CHAPTER 3

Ackermann, B. and Changnon, S.A., et al. (eds.): 1977/8, Summary of Metromex. Vol. I, II. Illinois State Water Survey, Urbana (cf. also J. Appl. Meteor. 15(1976), 544-570).

Ambs, A.: 1979, Dipl. Thesis Univ. Bonn.

Arlery, R.: 1970, The Climate of France, Belgium, The Netherlands and Luxemburg, World Survey of Climatology, Vol. 5, Ch. 4.

Baur, F.: Beilage Berliner Wetterkarte 77/75 vom 24.6.1975.

Bijl, W. van der: 1952, Statistische Methoden in der Klimatologie, Kon. Nederl. Meteor. Inst., Med. en Verh., no. 68.

Cehak, K.: 1977, Arch. Meteor. Geophys. Biokl., Ser. B 25, 209-219.

Chico, T. and Sellers, W.O.: 1979, Climatic Change 2, 139-148.

Craddock, J.M.: 1976, Quart. J. Roy. Meteor. Soc. 102, 823-840.

Craddock, J.M. and Wales-Smith, B.G.: 1977, Meteor. Mag. 106, 97-111.

Dettwiller, J.: 1978, La Météorologie, VIe Série, no. 13, 95-130.

Deutscher Wetterdienst: 1956-1975, Die Grosswetterlagen Europas. Deutscher Wetterdienst, Offenbach.

Dool, H.M. van den, Krijnen, H.J. and Schuurmans, C.J.E.: 1978, Climatic Change 1, 319-330.

Dronia, H.: 1967, Der Stadteinfluss auf den weltweiten Temperaturtrend, Meteor. Abhandl., 74, no. 4, F.U. Berlin.

Dronia, H.: 1974, Meteor. Rundsch. 27, 166-174.

Dupriez, D.L. and Sneyers, R.: 1978, Publ. Inst. Roy. Météor. Belgique, A 101.

Eriksson, B.: 1979, Temperature Fluctuations during the Latest 100 Years, Rapporter, Meteorologi och Klimatologi Sver. Meteor. Hydr. Inst., nr. RMK 14.

Garnier, M.: 1974, Mém. Météor. Nat. 53,Fasc. 1-4.

Gray, B.M.: 1976, Quart. J. Roy. Metor. Soc. 102, 627 f.

Hillebrand, C.: 1978, Dipl. Thesis Univ. Bonn.

Johannessen, Th.W.: 1970, The Climate of Scandinavia, World Survey of Climatology, Ch. 2, Vol. 5.

Jones, P.D., Wigley, T.M.L. and Kelly, P.M.: 1982, Monthly Weather Rev. 110, 59-70 .

Labrijn, A.: 1944, The Climate of the Netherlands over the Last Two and a Half Centuries, Kon. Nederl. Meteor. Inst., Med. en Verh., no. 49.

Labrijn, A.: 1945, Meded. Verhand. Kon. Nederl. Meteor. Inst. 49.

Lamb, H.H.: 1972,1977, Climate: Present, Past and Future, Methuen London, Vol. I.

Lamb, H.H.: 1982, Climate History and the Modern World. Methuen, London.

Lamb, H.H. and Johnson, A.I.: 1959, Geografiska Annaler 41, 94-134.

Lamb, H.H. and Johnson, A.I.: 1961, Geografiska Annaler 43, 363-400.

Landsberg, H.E.: 1962, Beitr. Phys. Atmos. 35, 184 f.

Ledger, D.C. and Thom, A.S.: 1977, Meteor. Mag. 106, 342-349.

Loon, H. van and Williams, J.: 1978, Monthly Weather Rev. 106, 1012-1017.

Manley, G.: 1970, The Climate of the British Isles, World Survey of Climatology, Vol. 5, Ch. 3.

Meyer zu Düttingdorf, A.: 1970, Klimaschwankungen im maritimen und kontinentalen Raum Europas seit 1871. Bochumer Geographische Arbeiten, Heft 32.

Neumann, Y.: 1977, Bull. Amer. Meteor. Soc. 58, 163-168.

Oke, T.R.: 1973, Atmospheric Environment 7, 769-779.

Palumbo, A. and Mazzarella, A.: 1980, Monthly Weather Rev. 108, 1041-1045.

Ratcliffe, R.A.S., Weller, J. and Collison, P.: 1978, Quart. J. Roy. Meteor. Soc. London 104, 243-255.

Reidat, R.: 1971, Wetter und Leben 23, 1-6.

Rudloff, H. von: 1967, Die Schwankungen und Pendelungen des Klimas in Europa seit dem Beginn der regelmässigen Instrumenten-Beobachtungen (1670), Vieweg, Braunschweig.

Schönwiese, C.D.: 1969, Spektrale Varianzanalyse klimatologischer Reihen im langperiodischen Bereich, Wiss. Mitteilungen Nr. 15, Meteor. Inst. Univ. München.

Schönwiese, C.D.: 1978, Beitr. Phys. Atmos. 51, 139-152.

Seibel, M.: 1980, Dipl. Thesis Univ. Bonn.

Shapiro, R.: 1975, Quart. J. Roy. Meteor. Soc. London 101, 679-681.

Tavakol, R.K.: 1979, Climate Monitor 8, no. 3, 76-81, CRU, Norwich.

Wacker, U.: 1981, Arch. Meteor. Geophys. Bioklim. Serie B, 29, 269-281.

Wales-Smith, B.G.: 1971, Meteor. Mag. 100, 345-362: l.c. 102 (1973), 157-171.

Wales-Smith, B.G.: 1977, Meteor. Mag. 106, 297-312.

Wallén, C.C.: 1953, Tellus 5, 157-178.

Wallén, C.C.: 1970,1977, Climates of Northern and Western Europe, Vol. 5 and Climates of Central and Southern Europe, Vol. 4, World Survey of Climatology, Amsterdam.

Wigley, T.M.L. and Atkinson, T.C.: 1977, Nature 265, 431-434.

Witter, J.V.: 1982, Landbouw Hogeschool Wageningen, Dept. of Hydraulics, Nota 55.

WMO/UNESCO: 1970, WMO/UNESCO, Climatic Atlas of Europe.

CHAPTER 4

Adams, J.A.S., Montovani, M.S.M. and Lundell, L.L.: 1977, Science 196, 54-56.

Augustsson, T. and Ramanathan, V.: 1977, J. Atmos. Sci. 34, 448-451.

Bacastow, R.B. and Keeling, C.D.: 1981, in Bolin (ed.): Carbon Cycle Modelling, Scope 16, John Wiley and Sons, Chichester, pp. 103-112.

Bach, W.: 1976, Rev. Geophys. Space Phys. 14, 429-474.

Bach, W.: in J. Williams (ed.), 1978, Carbon Dioxide, Climate and Society, IIASA Proceedings series Environment, Pergamon Press, Oxford, pp. 141-168.

Bach, W., Pankrath, J. and Kellogg, W. (eds.): 1979, Man's Impact on Climate, Elsevier, Amsterdam.

Bach, W., Pankrath, J. and Williams, J., (eds.): 1980, Interactions of Energy and Climate, D. Reidel, Dordrecht, Holland.

Bach, W.: 1982, Progress in Physical Geography 6, 549-560.

Bach, W.: 1982, Gefahr für unser Klima. Verslag C.F. Müller, Karlsruhe. English translation: 1984, Our Threatened Climate, Ways of averting the CO_2 -problem through rational energy use. D. Reidel, Dordrecht, Holland.

Bach, W., Crane, A., Berger, A. and Longhetto, A.: 1983, Carbon Dioxide, Current views and developments in energy/climate research. Proceedings of the 2nd International School of Climatology, D. Reidel, Dordrecht, Holland.

Baes Jr., C.F., Goeller, H.E., Olson, J.S. and Rotty, R.M.: 1976, ORNL-5194 (Oak Ridge National Laboratory, Oak Ridge, Tennessee.

Barry, R.G.: 1978, in J. Williams (ed.): Carbon Dioxide, Climate and Society, IIASA Proceedings series Environment, Pergamon Press, Oxford, 169-180 pp.

Barry, R.G.: 1980, Cold Regions Science and Technology 2, 133-150.

Baumgartner, A. and Reichel, E.: 1975, The World Water Balance, Elsevier, Amsterdam.

Baumgartner, A., Kirchner, M. and Mayer, H.: 1978, Promet 2/3 Meteorologische Fortbildung, Deutscher Wetterdienst, 32-42 pp.

Baumgartner, A. and Kirchner, M.: 1980, in W. Bach, J. Pankrath and J. Williams (eds.): Interactions of Energy and Climate, D. Reidel, Dordrecht, pp. 305-316.

Berger, A., (ed.): 1981, Climatic Variations and Variability: Facts and Theories, First Erice International School of Climatology, NATO ASI C72, D. Reidel, Dordrecht, 795 pp.

Bischof, W.: 1977, Tellus, 435-444.

Björkström A.: 1979, in B. Bolin et al. (eds.): The Global Carbon Cycle, SCOPE 13, J. Wiley and Sons, Chichester, pp. 1-53.

Bischof, W. and Bolin, B.: 1966, Tellus 18, 155-159.

Böhlen, T.: 1978, Meteorologische Rundschau 31, Heft 3, 87-91.

Böhm, R.: 1979, Archiv für Meteorologie, Geophysik und Bioklimatologie, Ser. B 27, 31-46 pp.

Bohn, H.L.: 1976, Soil Sci. Soc. Amer. J. 40, 468-470.

Bolin, B.: 1977, Science 196, 613-615.

Bolin, B. and Arrhenius, E.: 1977, Ambio 6, 96-105.

Bolin, B.: 1979, in Proc. World Climate Conf. Geneva, February 1979, World Meteorological Organization No. 537, pp. 27-50.

Bolin, B. and Keeling, C.D.: 1963, J. Geophys. Res. 68, 3899-3920.

Bolin, B., Degens, E.T., Duvigneaud, F. and Kempe, S.: 1979a, in B. Bolin et al. (eds.): The Global Carbon Cycle, SCOPE 13, John Wiley and Sons, Chichester, pp. 1-53.

Bolin, B., Degens, E.T., Kempe, S. and Ketner, P., (eds.): 1979b Global Carbon Cycle, SCOPE 13, John Wiley and Sons, Chichester.

Bolin, B., (ed.): 1981, Carbon Cycle Modelling, SCOPE 16, John Wiley and Sons, Chichester.

Brewer, P.G.: 1978, Geophys. Res. Lett. 5, 997-1000.

Broecker, W.S.: 1963, in M.N. Hill (ed.), The Sea, Vol. 2, Interscience Publishers, New York, pp. 88-108.

Broecker, W.S., Li, Y-H. and Peng, T-H.: 1971, in: D.H. Wood (ed.), Impingement of Man on the Oceans, John Wiley and Sons, New York, pp. 297-324.

Broecker, W.S. and Takahashi, T.: 1977, in: N.R. Andersen and A. Malahoff (eds.), the Fate of Fossil Fuel CO_2 in the Oceans, Plenum Press, New York, pp. 213-241.

Broecker, W.S., Takahashi, T., Simpson, H.J. and Pneg, T-H.: 1979, Science 206, pp. 409-418.

Bryan, K., Komro, F.G., Manabe, S. and Spelman, M.J.: 1982, Science 215, 56-58.

Bryson, R.A. and Coodrans: 1980, Science 207 (4435), 1041-1044.

CEQ: 1981, Global Energy Futures and the Carbon Dioxide Problem, Council on Environmental Quality, Government Printing Office, Washington D.C.

Cess, R.D.: 1976, J. Atmos. Sci. 33, 1831-1843.

Cess, R.D. and Goldenberg, S.D.: 1981, J. Geophys. Res. 86, 498-502.

Chan, Y.H., Olson, J.S. and Emanuel, W.R.: 1979, Simulation of Land Use Patterns Affecting the Global Carbon Cycle, Environmental Sciences Division, Publication no. 1273, Oak Ridge National Laboratory, Oak Ridge, Tennessee.

Changnon, S.A., Jr. and Semonin, R.G.: 1979, Rev. Geophys. Space Phys. 17. 1891-1900.

Charney, J., Quirk, W.J., Chow, S.H. and Kornfield, J.: 1977, J. Atmos. Sci. 34, 1366-1385.

Chen, G-T. and Millero, F.J.: 1979, Nature 277, 205-206.

Choudbury, B. and Kukla, G.: 1979, Nature 280, 668-671.

Chylek, P. and Coakley, J.A. Jr.: 1974, Science 183, 75-77.

Clark, W.C.: 1982, Carbon Dioxide Review, Oxford University Press.

Climate Research Board: 1979, Carbon Dioxide and Climate: A Scientific Assessment, National Academy of Sciences, Washington D.C., 22 pp.

Climate Research Board: 1980, Letter Report of the Ad Hoc Study Panel on Economic and Social Aspects of Carbon Dioxide Increase, National Academy of Sciences, Washington D.C., 11 pp.

CO_2- Climate Review Panel: 1982, Carbon Dioxide and Climate: A Second Assessment, National Research Council, National Academy Press, Washington D.C.

CSIRO: 1978, Atmospheric Chemistry Research, CSIRO Div. of Atm. Res., Mordialloc, Victoria, Australia.

Dalrymple, G.J. and Unsworth, M.H.: 1978, Quart. J. Roy. Meteor. Soc. 104, 989-998.

Dettwiller, J. and Changnon, S.A., Jr.: 1976, J. Appl. Meteor. 15, 517-519.

Dittberner, G.J.: 1978a, in Conference on Climate and Energy, Climatological Aspects and Industrial Operations, Asheville, North Carolina, American Meteorological Society, pp. 101-106.

Dittberner, G.J.: 1978b, in Conference on Climate and Energy, Climatological Aspects and Industrial Operations, Asheville, North Carolina, American Meteorological Society, pp. 125-132.

DOE: 1980, Climate Program Plan, Office of Health and Environment Research, Carbon Dioxide and Climate, Research Program, Washington, 80 pp.

DOE: 1982, Proceedings of the Workshop on First Detection of Carbon Dioxide Effects: Carbon Dioxide Effects, Research and Assessment Program, Office of Energy Research, Washington.

Dreiseitl, E. and Reiter, E.R.: 1978, Archiv für Meteorologie, Geophysik und Bioklimatologie, Serie B 26, 305-317.

Eddy, J.A.: 1977, Climatic Change 1, 173-190.

Elliott, W.F. and Machta, L., (eds.): 1977, Proceedings of the ERDA Workschop on Environmental Effects of Carbon Dioxide from Fossil Fuel Combustion, Department of Energy, Washington D.C.

Ellsaesser, H.W., Mac Cracken, M.C., Potter, G.L. and Luther, F.M.: 1976, Quart. J. Roy. Meteor. Soc. 102 (433), 655-667.

Fletcher, J.Ol, Mintz, Y., Arakawa, A. and Fox, R.: 1973, in Energy Fluxes over Polar Surfaces, Proceedings of IAMAP - IAPSO - SCAR - WMO Symposium, WMO no. 361, Tech. Note no. 129, Genèva, pp. 181-218.

Flohn, H.: 1973, Naturwiss. 60, 340-348.

Flohn, H.: 1975, in B. Bolin (ed.): The Physical Basis of Climate and Climate Modeling, GARP Publ. Ser. 16, pp. 106-118.

Flohn, H.: 1978, in J. Williams (ed.): Carbon Dioxide, Climate and Society, IIASA Proceedings series, Environment, Pergamon Press, Oxford, pp. 227-237.

Flohn, H.: 1979, Possible Climatic Consequence of a Man-Made Global Warming, IIASA working paper 79-86, Laxenburg (Austria).

Freyer, H.D.: 1978, in J. Williams (ed.), Carbon Dioxide, Climate and Society, Vol. 1, Pergamon Press, London, pp. 69-87.

Georgii, H.W.: 1979, in W. Bach et al. (eds.): Man's Impact on Climate, Elsevier, Amsterdam, pp. 181-192.

Gilchrist, A.: 1983, in Bach et al. (eds.), Carbon Dioxide, D. Reidel, Dordrecht, Holland, pp. 219-258.

Gilliland, R.L.: 1982, Climatic Change 4(2), 111-132.

Glantz, M.: 1977, Bull. Amer. Meteor. Soc. 58, 150-158.

Goudriaan, J. and Ajtay, G.L.: 1979, in B. Bolin et al. (eds.) The Global Carbon Cycle, SCOPE 13, J. Wiley and Sons, Chichester, New York, pp. 237-249.

Grams, G.W., Blifford, I.H., Jr., Gillette, D.A. and Russell P.B.: 1974, J. App. Meteor. 13, 459-471.

Groves, K.S. and Tuck, A.F.: 1979, Nature 280, 127-129.

Häfele, W., et al.: 1976, Second status report of the IIASA Project on energy systems, Rh-76-1, Int. Inst. for Appl. Systems Analysis, Laxenburg, Austria. Cited in J. Williams: 1978, Carbon Dioxide, Climate and Society, IIASA Proceedings series Environment, Pergamon Press, Oxford Vol. I., pp. 239-248.

Hahn, J.: 1979, in W. Bach et al. (eds.): Man's Impact on Climate Elsevier, Amsterdam, pp. 193-213.

Hall, C.A.S., Ekdahl, C.A. and Wartenberg, D.E.: 1975, Nature 255, 136-138.

Hameed, S., Cess, R.D. and Hogan, J.S.: 1980, J. Geophys. Res. 85, 7537-7545.

Hampicke, U.: 1979a, in B. Bolin et al. (eds.): The Global Carbon Cycle, SCOPE 13, J. Wiley and Sons, Chichester, pp. 219-236.

Hampicke, U.: 1979b, in W. Bach et al. (eds.): Man's Impact on Climate Elsevier, Amsterdam, pp. 139-159.

Hanna, S.R.: 1977, J. Appl. Meteor. 16, 880-887.

Hansen, J., Johnson, D., Lacis, A., Lebedeff, S., Lee, P., Rind, D. and Russell, G.: 1981, Science 213, 957-966.

Hare, F.K.: 1977, in Secretariat of the United Nations Conference on Desertification (ed.): Desertification: Its Causes and Consequences, Pergamon Press, Oxford, pp. 63-168.

Hollin, J. and Barry, R.G.: 1979, Environment International 2, 437-444.

Hollis, G.E.: 1978, Geograph. J. 144, 62-80.

Hoover, T.E. and Berkshire, D.C.: 1969, J. Geophys. Res. 74, 456-464.

Hosler, C.L. and Landsberg, H.E.: 1977, in Engergy and Climate: Outer Limits to Growth, Geophysics Study Committee, National Academy of Science, Washington D.C., pp. 96-105.

Hoyt, D.V.: 1979, Nature 282, 388-390.

Idso, S.B.: 1977, Quart. J. Roy. Meteor. Soc. 103, 369-370.

Idso, S.B.: 1982, Archiv. für Meteorologie, Geophysik and Bioklimatologie 31, 325-330.

Isaksen, I.S.A., Hesstvedt, E. and Stordal, F.: 1980, Nature 283, 189-191.

JASON: 1979, The Long Term Impact of Atmospheric Carbon Dioxide on Climate, Technical Report, JSR-78-07, SRI International, Arlington, Virginia.

Jones, P.D., Wigley, T.M.L. and Kelly, P.M.: 1982, Monthly Weather Rev. 110, 59-70.

Junge, C.: 1978, Promet 2/3, Meteorologische Forbildung, Deutscher Wetterdienstes, 21-32.

Kandel, R.S.: 1983, in Bach et al. (eds.): Carbon Dioxide, D. Reidel, Dordrecht, Holland, pp. 179-218.

Kanwishers, J.: 1963, Deep Sea Res. 10, 195-207.

Keeling, C.D.: 1980, in Bach et al. (eds.): Interactions of Energy and Climate, D. Reidel, Dordrecht, Holland, pp. 129-147.

Keeling, C.D. and Bacastow, R.B.: 1977, in Energy and Climate: Outer Limits of Growth, Geophysics Study Committee, National Academy of Science, Washington D.C., pp. 72-95.

Keeling, C. D., Mook, W.G. and Tans, P.P.: 1979, Nature 277, 121-123.

Keeling, C.D. and Stuiver, M.: 1978, Science 202, 1109.

Kellogg, W.W., Coakley, J.A., Jr. and Grams, G.W.: 1975, in Proceedings of the WMO-IAMAP Symposium on Long-Term Climatic Fluctuations, WMO no. 421, Genéva, pp. 323-330.

Kellogg, W.W.: 1977, Effects of Human Activities on Global Climate, World Meteorological Organization, Technical Note no. 156, Genéva.

Kellogg, W.W.: 1978, Bull. Atomic Sci. 34, 10-19.

Kellogg, W.W. and Schware, R.: 1981, Climate Change and Society, Westview Press, Boulder, Colorado.

Kukla, G.J., Angell, J.K., Korshover, J., Dronia, H., Hoshiai, M., Namias, J., Rodewald, M., Yamamoto, R. and Iwsahima, T.: 1977, Nature 270, 573-580.

Kukla, G.T., Gavin, T.: 1981, Science 214, 497-504.

Lacis, A., Hansen, J., Lee, P., Mitchell, T. and Lebedeff, S.: 1981, Geophys. Res. Lett. 8, 1035-1038 pp.

Lamb, H.H.: 1971, Earth Sci. Rev. 7, 87-95.

Landsberg, M.E.: 1970, in S. Teweles and J. Giraytys (eds.): Meteorological Observations and Instrumentation, Meteorological Monographs, pp. 11-91.

Landsberg, H.E.: 1981, The Urban Climate. Int. Geophys. Ser. 28, Academic Press, New York, 275 pp.

Lee, R.: 1973, J. Appl. Meteor. 12, 556-557.

Le Houerou, H.N.: 1979, La Recherche 10, 336-344.

Lemon, E.: 1977, in N.R. Andersen and A. Malahoff (eds.), The Fate of Fossil Fuel CO_2 in the Oceans, Plenum Press, New York, pp. 97-130.

Lettau, H., Lettau, K. and Molton, L.C.B.: 1979, Monthly Weather Rev. 107, 227-238.

Lian, M.S. and Cress, R.D.: 1977, J. Atmos. Sci. 34, 1058-1062.

Liou, K.N., Freeman, K.F. and Sasamori, T.: 1978, Tellus 30, 62-69.

Lorius, C. and Raynaud, D.: 1983, in Bach et al. (eds.): Carbon Dioxide, D. Reidel, Dordrecht, Holland, pp. 145-178.

Lovins, A.B., Lovins, L.H., Krause, F. and Bach W.: 1982, Least-Cost Energy: Solving the CO_2 Problem, Brick House Publ. Co., Andover, USA.

Lowe, D.C., Guenther, P.R. and Keeling, C.D.: 1979, Tellus 31, 58-67.

L'Vovitch, I.M.: 1977, Ambio 6, 13-22.

L'Vovitch, M.I.: 1979, World Water Resources and Their Future, Publ. by American Geophysical Union, Washington D.C.

MacDonald, G.J., (ed.): 1982, The Long-Term Impact of Increasing Atmospheric Carbon Dioxide Levels, Ballinger Publ. Com., Cambridge Mass.

Machta, L., Hanson, K. and Keeling, C.D.: 1979, in N.R. Andersen and A. Malahoff (eds.), The Fate of Fossil Fuel CO_2 in the Oceans, Plenum Press, New York, pp. 131-144.

Madden, R.A. and Ramanathan, V.: 1980, Science 209, 763-768.

Manabe, S. and Stouffer, R.J.: 1980, J. Geophys. Res. 85, 5529-5554.

Manabe, S. and Wetherald, R.T.: 1975, J. Atmos. Sci. 32(1), 3-15.

Manabe S. and Wetherald, R.T.: 1980, J. Atmos. Sci. 37, 99-118.

Manabe, S., Wetherald, R.T. and Stouffer, R.J.: 1981, Climatic Change 3, 347-386 pp.

Marland, G. and Rotty, R.M.: 1979, Rev. Geophys. Space Phys., 17(7), 1813-1824 and Consensus, 41-52 (1980/2-3).

Mason, B.J.: 1979, in World Climate Conference, WMO no 537, Genéva, pp. 210-242.

Mass, C. and Schneider, S.H.: 1977, J. Atmos. Sci. 34(12), 1995-2004.

Michael, F., Hoffert, M., Tobias, M. and Tichler, J.: 1981, Climatic Change 3, 137-153.

Miles, M.K. and Gildersleeves, P.B.: 1978, Nature 271, 735-736.

Mitchell, J.M., Jr.: 1975, in S.F. Singer (ed.): The Changing Global Environment, D. Reidel, Dordrecht, Holland, pp. 149-175.

Mitchell, J.M., Jr.: 1977, Carbon Dioxide and Future Climate, Environment Data Service, National Oceanic and Atmospheric Administration, U.S. Department of Commerce.

Mitchell, J.M., Jr.: 1982, Weatherwise 35, 252-262.

Mitchell, J.F.B.: 1983, Quart. J. Roy. Meteor. Soc. 109, 113-152.

NAS: Halocarbons: 1976, Environmental Effects of Chlorofluoromethane Release, Committee on Impacts of Stratospheric Change, National Academy of Sciences, Washington D.C.

Newell, R.E., Herman, G.F., Dopplinck, T.G. and Boer, G.J.: 1972, J. Appl. Meteor. 11, 864-867.

Newell, R.E. and Dopplick, R.G.: 1979, J. Appl. Meteor. 18, 822-825.

Niehaus, F.: 1979, in W. Bach et al. (eds.), Man's Impact on Climate, Elsevier, Amsterdam, pp. 285-298.

Nriagu, J.O.: 1979, Nature 279, 409-411.

Nye, P.H. and Greenland, D.J.: 1965, Tech. Comm. No. 51, Commonwealth Bureau of Soils, Harpenden.

Obeng, L.: 1977, Ambio, 6, 46-50.

Oke, T.R.: 1974, Review of Urban Climatology 1968-1973. World Meteorological Organization, WMO 383 Technical Note No. 134, Genéva.

Olson, J.S.: 1970, in D.E. Reichle (ed.), Analysis of Temperate Forest Ecosystems, Springer-Verlag, Berlin, pp. 226-241.

Olson, J.S.: 1974, in H. Hemingway Benton (Publisher) Encyclopaedia Britannica, 15th ed. pp. 144-149.

Olson, J.S.: 1975, in D.E. Reichle, J.F. Franklin, D.W. Goodall (eds.), Productivity of World Ecosystems, National Academy of Sciences, Washington DC, pp. 33-43.

Olson, J.S., Pfuderer, H.A. and Chan, J.H.: 1973, Changes in the Global Carbon Cycle and The Biosphere, Environmental Sciences Division, Publication no. 1050, Oak Ridge National Laboratory, Oak Ridge, Tennessee.

Parkinson, C.L. and Kellogg, W.W.: 1974, Climatic Change 2(2), 149-162.

Peng, T.H., Broecker, W.S., Kipphut, G. and Shackleton, N.: 1977, N.R. Andersen and A. Malahoff (eds.), The Fate of Fossil Fuel CO_2 in the Oceans, Plenum Press, New York, pp. 355-373.

Perry, H. and Landsberg, H.H.: 1977, in Energy and Climate: Outer Limits to Growth, Geophysics Study Committee, National Academy of Science, Washington D.C., pp. 35-50.

Porch, W.M. and MacCraken, M.C.: 1982, Atmospheric Environment 16, 1365-1372.

Potter, G.L., Ellsaesser, H.W., Mc. Cracken, M.C. and Luther F.M.: 1975, Nature 258, 697-698.

Potter, G.L., Ellsaesser, H.W., Mac Cracken, M.C., Ellis, J.S. and Luther, F.M.: 1980, in W. Bach, J. Pankrath and J. Williams (eds.): Interactions of Energy and Climate, D. Reidel, Dordrecht, Holland, 317-326 pp.

Quinn, J.A. and Otto, N.C.: 1971, J. Geophys. Res. 76. 1539-1549.

Ramanathan, V.: 1975, Science 190, 50-52.

Ramanathan, V. and Dickinson, R.E.: 1979, J. Atmos. Sci. 36, 1084-1104.

Ramanathan, V.: 1981, J. Atmos. Sci. 38, 918-930.

Rampino, M.R. and Self, S.: 1982, Quatenary Res. 18, 127-143

Rebello, A. and Wagener, K.: 1976, 4th Intern. Symp. Chemistry of the Mediterranean, Rovinj, Yugoslavia.

Reck, R.A.: 1976, Science 192, 557-559.

Reiners, W.A., et al.: 1973, in G.M. Woodwell and E.V. Pecan (eds.), Carbon and the Biosphere, conf-720510, AEC Symp. Ser. 30, Tech. Information Center, Oak Ridge, Tennessee.

Revelle, R.: 1982, Scientific American 247, 33-41.

Roads, J.O.: 1978, J. Atmos. Sci. 35, 753-773.

Robock, A. and Mass, C.: 1982, Science 216, 628-630.

Rodin, L.E., Bazilevich, N.I. and Rozov, N.N.: 1975, D.E. Reichle J.F. Franklin and D.W. Goodall (eds.), Productivity of World Ecosystems, National Academy of Sciences, Washington D.C. pp. 13-26.

Rotty, R.M.: 1977, in N.R. Andersen and A. Malahoff (eds.), The Fate of Fossil Fuel CO_2 in the Oceans, Plenum Press, New York, pp. 167-181.

Rotty, R.M.: 1979, in W. Bach et al. (eds.), Man's Impact on Climate, Elsevier, Amsterdam, pp. 269-283.

Rotty, R.M.: 1980, Experentia 36, 781-783.

Rotty, R.M.: 1981, in Analysis and Interpretation of Atmospheric CO_2 Data, WCP-14, World Meteorological Organization, Geneva.

Rowntree, P.R. and Walker, J.: 1978, in J. Williams (ed.): Carbon Dioxide, Climate and Society, IIASA Proceedings series, Environment, Pergamon Press, New York, 181-191 pp.

Ryabchikov, A.: 1975, The Changing Face of Earth. The Structure and Dynamics of the Geosphere, Its Natural Development and the Changes Caused by Man, Progress Publishers, Moscow, 200 pp.

Ryther, J.H.: 1969, Science 166, 72.

SCEP: 1970, Man's Impact on the Global Environment, Assessment and Recommendations for Action, Study of Critical Environment Problems, M.I.T. Press.

Schlesinger, M.E.: 1982, Proceedings of the U.S. Department of Energy CO_2 , Research Conference: Carbon Dioxide, Science and Consensus, Coolfont Conference Center, Berkeley Springs, West Virginia, 19-23 September 1982.

Schneider, S.H. and Mesirow, L.E.: 1976, The Genesis Strategy, Plenum Press, New York and London.

Schneider, S.H. and Thompson, S.L.: 1981, J. Geophys. Res. 86, 3135-3147.

Schoeberl, M.R. and Strobel, D.F.: 1978, J. Atmos. Sci. 35, 1751-1757.

Schuurmans, C.J.E.: 1983, in W. Bach et al. (eds.), Carbon Dioxide, D. Reidel, Dordrecht, Holland, pp. 337-352.

Sellers, W.D.: 1965, Physical Climatology, University of Chicago Press, Chicago.

Sellers, W.D.: 1977, Ambio 6, 10-12.

Siegenthaler, U. and Oeschger, H.: 1978, Science 199, 388-395.

SMIC: 1971, Inadvertent Climate Modification, MIT Press, Cambridge.

Strickland, J.D.H.: 1965, in J.P. Riley and G. Skirrow (eds.), Chemical Oceanography, Vol. 1, Academic Press, London, pp. 477-610.

Stuiver, M.: 1978, Science 199, 253-258.

Thompson, S.L. and Schneider, S.H.: 1982, Nature 295, 645-646.

Thrush, B.A.: 1978, Nature 276, 345-348.

U.N.: 1977, Desertification, Its Causes and Consequences, United Nations conference on Desertification, Nairobi, Kenya, Pergamon Press, Oxford.

Untersteiner, N.: 1975, in B. Bolin (ed.), The Physical Basis of Climate and Climate Modeling, GARP Publ. Ser. 16, 206-224.

Volz, A.: 1983, in W. Bach et al. (eds.), Carbon Dioxide, D. Reidel, Dordrecht, Holland, pp. 353-378.

Wade, N.: 1979, Science 206, 912-913.

Wang, W.C., Yung, Y.L., Lacis, A.A., Mo, T. and Hansen, J.E.:1976, Science 194, 685-691.

Weiss, R.E., Wagoner, A.P., Charlson, R.J. and Ahlquist, N.C.: 1977, Science 195, 979-981.

Wetherald, R.T. and Manabe, S.: 1975, J. Atmos. Sci. 32, 2044-2059.

Wetherald, R.T. and Manabe, S.: 1981, J. Geophys. Res. 16, 1194-1204.

Whittaker, R.H. and Likens, G.E.: 1973, in G.M. Woodwell and E.V. Fecan (eds.), Carbon and the Biosphere, Conf-720510, AEC Symp. Ser. 30, Techn. Information Center, Oak ridge, Tennessee.

Whitten, R.C. and Borucki, W.J., Capone, L.A., Reigel, C.A. and Turco, R.P.: 1980, Nature 283, 191-192.

Wigley, T.M.L., Jones, P.D. and Kelly, P.M.: 1980, Nature 283, 17-21.

Wigley, T.M.L. and Jones, P.D.: 1981, Nature 292, 205-208.

Williams, J. and Loon H. van: 1976, Monthly Weather Rev. 104, 1591-1596.

Williams, J., Krömer, G. and Weingart, J., (eds.): 1977, Climate and Solar Energy conversion, Proceedings of a IIASA Workshop. CP-77-9, IIASA, Laxenburg, Austria.

W.M.O.: 1979, Report of the first Session of the CAS Working Group on Atmospheric Carbon Dioxide, WMO Project on Research and Monitoring of Atmospheric CO_2, Report no. 2 CAS, WMO, Geneva, 49 pp.

W.M.O.: 1979, Proceedings of the World Climate Conference, WMO Report no 537, World Meteorological Organization, Geneva.

W.M.O.: 1981, Joint WMO/ICSU/UNEP Meeting of Experts on the Assessment of the Role of CO_2 on Climate Variations and Their Impact. Joint Flanning Staff, World meteorological Organization, Geneva.

Wong, C.S.: 1978, Science 200, 197-200.

Wong, C.S.: 1978, Mar. Pollution Bull. 9, 257-264.

Woodwell, G.M.: 1978, Scientific American 238, 34-43.

Woodwell, G.M., Whittaker, R.H., Reiners, W.A., Likens, G.E., Delwiche, C.C. and Botkin, D.B.: 1978, Science 199, 141-146.

Zimen, K.E.: 1979, in W. Bach et al. (eds.): Man's Impact on Climate, Elsevier, Amsterdam, pp. 129-139.

See also Several authors in: WMO/ICSU/UNEP Conference, Analysis and Interpretation of CO_2 data, World Climate Research Programme, Bern 14-18 September (1981) (WCP-14).

CHAPTER 5

Aagard, K.: 1975, L.K. Coachman: EOS (Transact. Amer. Geophys. Un.) 56, 484-487.

Andersen, S.T.: 1960, Geol. Survey of Denmark, Ser. II, No. 75, pp. 1-175.

Andersen, S.T.: 1969, Medd. Dansk Geol. Foren. 15, 90-102.

Andrews, J.T., Barry, R.G., Bradley, R.S., Miller, G.H. and Williams, L.D.: 1972, Quat. Res. 2, 303-314.

Andrews, J.T., et al.: 1972, Nature 239, 147-149.

Barkow, N.I., Gordienko, F.G., Korothevich, E.S. and Kotlyakov, V.M.: 1975, in I.U.G.G. (ed.), Symposium on Isotopes and Impureties in Snow and Ice, Grenoble.

Barry, R.G., et al.: 1977, Arctic and Alpine Research 9, 193-210.

Bé, A.W.H. and Duplessy, J-C.: 1976, Science 194, 419-422.

Berger, A.L.: 1976, Astron. Astrophys. J.51, 127-135.

Berger, A.L.: 1978, Quat. Res. 9, 139-167.

Berger, A.L., Guiot, J., Kukla, G. and Pestiaux, P.: 1981, Geol. Rundsch. 70, 748-758.

Berger, A.L.: 1980a, Geophys. Surveys 3, 351-402.

Berger, A.L.: 1980b, Vistas in Astronomy 24, 103-122.

Berger, W.H.: 1971, Mar. Geol. 11, 325-358.

Berger, W.H.: 1977, in N.R. Andersen and A. Malahoff (eds.), The Fate of Fossil Fuel CO_2 in the Oceans, Plenum Press, New York, pp. 505-542.

Berger, W.H. and Heath, G.R.: 1968, J. Mar. Res. 26, 135-143.

Berkofsky, L.: 1977, Contrib. Phys. Atmos. 50, 312-320.

Bloom, A.L., Broecker, W.S., Chappell, J.M.A., Matthews, R.K. and Mesolella, K.J.: 1974, Quat. Res. 4, 185-205.

Brady, H. and Martin, H.: 1979, Science 203, 437-438.

Bray, J.R.: 1976, Nature 260, 414-415.

Bray, J.R.: 1977, Science 197, 251-254.

Buchardt, B.: 1978, Nature 275, 121-123.

Budd, W., Jenssen, D. and Radok, U.: 1971, Univ. of Melbourne, Dept of Meteorol., Publ. No. 18.

Budyko, M.J.: 1962, Izvest. Ak. Nauk USSR, Ser. Geogr. No. 6, 3-10.

Budyko, M.I.: 1969, Tellus 21, 611-619.

Budyko, M.J.: 1972, EOS (Transact. Amer. Geophys. Un.) 53, 868-874.

Budyko, M.J.: 1977, Climate Changes. Amer. Geophys. Union, V + 261.

Chappell, J.M.A.: 1974, J. Geophys. Res. 79, 390-398.

Charney, J.G.: 1975, Quart. J. Roy. Meteor. Soc. 101, 193-202.

Clark, D.L.: 1971, Geol. Soc. Am. Bull. 82, 3313-3324.

Cline, R.M. and Hays, J.D.: 1976, Investigation of Late Quaternary Paleoceanography and Faleoclimatology, the Geological Society of America, Inc., Memoir 145.

Coppens, Y.: 1978, in W.W. Bishop (ed.): Geological Background to Fossil Man.

Croll, J.: 1875, Climate and Time, Appleton, New York.

Dalrymple, G.B.: 1972, in Bishof and Miller (eds.), Potassium Argon Dating of Geomagnetic Reversals.Scott. Academic Press.

Dansgaard, W.: 1964, Tellus 16, 436-468.

Dansgaard, W. and Tauber, H.: 1969, Science 166, 499-502.

Dansgaard, W., Clausen, H.B., Gundestrup, N., Hammer, C.U., Johnsen, S.J., Kristinsdottir and Reeh, N.: 1982, Science 218, 1273-1277.

Dansgaard, W., Clausen, H.B., Gundestrup, N., Johnsen, S.J. and Rygner, C.: 1983, Proc. G.I.S.P. Symp., Philadelphia, June 1982, Am. Geophys. Union, Spec. Volume (in press).

Dansgaard, W., Johnsen, S.J., Clausen, H.B. and Gundestrup, N.: 1973, Medd. om Grønland 197, 1-52.

Dansgaard, W., Johnsen, S.J., Clausen, H.B. and Langway, C.C.,Jr : 1971, in K.K. Turekian (ed.): Late Cenozoic Glacial Ages, Yale University Press, New Haven, pp. 37-56.

Delmas, L.J., et al.: 1980, Nature 284, 155-157.

Dennison, B., and Mansfield, V.N.: 1976, Nature 261, 32.

Denton, G.H., Armstrong, R.L. and Stuiver, M.: 1971, in K.K. Turekian (ed.), Late Cenozoic Glacial Ages, Yale University Press, pp. 267-306.

Denton, G.H. and Hughes, T.J., (eds.): 1981, The Last Great Ice Shields. J. Wiley and Sons, New York Vol. I,II.

Dorf, E.: 1960, American Scientist, 341-364.

Duplessy, J-C.: 1978, in J. Gribbin (ed.): Climatic Change, Cambridge University Press, Cambridge, pp. 46-67.

Duplessy, J-C., Lalou, C. and Vinot, A-C.: 1970, Science 168, 250-251.

Duplessy, J-C., Chenouard, L. and Vila, F.: 1975, Science 188, 1208-1209.

Duplessy, J.C., et al.: 1981, Palaeogeogr. Palaeoclimat. Palaeoecol. 35, 1981, 121-144.

Emiliani, C.: 1955, J. Geol. 63, 538-578.

Emiliani, C.: 1966, J. Geol. 74, 109-124.

Emiliani, C.: 1972, Science 178, 398-401.

Findlater, J.: 1977, Pure Appl. Geophys. 115, 1251-1262.

Fletcher, J.O., et al.: 1973, WMO Technical Note 129, 181-218.

Flint, R.F.: 1971, Glacial and Quaternary Geology, J. Wiley and Sons, New York.

Flohn, H.: 1964, Geol. Rundschau 54, 504-515.

Flohn, H.: 1964a, Bonner Meteor. Abhandl. 4.

Flohn, H.: 1966, Z.f. Meteor. 17, 316-320.

Flohn, H.: 1968, Colorado State University: Atmos. Sci. Paper 130, VI + 120.

Flohn, H.: 1974, Quat. Res. 4, 385-404.

Flohn, H.: 1978, in A.B. Pittock et al. (eds.), Climatic Change and Variability - A Southern Perspective, Cambridge University Press, pp. 124-134.

Flohn, H.: 1978, in J. Williams (ed.), Carbon Dioxide, Climate and Society, Pergamon Press, pp. 227-237.

Flohn, H.: 1978, in E.M. van Zinderen Bakker (ed.), Antarctic Glacial History and World Palaeoenvironment, Rotterdam, VIII + 171 pp.

Flohn, H.: 1981, Geol. Rundschau 70, 725-736.

Flohn, H. and Nicholson, Sh.: 1980, Palaeoecology of Africa 12, 3-21.

Frakes, L.A.: 1978, in Climatic Changes and Variability - A Southern Perspective, Cambridge University Press, pp. 53-69.

Frakes, L.A.: 1979, Climates Throughout Geologic Time. Elsevier, Amsterdam, XII + 310 pp.

Frenzel, B.: 1968, Science 161, 637-649.

Frenzel, B.: 1973, Climatic Fluctuations of the Ice Age, The Press of Case Western Reserve University, Cleveland.

Gabriel, B.: 1977, Berliner Geogr. Abhandl. 27.

Goldthwait, R.P., Dreimanis, A., Forsyth, J.L., Karrow, P.F. and White, G.W.: 1965, in H.E. Wright, Jr and D.G. Frey (eds.), The Quaternary of the United States, Princeton, pp. 85-97.

Hammer, C.U.: 1980, J. Glaciol. 25, 359-372.

Hammer, C.U., Clausen, H.B., Dansgaard, W., Gundestrup, N., Johnsen, S.J. and Reeh, N.: 1978, J. Glaciol. 20, 3-26.

Hammer, C.U., Clausen, H.B. and Dansgaard, W.: 1980, Nature 288, 230-235.

Hare, F.K.: 1977, Climate and Desertification. In: Desertification its causes and consequences. Pergamon, London, pp. 63-120.

Hays, J.D.: 1978, in E.M. van Zinderen Bakker (ed.): Antarctic Glacial History and World Palaeoenvironments, Balkema, Rotterdam pp. 57-71.

Hays, J.D., Imbrie, J. and Shackleton, N.J.: 1976, Science 194, 1121-1132.

Herman, Y.: 1969, Palaeogeogr. Palaeoclimatol., Palaeoecol. 6, 251-276.

Hibler, W.D.: 1979, III: Phys. Oceanogr. 9, 815-846.

Hollin, J.T.: 1980, Nature 283, 629-633.

Hopkins, D.M., (ed.): 1967, The Bering Land Bridge, XIII + 495 pp.

Hopkins, D.M., et al.: 1971, Palaeogeogr., Palaeoclim., Palaeoecol. 9, 211-231.

Hsü, K.J., et al.: 1973/1977, Nature 242, 240-244 and 267, 399-403.

Hughes, T.: 1973, Geophys. Res. 78, 7884-7910.

Humphrey, W.J.: 1940, Physics of the Air, McGraw-Hill, New York.

Imbrie, J. and Kipp, N.G.: 1971, in K.K. Turekian (ed.), The Late Cenozoic Glacial Ages, Yale Univ. Press, New Haven Con., pp. 71-82.

Jenssen, D.: 1977, J. Glaciol. 18, 372-390.

Johnsen, S.J., Dansgaard, W., Clausen, H.B. and Langway, C.C.: 1972, Nature 235, 429-434.

Kellogg, T.B.: 1975, in G. Weller and S.A. Bowling (eds.), Climate of the Arctic, Proc. 24th Alaska Sci. Conf., Geophys. Inst. Univer. Alaska, Fairbanks, pp. 3-36.

Kellogg, T.B., Duplessy, J-C. and Shackleton, N.J.: 1978, Boreas 7, 61-73.

Kemp, E.M.: 1978, Palaeogeogr., Palaeoclim., Palaeoecol. 24, 169-203.

Kempe, S.: 1977, Mitt. Geol. Paläont. Inst. Univ. Hamburg 47, 125-228.

Kennett, J.P.: 1977, J. Geophys. Res. 82, 3843-3860.

Kennett, J.P. and Thunell, R.C.: 1975, Science 187, 497.

Kennett, J.P. and Thunell, R.C.: 1977, Science 196, 1231-1234.

Kennett, J.P., et al.: 1977, J. Volc. Geotherm. Res. 2, 145-163.

Koch, L.: 1945, Meddelelser om Grønland, 130, Nr. 3.

Korff, H.C. and Flohn, H.: 1969, Ann. Meteor. N.F. 4, 163-164.

Ku, T.L., Kimmel, M.A., Easton, W.H. and O'Neil, T.J.: 1974, Science 183, 959-961.

Kukla, G.J. and Berger, A.: 1979, Proc. Int. Sci. Assembly Symp. on Milankovitch's Life and Work, Serbian Academy of Sciences and Art, Belgrado.

Kutzbach, J.E. and Otto-Bliesner, B.L.: 1982, J. Atmos. Sci. 33, 1177-1188.

Lalou, C., Duplessy, J-C. and Nguyen Huu Van: 1971, Rev. Géogr. Phys. Géol. Dyn. 13, 447-462.

LaMarche, V.C.: 1974, Science 183, 1043-48.

Lamb, H.H.: 1968, The Changing Climate, Methuen, London 186 pp.

Lamb, H.H.: 1970, Phil. Trans. Roy. Soc. London, Ser. A., 266, 425-533; Amendments in Climate Monitor 6 (1977), No. 2.

Lamb. H.H.: 1977, Climate: Present, Past and Future, Vol. 2, Methuen, London.

Lorenz, E.N.: 1968, Meteor. Monogr. 8,1.

Lorius, C., et al.: in CNES, Evolution des atmosphères planétaires et climatologie de la terre. Coll. Internat. Nice, 16-20 Oct. 1978, 71-82.

Lotze, F.: 1963, in A.E.M. Nairn (ed.), Problems on Paleoclimatology, London.

Mägdefrau, K.: 1968, Paläobiologie der Pflanzen. 4. Auflage. Stuttgart, 549 pp.

Maley, J.: 1977, Nature 269, 573-577.

Maley, J.: 1980, in H. Faure and M.E.J. Williams (eds.), The Sahara and the Nile, pp. 63-86.

Manabe, S. and Stouffer, R.J.: 1980, J. Geophys. Res. 85, 5529-5554.

Manabe, S. and Wetherald, R.T.: 1976, J. Atmos. Sci. 32, 3-15; and 1980, J. Atmos. Sci. 37, 99-118.

Manabe, S., Wetherald, R.T. and Stouffer, R.J.: 1981, Climatic Change 3, 347-385.

Mangerud, J., Sønstegaard, E. and Sejrup, H-P.: 1979, Nature 277, 189-192.

Mangerud, J., Sønstegaard, E., Sejrup, H-P. and Holdorsen, S.: 1981, Boreas 10, 138-208.

Marchesoni, V. and Paganelli, A.: 1960, Rendiconti degli Ist Sci. Univ. Camerino 1, 47-54.

Mason, B.J.: 1976, Quart. J. Roy. Meteor. Soc. 102, 473-498.

Maykut, G.A. and Untersteiner, N.: 1971, J. Geophys. Res. 76, 1550-1575.

McClure, H.A.: 1976, Nature 263, 755-756.

Mercer, J.H.: 1968, Quat. Res. 6, 125-166.

Mercer, J.H.: 1978, Nature 271, 321-325.

Milankovitch, M.M.: 1930, Handb. der Klimatologie I, Teil A, Köppen and Wegener, Geiger, Berlin.

Milankovitch, M.M.: 1941, "Canon of Insolation and the Ice Age Problem". Königl. Serb. Akad. Beograd. Engl. Transl.: Israël Progr. for Sc. Transl. Publisher: U.S. Nat. Sci. Found., Wash., D.C.

Mitchell, J.M., Jr.: 1965, in The Quaternary of the United States H.E. Wright Jr. and D.G. Frey (eds.), Princeton Univ. Press, Princeton, N.J. 1965, pp. 881-901.

Mörner, N.A.: 1976, Palaeogeogr., Palaeoclim., Palaeoecol. 19, 63-85.

Müller, H.: 1974, Geologisches Jahrbuch, Heft A 21, 107-140.

Müller, H.: 1977, Science 196, 489-494.

Munn, R.E. and Machta, L.: 1979, in Proceedings of the World Climate Conference, Genéva, 12-23 Febr. 1979, WMO-No. 537, 170-209.

Neftel, A., et al.: 1982, Nature 295, 220-223.

Newell, R.E.: 1974, Quat. Res. 4, 117-127.

Nichols, H.: 1975, Inst. of Arctic and Alpine Research, Univ. of Colorado Paper 15.

Oerlemans, J., and van den Dool, H.M.: 1978, J. Atmos. Sci. 35, 371-381.

Oeschger, H., Siegenthaler, U., Schotterer, U. and Gugelmann, A.: 1975, Tellus 27, 168-192.

Olausson, E.: 1965, in Mary Sears (ed.), Progress in Oceanography 3, Pergamon Press, Oxford, pp. 221-252.

Parkinson, C.L. and Washington, W.M.: 1979, J. Geophys. Res. 84, 311-332.

Parkinson, C.L. and Kellogg, W.W.: 1979, Climatic Change 2, 149-162.

Pastouret, L. et al.: 1978, Oceanologia Acta 1, 217-232.

Paterson, W.B.S., Koerner, R.M., Fisher, D., Johnsen, S.J., Clausen, H.B., Dansgaard, W., Bucher, P. and Oeschger, H.: 1977, Nature 266, 508-511.

Patzelt, G.: 1973, Z. Geomorph. N.F. Suppl. Band 16, 25-72.

Peng, T.H., Broecker, W.S., Kipphut, G. and Shackleton, N.: 1977, in N.R. Andersen and A. Malahoff (eds.), The Fate of Fossil Fuel CO_2 in the Oceans, Plenum Press, New York, pp. 355-373.

Plass, G.N.: 1956, Tellus 8, 140.

Reeh, N.: 1984, Annals of Glaciol. 5 (in press).

Richards, H.G. and Judson, S.: 1965, in H.E. Wright Jr. and D.G. Frey (eds.): The Quaternary of the United States, Princeton University Press, Princeton, 129-136.

Robin, G. de Q. and Adie, R.J.: 1964, in R. Priestley, R.J. Adie and G. de Q. Robin (eds.), Antarctic Research, Butterworths London, 100 p.

Rognon P, Williams, M.A.J.: 1977, Palaeogeogr., Palaeoclim., Palaeoecol. 21, 285-327.

Ruddiman, W.F. and McIntyre, A.: 1977, J. Geophys. Res. 82, 3877-3887.

Ruddiman, W.F. and McIntyre, A.: 1979, Science 204, 173-175.

Saito, T., et al.: 1975, in Late Neogene Epoch Boundaries, pp. 226-244.

Sancetta, C., Imbrie, J., Kipp, N.G., McIntyre, A. and Ruddiman, W.F.: 1972, Quat. Res. 2, 363-367.

Sarnthein, M.: 1978, Nature 272, 43-46.

Savin, S.M. and Douglas, R.G.: 1973, Bull. Geol. Soc. Am. 84, 2327-2347.

Schwarzbach, M.: 1974, Das Klima der Vorzeit. 3. Auflage. Enke, Stuttgart, VIII + 380 pp.

Shackleton, N.J.: 1967, Nature 215, 15-19.

Shackleton, N.J.: 1977a, in N.R. Anderson and A. Malahoff (eds.) The Fate of Fossil Fuel CO_2 in the Oceans, Plenum Fress, New York, pp. 401-427.

Shackleton, N.J.: 1977b, Phil. Trans. R. Soc. Lond. B 280, 169-182.

Shackleton, N.J. and Opdyke, N.D.: 1973, Quat. Res. 3, 39-55.

Shackleton, N. and Opdyke, N.D.: 1977, Nature 270, 216-219.

Shaw, D.M. and Donn, W.L.: 1968, Science 162, 1270-72.

Singh, G., et al.: 1974, Philos. Transact. Roy. Soc. London B 267, 467-501.

Smagorinsky, J. : 1963, Monthly Weather Rev. 91, 99-165.

Steinen, R.P., Harrison, R.S. and Matthews, R.K.: 1973, Geol. Soc. Am. Bull. 84. 63-70.

Street-Perrott, F.A. and Grove, A.T.: 1976, Nature 261, 385-390.

Street-Perrott, F.A. and Roberts, N.: 1979, Quat. Res. 12, 83-116.

Thomas, R.H.: 1976, Nature 259, 180-183.

Thomas, R.H., Sanderson, T.J.O. and Rose, K.E.: 1979, Nature 277, 355-358.

Thorndike, A.S., et al.: 1975, J. Geophys. Res. 80, 4501-4513.

Tolstikov, Y.I., (ed.): 1966, Atlas Antartiki 1, 76 Glavnoye Upravleniye Geodezi i Kartografii, Moscow and Leningrad.

Tooley, M.J.: 1974, Geogr. J. 140, 18-24.

Ulrych, T.J. and Bishop, T.N.: 1975, Rev. Geophys. 13, 183-200.

Van Loon, H. and Rogers, J.C.: 1978/1979, Monthly Weather Rev. 106, 296-310.; 107, 509-519.

Veeh, H.H. and Chappell, J.: 1970, Science 167, 862-865.

Vernekar, A.D.: 1972, Meteor. Monogr. 12, No. 34.

Wang, W.C., et al.: 1976, Science 194, 685-690.

Wade, N.: 1979, Science 206, 912-913.

Vowinckel, E. and Orvig, Sv.: 1970, World Survey of Climatology Vol. 14, 129-252.

Weertman, J.: 1964, J. Glaciol. 38, 145.

Weertman, J.: 1976, Nature 260, 284: 261, 17.

Weidick, A.: 1976 , in A. Escher and W.S. Watt, (eds.), Geology of Greenland, Geological Survey of Greenland, Copenhagen, pp. 431-458.

Wetherald, R.T. and Manabe, S.: 1980, J. Geophys. Res. 86, 1194-1204 .

Whillans, I.M.: 1977, J. Glaciol. 18, 359.

Wigley, T.M.L.: 1976, Nature 264, 629-631.

Wilson, A.T.: 1964, Nature 201, 147-149.

Woillard, G.: 1975, Thesis, Univ. Catholique de Louvain.

Woillard, G.: 1978, Quat. Res. 9, 1-21.

Woillard, G.: 1979, Nature 281, 558-562.

Woillard, G. and Mook, W.G.: 1982, Science 215, 159-161.

Wollin, G., Ericson, D.B. and Ryan, W.B.F.: 1971, Nature 232, 549

Wollin, G., Ericson, D.B. and Ryan, W.B.F.: 1978, Earth Planet. Sci. Lett. 41, 395

Wijmstra, T.A.: 1969, Acta Bot. Neerl. 18, 511-527.

Yapp, C.J. and Epstein, S.: 1977, Earth Planet. Sci. Lett. 34, 333-350.

Zagwijn, W.H.: 1977, X INQUA Congress, Birmingham.

CHAPTER 6

Anon: 1971, British Farmer and Stockbreeder 1, 63 pp.

Azzi, G.: 1921, Reale Accademia dei Lincei, Vol. XXIX, serie V, Fasc. II - Roma.

Bach, W., Pankrath, J. and Schneider, S.H., (eds.): 1981, Food-Climate Interactions. D. Reidel, Dordrecht, Holland, XXXI + 504 pp.

Bergthorsson, P.: 1967, Jökull 19, 94-101.

Boate, G.: 1755, A Natural History of Ireland, Dublin.

Bourke, P.M.A.: 1964, Nature 203, 805-808.

Drake, M.: 1968, in T.W. Mooney (ed.), Historical Studies VI, London, 101-124.

Eythorsson, J. and Sigtryggsson, H.: 1971, The Zoology of Iceland 1 (3), 1-62.

Feekes, W.: 1967, Neth. J. Flant Pathol. 73, Supplement 1, 97-115.

Flohn, H.: 1978, Bild der Wissenschaft 12, 136-139.

Food and Agriculture Organization (FAO, Rome) - Production Year Books.

Fridriksson, S.: 1973, Intern. J. Biometeor. 17, 359-362.

Fukui, H.: 1979, Climatic Variability and Agriculture in Tropical Moist Regions. WMO, World Climate Conference, WMO-No. 537, Genéva, 1979.

Gerbier, N. and Rémois, P.: 1977, Météorologie nationale Mon. no 106, Paris.

Hoskins, W.G.: 1968, Agricultural History Rev. 12, 28-46; 16 (1), 15-31.

Instituto Centrale di Statistica of Italy, Roma, Annali di Statistica Agraria.

Jones, E.L.: 1964, Seasons and Prices: The Role of The Weather in English Agricultural History, London.

Landsberg, H.E. and Albert, J.M.: 1974, Weatherwise 27, 63-66.

Legrand, J.P.: 1977, "La Météorologie" VI serie no 8, Paris, 73 pp.

Le Roy Ladurie, E.: 1972, Times of Feast, Times of Famine: A History of Climate Since the Year 1000, London.

Le Roy Ladurie, E. and Baulant, M.: 1980, J. Interdiscipl. History 10, 839-849.

Mangianti, F. and Bonelli, G.: 1973, Rivista Italiana di Geofisica Vol. XXII no 5/6 Genova, 415 pp.

Manley, G.: 1958, Weather 13, 11-17.

Manley, G.: 1959, Archiv für Meteorologie, Geophysik und Bioklimatologie, Serie B 9, 360-412.

Manley, G.: 1974, Quart. J. Roy. Meteor. Soc. 100, 389-405.

Milham, W.I.: 1924, Monthly Weather Rev. 52, 563-570.

Neumann, J.: 1977, Bull. Amer. Meteor. Soc. 58, 1163-1168.

Neumann, J. and Lindgren, S.: 1979, Bull. Amer. Meteor. Soc. 60, 775-787; see also 1981, Climatic Change 3, 173-187.

Osvald, H.: 1952, Swedish Agriculture, Stockholm.

Post, J.D.: 1977, The Last Great Subsistence Crisis in the Western World, Baltimore and London.

Roncali, G.: 1955, Sui danni della grandine in Italia, Ufficio Centrale di Ecologia Agraria, Roma.

Rosini, E.: 1973, Prima Relazione Ambientale, Urbino (Italy).

Seemann, J., Chirkov, Y.I., Lomas, J. and Primault, B.: 1979, Agrometeorology, Springer-Verlag, Berlin.

Slichter van Bath, B.H.S.: 1963, The Agrarian History of Western Europe: AD 500-1850, London.

Smith, L.P.: 1965, in C.G. Johnson and L.P. Smith (eds.), The Biological Significance of Climatic Change in Britain, London and New York, pp. 187-191.

Takahashi, K. and Yoshino, M.M.: 1978, Climate Change and Food Production, University of Tokyo.

Ten Houten, J.G.: 1974, Ann. Rev. Phytopath. 12, 3-11.

U.S. Department of Agriculture, Economic Research Service: 1975, Foreign Agricultural Economic Report No. 98, Washington.

Utterström, G.: 1961, Scandinavian Economic History Rev. 9, 176-194.

Vargo, U.: 1973, Intern. J. Biometeor. 17, 363-369.

Wallén, C.C. (ed.): 1970, World Survey of Climatology Vol. 5, and 6, Amsterdam.

Woodham-Smith, C.: 1962, The Great Hunger: Ireland 1845-1849, London.

World Meteorological Organization: 1975, WMO Publication no 396 and Supplement, Genéva.

Zadoks, J.C.: 1965, FAO Plant Protection Bulletin 13, 97-108.

INDEX